Manual of Economic Analysis of Chemical Processes

Manual of Economic Analysis of Chemical Processes

Feasibility Studies in Refinery and Petrochemical Processes

Institut Français du Pétrole

Alain Chauvel
Pierre Leprince
Yves Barthel
Claude Raimbault
Jean-Pierre Arlie

Translated from the French by
Ryle Miller and
Ethel B. Miller

McGraw-Hill Book Company

New York St. Louis San Francisco Auckland Bogotá Hamburg
Johannesburg London Madrid Montreal New Delhi Panama
Paris São Paulo Singapore Sydney Tokyo Toronto Mexico

Library of Congress Cataloging in Publication Data

Main entry under title:
Manual of economic analysis of chemical processes.

Translation of Manuel d'évaluation économique des procédés.
Bibliography: p.
Includes index.
1. Petroleum—Refining. 2. Petroleum chemicals.
I. Chauvel, Alain. II. Rueil-Malmaison, France.
Institut français du pétrole.
TP690.M2913 338.4'566'550944 79-25247
ISBN 0-07-031745-3

Manuel D'évaluation Économique Des Procédés
Avant-projets en raffinage et pétrochimie
(Collection Pratique du Pétrole N° 6)

1234567890 KPKP 8987654321

The editors for this book were Jeremy Robinson, Robert L. Davidson, and Susan Thomas, the designer was Mark E. Safran, and the production supervisor was Paul A. Malchow. It was set in Baskerville by Haddon Craftsmen, Inc.

Printed and bound by The Kingsport Press.

Contents

Preface

There are times when bringing together known information can result in a synergistic association that affords unprecedented effectiveness for that information. The authors have achieved such a synergism with this book.

Profitability calculations, market research, chemical engineering cost estimating, and shortcut process design methods, which are all discussed here, are well-known skills. However, anyone who has ever faced a feasibility study for a refining, a petrochemical, or a chemical plant will have discovered that finding reliable answers is something like following a will-o'-the-wisp. Profitability depends on revenues, operating costs, and investment, which in turn depend on market conditions, manufacturing efficiencies, and cost estimates, which in turn depend on the process design, so that the best efforts are frequently frustrated by one uncertain link in a long chain of calculations.

This single book spans that entire gamut of information, from discounting rate to heat-transfer coefficient, while maintaining perspective on the relative error potential so that an inordinate amount of time need not be invested in one aspect of a project only to have that time wasted by uncertainty introduced somewhere else. Perhaps this perspective is best illustrated by a paraphrase of the discussion with which the authors introduce methods for process design estimating in Section 43.3.3.:

> Even when a detailed estimate is accompanied by information on sizes for a unit, it is best to recalculate the sizes according to a consistent estimating method,

> since use of a consistent method puts everything into the same framework for making comparisons.
>
> An extrapolated estimate based on a plant known in detail will differ from an estimate based on the consistent sizing method, and the extrapolated estimate should be more reliable. If the difference is large, an error or false information is indicated. If the difference is not too great and the basic data are reliable, the difference between the two should be converted into coefficients that can be applied to the consistent method for future estimates.

One primary source of uncertainty in feasibility studies has been the accuracy of the process design on which the estimates were based. In the past, we have carried a personal pocket notebook from which we would extract critical numbers and quick-design rules for estimating equipment variations at a proposal meeting. Many design engineers have carried such a book. However, there are very few, if any, companies that have lifted the numbers and rules of those personal notebooks to the level of formalized methods that could be tested against known accurate values and thus tuned and updated. Detailed design procedures have been thus formalized, but not estimating procedures. In this book, the authors have given us a collection of such formalized methods in their Appendixes.

Some readers may wonder, perhaps cynically, how a company happens to publish such information rather than to keep it proprietary. If so, those readers should look further into Institut Français du Pétrole (IFP). IFP was formed near the end of World War II to help close the technological gap by which France lagged behind such countries as the United States. A subsidy was established in the form of a small fraction of the French national tax on gasoline. As the sales of gasoline have grown, so has the subsidy. Also, much of IFP's research and many of its enterprises have literally paid off, so that some 30% of its present income is accounted for by its achievements, specifically licensing, royalties, and research contracts. This holds true even though, being a not-for-profit institution, IFP cannot have conventional commercial activities. Value analysis and industrialization of some of its results have thus resulted in the creation and spin-off of newly formed companies—Technip, Procatalyse, Franlab, Coflexip, etc. Result: the IFP group is one of the world's more important research-based groups.

In addition to its obligation to perform research and industrial development, IFP also has a statutory mandate to disseminate information through adequate means, such as by maintaining an information and documentation center and by promoting the transfer of knowledge and know-how through publications. An IFP subsidiary, Société des Editions Technip, publishes books, periodicals, and technical papers.

Perhaps the first of its books to attract widespread attention outside of France was *L'industrie pétrochimique et ses possibilités d'implantation dans les pays en voie de développement,* which was presented as a paper at the first United Nations conference on agricultural chemicals in Teheran in 1964 and published afterwards simultaneously in French and English (1966). Preprints of the English version were snatched up by attendees of the conference. One senior engineer with a major U.S. research company was overheard to remark, "A consultant could make a living out of that book." Continuing this publishing policy, another book, *Procédés de pétrochimie,* was published in 1971. However this edition, which is in French, has failed to find its way to many American users.

We have tried to avert a similar fate for the present book by translating it. However, we are inclined to think that, largely because of IFP, American process engineers would do well to make French a second language—or at least to gain a reading acquaintance with French, as they used to do with German. Otherwise they are apt to miss not only synergistic combinations of already available material but also information about new technology that is only now being developed.

We publicly want to thank René E. G. Smith of Mobil Research and Development Corporation for taking precious leisure time from a North Sea drilling platform to read and criticize parts of our translation. Also, we want to caution the reader that this is a liberal translation, in which we have tried to capture the authors' intent rather than their precise expression. In doing this and in converting francs to dollars for the user's convenience, we may inadvertently have let some minor errors slip by us. We would be grateful to the reader for bringing them to our attention.

Ryle Miller
Ethel B. Miller

Introduction

A new manufacturing plant can be put into perspective quantitatively by a technical and economic study. If the plant involves new technology, this study should involve not only the engineering and operating companies responsible for designing the commercial plant, constructing it, and putting it into operation but also the organization that invents and develops the technology.

Any technical and economic study more accurately reveals the potentials for a new project—whether new plant, new process, or new product—when the data, the basic assumptions, and the calculated results are thoroughly understood. The evaluation should cover technical correlations, operating feasibility for purchased equipment, properties of the manufactured product, and the financial requirements for purchasing and maintaining the equipment. Also, the needs of the industrial environment are constantly changing, and the changes should be continually incorporated into new studies that thus reexamine earlier studies and improve the accuracy of the forecasts.

The evaluation of a manufacturing project must begin by examining the economic context in which the plant will exist throughout its lifetime; e.g., it must forecast the availability of feedstocks and utilities as well as the market for the product. Next, the study must define the technology—the types and sizes of required equipment, its prices, its efficiencies, its energy consumptions, and the work force needed for its operation. The effort placed in this technical study, which is normally made by specialized departments, varies according to the state of the development of the project. A limit is reached in "control estimates" made by engineering and construction contractors from engineering drawings developed as part of a contract between the contractor and the client.

In the last stage of evaluation, the economic forecasts are combined with the technical estimates according to selected operating criteria in order to establish an optimum profitability, which guides the final decision.

In this book, when conventional methods for making economic evaluations are assembled, the authors have tried to point out the relative importance of the different aspects of the calculations, whether they relate more or less directly to the technical nature of the process, to the properties of the product in question, or to the financial means and management methods of the company under consideration.

Accordingly, the first part of the book offers a description of the principles of economic calculations, including market analysis, procedures of accounting and economic evaluation, and cost estimation, while in the second part these principles are applied to both industrial and research projects. In addition, the appendixes present a group of methods for selecting, sizing, and pricing equipment for purposes of establishing cost estimates.

All the calculations presented here were made in French francs pegged to a common base, but have been converted to dollars where necessary for the reader's convenience (conversions are given in Fig. 4.6). All of the monetary transactions are in terms of the year when the study was made, taken as the year of reference. One cannot estimate the effects of monetary depreciation over a long period. The type of calculation using fluctuating francs, in which the cash flow is corrected for the value of currency at the year in question, would be too uncertain to be of any use in the evaluation of a project.

Metric units of measurement are used in this manuscript, along with the appropriate calculation constants in the equations and calculation forms (see Appendix 13, p. 421),* so that American analysts and engineers might use these for a general reference for calculations in metric units rather than U.S. customary units. Accordingly, the metric ton of 1,000 kg (2,200 lb) is intended wherever the word ton or the abbreviation t is used.

*We have not translated the term *cheval-vapeur* (cv) into its English equivalent, *horsepower,* because a cheval is actually 0.986 horsepower; however, the terms are frequently interchanged when an approximation is acceptable.

Part 1

Principles of Economic Evaluation

Chapter 1

Market Research

Before undertaking the economic evaluation of a manufacturing project, it is necessary to have a certain amount of information about the marketing prospects of the product to be manufactured. In order to make a product, it is necessary first to buy the raw materials and chemicals and then to realize a profit in selling the product. It is therefore indispensable, on the one hand, to know that the raw materials are available and can be bought for a price and, on the other hand, to estimate the ability of a user to pay for the product. Consequently, availability-to-price relationships of both raw materials and products should be constantly kept in mind.

1.1 AVAILABILITY OF RAW MATERIALS

For raw materials, quantity-price relationships are illustrated by an example in which several parameters govern. Suppose that a company wants to build in France a plant that would use butene-2 as feedstock. There are four ways of procuring butene-2:

1. Purchase from an existing plant.
2. Separation from the C_4 distillate fraction resulting from catalytic cracking.

3. Separation from the C_4 distillate fraction resulting from steam pyrolysis for the production of ethylene and propylene.
4. Catalytic dehydrogenation of normal butane.

The first method assumes obtaining surplus quantities from a manufacturer whose price is fixed by his principal use for butene-2. The following two methods involve the by-products of existing plants, and the last implies construction of a new plant. Each of these different methods thus imposes its own restrictions.

If it is decided to buy the product, the cost of shipping is added to its price at the manufacturing plant. If the C_4 fractions are used, the unit to separate the butene-2 will have to be connected in some manner to the catalytic cracking or steam pyrolysis plants. If catalytic dehydrogenation is used, the supply of normal butane (refinery or natural gas field) will be of concern.

In the event that any one of these sources alone is not enough, some combination of two or more sources will be necessary. Consequently, it will be necessary to take into account the particular limitations of each source, as reflected in a specific cost.

Added to this problem of supply is the fact that C_4 distillate fractions from catalytic cracking and steam pyrolysis contain concentrations of butene-2 that vary from one installation to another. As shown in Table 1.1, the percentage can range from 10 to 22%. It follows that the cost of separation will be different.

To help relate such disparate conditions, the amounts of available material can be classed as a function of price at the plant, starting with those available at the lowest price. Thus, in the case of butene-2, the following case might typify a given location:

Quantity available, Q_1 = 20,000 tons/yr, at price P_1 = 560 fr/ton
Quantity available, Q_2 = 20,000 tons/yr, at price P_2 = 680 fr/ton
Quantity available, Q_3 = 10,000 tons/yr, at price P_3 = 720 fr/ton
Quantity available, Q_4 = 40,000 tons/yr, at price P_4 = 1,000 fr/ton

The price for each of these quantities includes

Cost of the initial raw material (C_4 fraction or *n*-butane)
Cost of separation and purification
Cost of transporting the product from place of production to point of use

Using these prices, a curve of price versus quantity can be plotted (Fig. 1.1) to determine the average price P_m:

$$P_m = \frac{\Sigma Q_i P_i}{\Sigma Q_i}$$

Such data enable one to calculate an economic optimum for the butene-2 transformation unit by assigning various feedstock prices to corresponding capacities.

A variation of this situation can be analyzed by studying the case in which several C_4 fractions are brought together at one site, so as to take advantage of an increase in size of the butene-2 separation unit and thus obtain a lower average price (provided sales of the design plant output are assured).

When the location of the plant is open, an optimum site includes consideration of the raw material costs. Treated abstractly, such problems are handled by a large number of equations whose optimum solution is found with the aid of a computer. This case is rare in the chemical industry, however, because

TABLE 1.1 Typical Compositions of C_4 Fractions, wt%

Compounds	From Steam Pyrolysis for Ethylene	From Steam Pyrolysis after Butadiene Extraction	From Refinery Catalytic Cracking
C_3 hydrocarbons	—	—	2.3
n-Butane	1.3	2.1	16.0
Isobutane	0.3	0.5	25.1
Butene-1	16.5	26.2	13.3
Butene-2	10.0	15.9	22.1
Isobutene	34.2	54.2	20.0
1,3-Butadiene	37.7	1.1	0.4
C_5^+	—	—	0.8
Recovery based on plant feedstock (wt%)	9.5–10.3	5–8	8–10

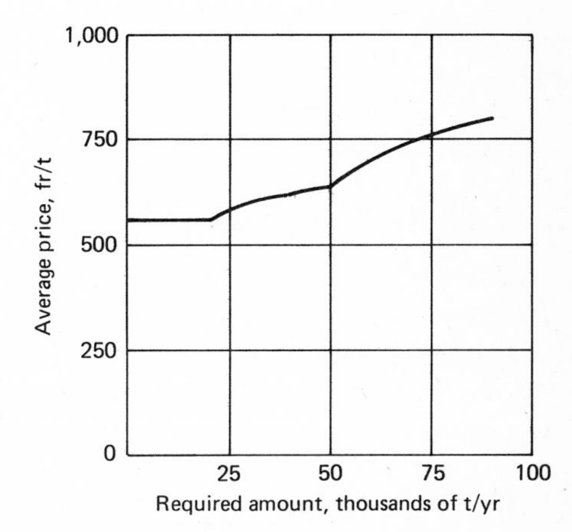

Fig. 1.1 Effect of plant consumption on the price of butene-2.

plant sites are often determined by an existing installation, or by the special advantages of a nearby port, highway, railroad, or pipeline.

For a competent analysis of raw materials availability, it is necessary to make inquiries of the producers and the shipping companies and to estimate the cost of transportation.

1.2 ESTIMATING THE POSSIBLE PRICE FOR A MANUFACTURED PRODUCT

A markedly different type of analysis from the preceding one is needed for estimating the market for the product of an installation under study. The purpose, here, is to determine the amount of a given product that can be absorbed by the market. When this is known, it is possible to fix the size of the new unit. However, the size of the market can depend on the selling price of the product, and two cases present themselves: that of an established product and that of a new product.

1.2.1 ESTABLISHED PRODUCTS

An established product will already be widely used and involved in large contracts; its price will be known; and it will be relatively independent of the quantities produced, although it can undergo sensitive modifications during the course of time. Since the plant under study will not be coming on stream before 3–5 years, a forecast of the future prices of this established product becomes a necessary part of the economic evaluation. The task is particularly delicate, because it is difficult to imagine, even in the near future, the direction and the rate of change that the price will follow. Past price variations of chemical products demonstrate this (Figs. 1.2 and 1.3).

The figures reflect a general tendency in the United States for progressively lowering prices for most organic chemical products, due to a steady growth of markets that makes it possible to achieve lower and lower operating costs through increased plant capacities and improved production methods. Certain economists have even evolved a rule for extrapolating this steady growth tendency. It is said, for example, that if the volume of sales increases by a multiple of 2 or 3, the price will decline by a factor of approximately $\frac{2}{3}$. Another method consists of charting the price of a product from the moment it begins to decline, i.e., from the time its manufacturing plants begin to improve production efficiencies. Figure 1.4 shows that the average decline of prices for organic chemicals has been steady and amounts to more than 50% for the past 15 years.

However, extrapolation of price trends can give false results if not made judiciously. The method should be complemented by a detailed study of

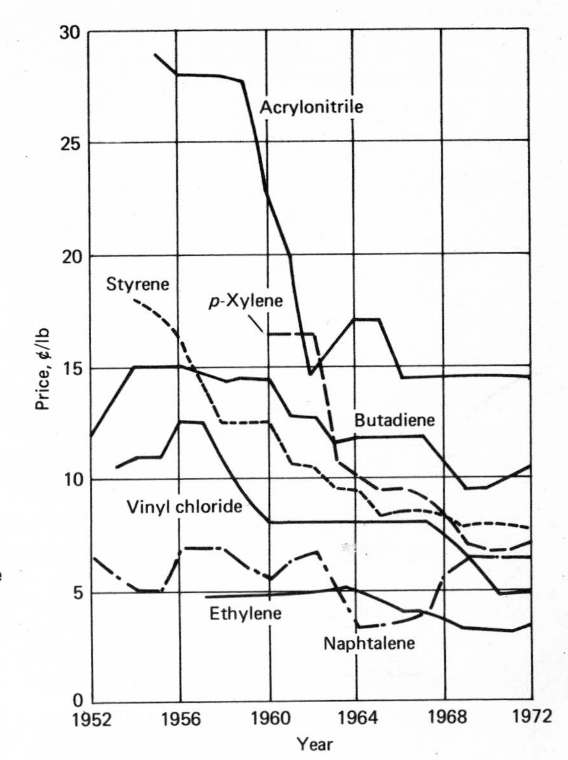

Fig. 1.2 Price histories of some chemicals in the United States.

the product under consideration. Useful information is drawn from the following:

- A comparison between actual production and the installed capacity; if capacity is much higher than demand, the overcapacity induces price stability, or may even provoke depressed prices.
- The profit margins of existing plants that set the current lowest price.
- Foreseeable changes in costs of production (increase in labor costs, costs of energy, costs of cooling water, etc.).
- The general trend of the economy (production growth, inflation, etc.).

Since there is no sure method of predicting the price of a product, it is wise to adopt the least favorable case, in order to assure that the investment makes a profit.

1.2.2 A NEW PRODUCT

A new product must enter a market where several others already have prices established according to quality and performance. Such an established market

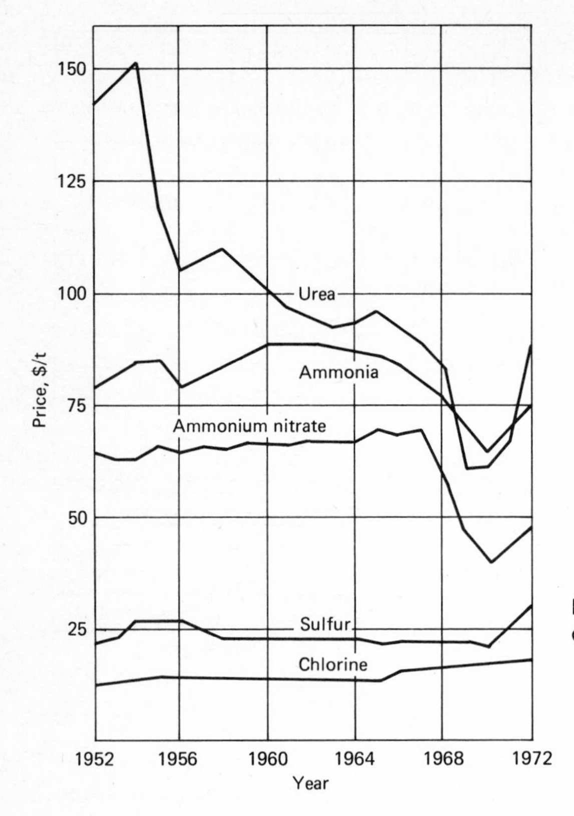

Fig. 1.3 Price histories of some chemicals in the United States.

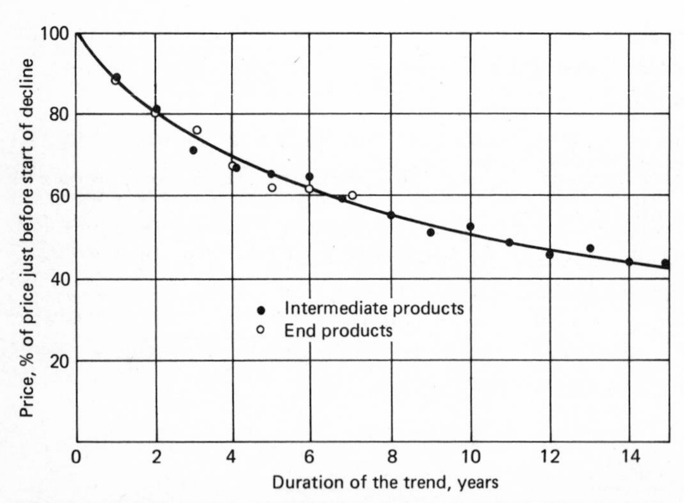

Fig. 1.4 Price trends in the United States for chemicals with declining production costs.

(plastics, rubbers, etc.) can often be characterized by a relationship between sales volume and price for each product. If the characteristics of the new product place it positively within one such group, it can enter into competition with the other products in the group, and its price alone will determine the accessible market.

However, this situation rarely arises, because each product inevitably exhibits certain unique properties that make it more or less easy to use. A polymer, for example, might be easier to form than its competitors; or an extremely desirable property might open up a new market where the new product can replace an entirely different material such as wood or metal. Instead of merely estimating the part of a given market made available by price, therefore, the actual market accessible to a new product should be determined by a more careful study.

Such a careful study may produce distinct prices associated with specific markets. Suppose the product under consideration can be used to make objects that would compete with the following kinds of materials:

1. Laboratory glassware
2. Plastic for medical syringes
3. Porcelain dishes
4. Transparent flat glass

Furthermore, suppose that these outlets represent the following sales and prices for a geographically defined market in a given industrial context:

1. 20,000 tons/yr at 50 price units
2. 10,000 tons/yr at 100 price units
3. 15,000 tons/yr at 70 price units
4. 50,000 tons/yr at 30 price units

The volume of sales from these several market outlets can be accumulated as a function of price (Fig. 1.5), so as to provide the elements for a calculation to determine the most favorable economic situation for this example. With such information it is even possible to analyze for the best return on investment (ROI). Thus if the chosen criterion is the maximum cash flow (see Sec. 2.6.3.1), the problem might be treated in the following manner.

A maximum is sought for the following function:

$$F = M(P - V)$$

where M = potential market
P = sales price
V = operating costs (raw materials, utilities, etc.)

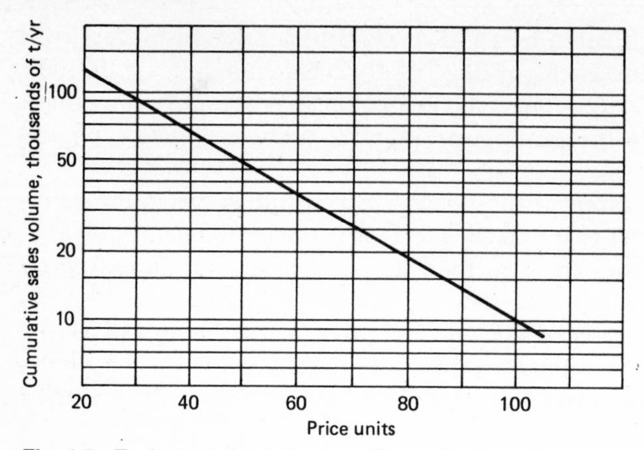

Fig. 1.5 Typical relation between price and sales volume.

This maximum can be found by differentiating:

$$\frac{dF}{dP} = M + (P - V)\frac{dM}{dP} = 0 \tag{1.1}$$

$$\frac{d^2F}{dP^2} = 2\frac{dM}{dP} + (P - V)\frac{d^2M}{dP^2} < 0 \tag{1.2}$$

If the elasticity of this market* is defined by the variation of its volume with its price according to the function

$$\alpha = -\frac{d(\ln M)}{d(\ln P)} = -\frac{dM}{M} \cdot \frac{P}{dP}$$

the equation for the optimum price of the product is

$$P_0 = \frac{\alpha}{\alpha - 1}V$$

Thus, if the variation of price P with the market M is known $[P = f(M)]$, the elasticity can be calculated and the optimum price obtained. In the example of the glasslike polymer, the relation plotted in Fig. 1.5 can be identified as a curve of the form

$$M = Ae^{-BP}$$

where A and B are constants. Substituting this expression for M in the equation for α, it is possible to determine α as equal to BP. Furthermore, substituting values from Fig. 1.5 for M and P, it is possible to determine constants A and B:

$$M = 2.3 \times 10^5 e^{-0.0315P}$$

*If $\alpha < 1.0$, the market is not elastic; if $\alpha > 1.0$, the market is elastic.

From this,

$$\alpha = BP = 0.0315P$$

This value of α can be substituted into the expression for optimum price P_0 to give

$$P - V = \frac{1}{B} = 31.6$$

indicating that in this case the selling price of the polymer should equal the operating costs plus 31.6 price units. When the operating costs can be identified independently of the fixed costs, without fixing the size of the unit, this method becomes extremely valuable, for it furnishes both the optimum price and its corresponding market volume.

1.3 ESTIMATING SALES VOLUME

The preceding studies have shown the importance of the estimated future sales of a product. This estimate can be arrived at in various ways, including consultation, projection, historical comparisons, and correlations, each of which has distinctive features worth noting.

First, the field of the study must be determined, both technically and geographically.

From the technical point of view, it is necessary to research the properties of the product, its characteristics, its features, qualities, faults, and its uses. Specialized books offer thorough descriptions of the properties, uses, and end products of chemical intermediates. For a new product, the study must begin with information furnished by the research laboratories that have processed the trials of the product.

On the geographic level, it is useful to take a census of the countries, first, where it is possible to manufacture the product (developed countries possessing the necessary basic industries, developing countries possessing the energy sources or raw material) and, second, where outlets for the product are to be found either as indigenous needs or as demands for exports. Allowance is made for geographic zones where the market is saturated, or where a monopoly exists, or where competition is particularly active for one reason or another.

1.3.1 CONSULTATION WITH EXPERTS

The market estimate for a product can be based on user surveys, a method widely used by commercial groups which specialize in launching a new commercial product and in studying the replies made by a selected group of users to a questionnaire. It is generally possible to follow the same procedures for industrial products, but the replies often represent a situation that will endure only for a short time, so they are not too useful for making decisions involving several years. In such cases, it is better to turn to experts who, because of their experience and function, can furnish an opinion that will be valid for future demands and especially for the factors that can influence the demands. Such information enables the engineer in charge of the study to make a qualitative opinion of the situation. It is disputable whether a study should start with this inquiry or with a more quantitative analysis and test those conclusions against the opinions of the consultants.

1.3.2 PROJECTION FROM THE PAST

The problem here is to present the past so that data for the future can be extrapolated from it. It is generally accepted that the growth of consumption for a product passes through three principal phases:

1. Increasing rate of growth, characterized by an exponential curve
2. Constant rate of growth, characterized by a straight line with the same slope as the tangent from the end of the exponential growth curve
3. Declining rate of consumption, characterized by a curve whose slope is always smaller than the preceding straight-line growth and may eventually become negative

These three stages are illustrated by acetylene consumption in the United States since 1935 (Fig. 1.6).* Thus, past data can be extrapolated as (1) an exponential progression with a constant coefficient of growth, (2) a linear progression, or (3) a more or less rapid decrease.

However, if the extrapolation is made over a long period, the results can be very clearly wrong and can even rob other more valid results of the study of their significance. Except for certain particular situations (e.g., the consumption of energy, steel, and fertilizers) that have many outlets and exhibit even rates of growth over long periods, extrapolations that have an exponential character should be treated with suspicion; they could lead to ridiculous conclusions. Certain extrapolations of this sort are famous. If the number of members of the Chemical Marketing Research Association (CMRA) from 1950 to 1963 is extrapolated to the end of the century, the CMRA is predicted to have more members than the entire U.S. chemical industry has workers. By the same method, it can be shown that half of the world population will be theoretical physicists by the end of this century.

Various techniques can lend extrapolation a greater degree of certainty. W.

*A closer analysis shows that the decline in acetylene consumption is due largely to its displacement by ethylene in the synthesis of vinyl chloride (lower curve in Fig. 1.6).

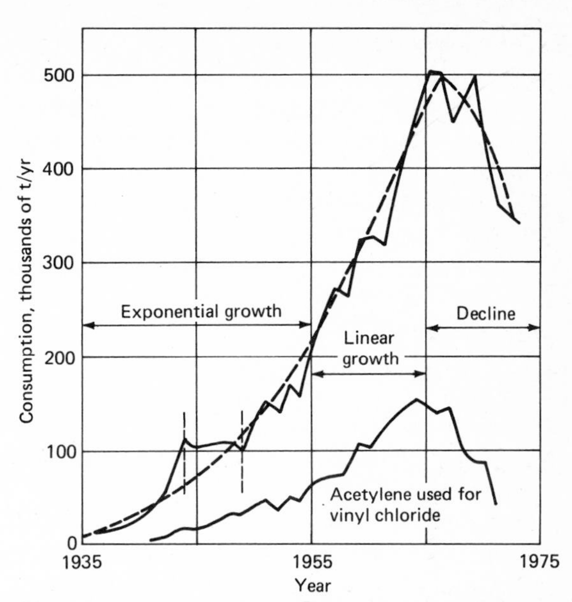

Fig. 1.6 Acetylene consumption in the United States—a study in the life of a chemical product.

W. Twaddle and J. B. Malloy have suggested that the exponential model (phase 2 above) can be perfected, first by introducing a correction that takes into account a slower growth, and then by fixing a limiting growth rate corresponding to the equilibrium between the consumption of the product in question and a characteristic size of the economy, for example, the gross national product (GNP). This calculation can be done in the following manner.

The rate of real growth at a given date is made up of two terms, r_∞ and $(r_0 - r_\infty)e^{-kt}$, so that

$$r = r_\infty + (r_0 - r_\infty)e^{-kt} \tag{1.3}$$

where r = rate of real growth
r_0 = rate of growth during the exponential period
r_∞ = limiting rate of growth
k = constant
t = time

Employing the exponential form for the increase that has occurred in demand, one has

$$D = D_0 \exp r_m t \tag{1.4}$$

where D = demand
D_0 = demand at time zero
r_m = average rate of growth

The average rate of growth, r_m can be expressed as a function of time t, so that

$$r_m = \frac{1}{t}\int_0^t r\,dt \tag{1.5}$$

Substituting the value of r from Eq. (1.3) into this equation gives

$$r_m = \frac{1}{t}\int_0^t [r_\infty + (r_0 - r_\infty)e^{-kt}]\,dt$$

which on integration gives

$$r_m = r_\infty + \frac{r_0 - r_\infty}{kt}(1 - e^{-kt}) \tag{1.6}$$

Accordingly, the expression for the demand is written

$$D = D_0 \exp(r_\infty + \frac{r_0 - r_\infty}{kt}(1 - e^{-kt}))t \tag{1.7}$$

In the particular case where $r_\infty = 0$, one has

$$r = r_0 e^{-kt}$$

and

$$D = D_0 \exp\left[\frac{r_0}{k}(1 - e^{-kt})\right]$$

Application of this model requires, first, that there be enough data to calculate the value of r and r_o and, second, that r_∞ be fixed. The value of k is then calculated by Eq. (1.3). Equation (1.7), where all the constants are known, permits extrapolation of the demand. However, the statistical data used to determine r should be corrected for variations due to external events (war, economic crises, inflation, devaluations, etc.). Figure 1.6 illustrates one such case, since acetylene production was very high between 1942 and 1944 because of World War II.

When applied to intermediate products, the method of projecting from the past can be perfected by examining the market for each of the end products separately. A more coherent generalization is obtained by adding all the results, thus reinforcing the value of the study.

1.3.3 HISTORICAL COMPARISONS

The consumptions of certain products are sometimes parallel over a period of time. For example, consumptions of nylon and acrylic fibers in the United States follow two parallel curves separated by a period of about 8 years (Fig. 1.7). This suggests a means of forecasting the consumption of acrylic fibers eight years ahead, as long as no new process or new application appears to disturb the situation. Polyolefin fibers, whose consumption is plotted since 1957 in Fig. 1.7, show an evolution that is not so nearly parallel as that of the others, so that the extrapolation of this curve parallel to the others is rather uncertain, expecially when taken to beyond 1980. If a comparison with polyester fibers is made, the expected similarity does not appear, since the consumption of this type of fiber has grown at an average rate of around 20% per year since 1950. Thus, application of this method is limited to cases where there is reason to believe that a similarity of markets exists.

1.3.4 MARKET CORRELATIONS

Generally preferred and widely used in various ways, this method consists of establishing a correlation between the volume or growth of a market and a convenient economic or technical indicator. The correct choice of this indicator is essential and comprises the real work of the analyst.

Use of these correlations assumes that studies made with the help of global economic models have furnished quantitative data on the future behavior of the chosen index. For this reason the economic indicators are usually limited to a few that can be established by general, independent studies (Fig. 1.8): total

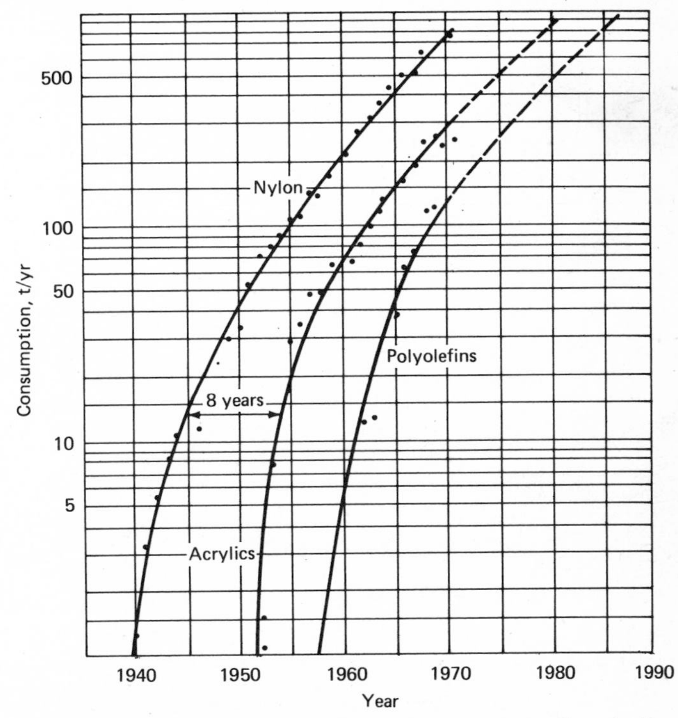

Fig. 1.7 Historical comparison of fiber consumptions in the United States.

GNP for a country or group of countries, GNP per capita, industrial production index, energy consumption, etc.

For example, a comparison of the consumption of plastics in Europe and in the United States as a function of the GNP (Fig. 1.8) shows that the correlation is good for Europe but not for the United States, where production of plastics is growing a little less rapidly. If the GNP of Europe in 1980 is on the order of 1,100 × 10^9 dollars (annual growth rate of 4.5%), the consumption of plastic materials can be forecast at about 25 × 10^6 tons/yr, for an average growth rate from now till then of 10%. An extrapolation of the same type for the United States would give a consumption of about 22 × 10^6 tons/yr.

This method can be applied to a more narrow group of products, for example, the polyolefins (Fig. 1.8), for which production in the United States in 1980 is forecast at about 8 × 10^6 tons/yr.

However, these extrapolations do not take into account the effects of market saturation, which could mean a lower consumption than anticipated. Even if the GNP continues to grow, the resulting wealth can be used by society to procure end products that do not necessarily entail consumption of such and such a raw material. For this reason a decreasing relationship between the rate

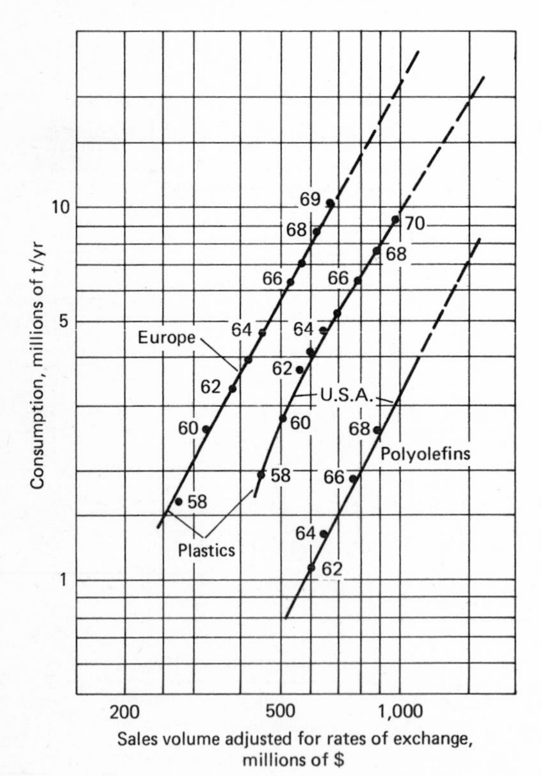

Fig. 1.8 Consumption of plastics in Western Europe and the United States.

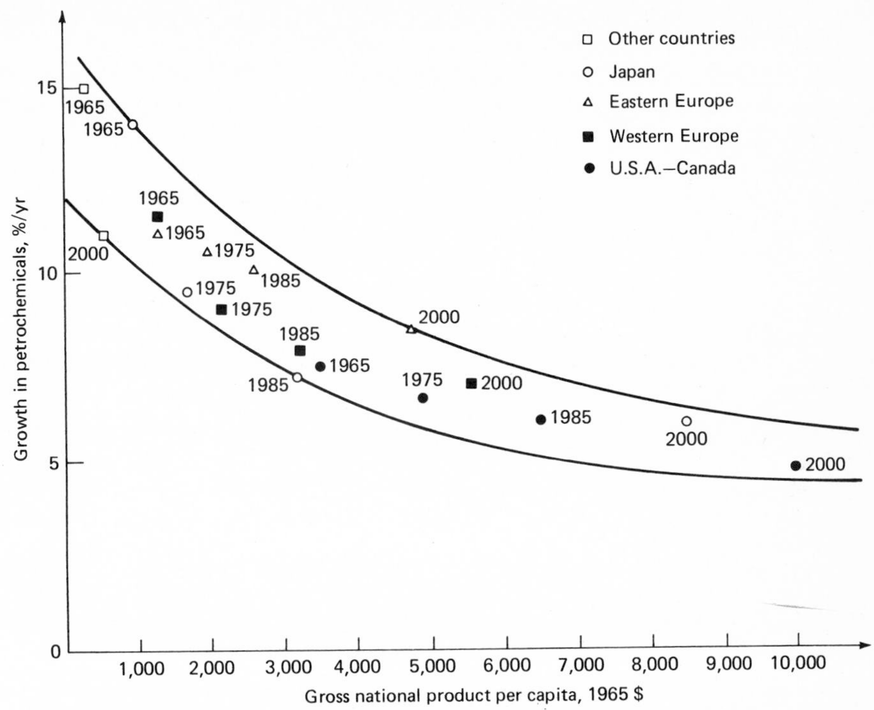

Fig. 1.9 The relation of GNP to demand—growth of petrochemicals in various countries.

of growth of chemical products and the per capita GNP is recommended for long-term forecasts, since such a relationship allows for progressive saturation of the market (Fig. 1.9).

1.4 CONCLUSION

Descriptions of the various available techniques show there is no truly quantitative method for a satisfying forecast of chemicals consumption. If by taking several of the methods simultaneously a satisfactory convergence is obtained, the result can of course be taken as relatively probable. However, it is impossible to forecast the different circumstances that can change the evolution of consumption. Even the most sophisticated mathematical models cannot arrive at reliable solutions for such complex problems.

However, market evaluation is necessary; even if the results cannot be taken as absolute, they do allow for determining the tendencies and thus the direction in which the manufacturer should turn.

Example 1

The Market for Benzene

The first part of this example consists of a study to determine world benzene production and consumption, with 1973 taken as the year of reference. In the second part, a projection of benzene demand is made over the period 1973–1980.

E1.1 ORGANIZING THE STUDY

A study of benzene in 1973 includes an identification of existing production capacity, plus an estimate of the demand.

E1.1.1 EXISTING BENZENE CAPACITY

The results are shown in Table E1.1. A distinction is made according to whether the raw material is coal or petroleum distillates. Subsequent categories of production depend on the process, for example,

Catalytic reforming of petroleum middle distillates

Treating pyrolysis gasoline by-produced in cracking naphtha for olefin production

Hydrodealkylating aromatic distillates

TABLE E 1.1 World Production Capacity for Benzene in 1973, 1,000 t/yr

Region	From Petroleum					
	Reforming	From Steam Pyrolysis	Hydrode-alkylation	Total	From Coal	Total
North America	3,536	710	1,932	6,178	398	6,576
South America	≃213	80	>0	>293	23	>316
TOTAL AMERICAN	≃3,749	790	>1,932	>6,471	421	>6,892
Western Europe	1,216	1,688	1,425	4,329	892	5,221
Eastern Europe	≃861	≃0	>72	>933	>1,593	>2,526
TOTAL EUROPEAN	2,077	≃1,688	>1,497	>5,262	>2,485	>7,747
Asia	≃402	>1,202	656	>2,260	612	>2,872
Africa	—	—	—	—	47	47
Australia	45	—	—	45	—	45
WORLD TOTAL	≃6,273	>3,680	>4,085	>14,038	3,565	>17,603

Benzene shown under the heading of pyrolysis gasoline is *primary,* i.e., already present and independent of subsequent treatment, whether that treatment is solvent extraction or hydrodealkylation of the total mixture. Consequently, the heading hydrodealkylation includes only *secondary* benzene, i.e., benzene obtained by removing the alkyl radicals from toluene, xylenes, or higher aromatics, either isolated from benzene or in a mixture with it. The source of these alkylated benzenes can be pyrolysis gasoline or gasoline produced by catalytic reforming.

E1.1.2 ESTIMATING THE BENZENE DEMAND

The growth in benzene depends on changes in the markets for its derivatives—cumene, cyclohexane, ethylbenzene, maleic anhydride, etc. (Figs. E1.1, E1.2, and E1.3). When data from such independent studies are summarized to show the trends in percentage of total consumption for the various applications over the years 1973, 1975, and 1980 for the United States, Western Europe, and Japan (Table E1.2), the worldwide growth in benzene needs is determined.

This method furnishes the primary approach to forecasting demand; its value rests on the validity of the detailed studies undertaken independently for each of the derivatives.

As was described in Sec. 1.3.2, it is possible to study the past growth for a given country and extrapolate this on logarithmic coordinates (Fig. E1.4). This method was shown to be dangerous, for it implies that an exponential growth continues; and the method is improved by modifying the growth

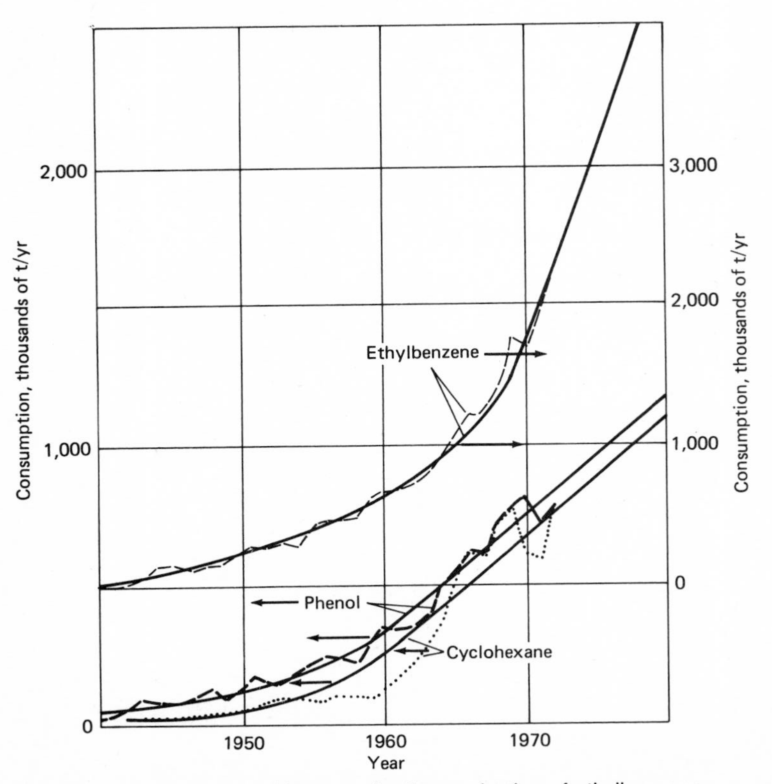

Fig. E1.1 Consumption of benzene for the production of ethylbenzene, phenol, and cyclohexane.

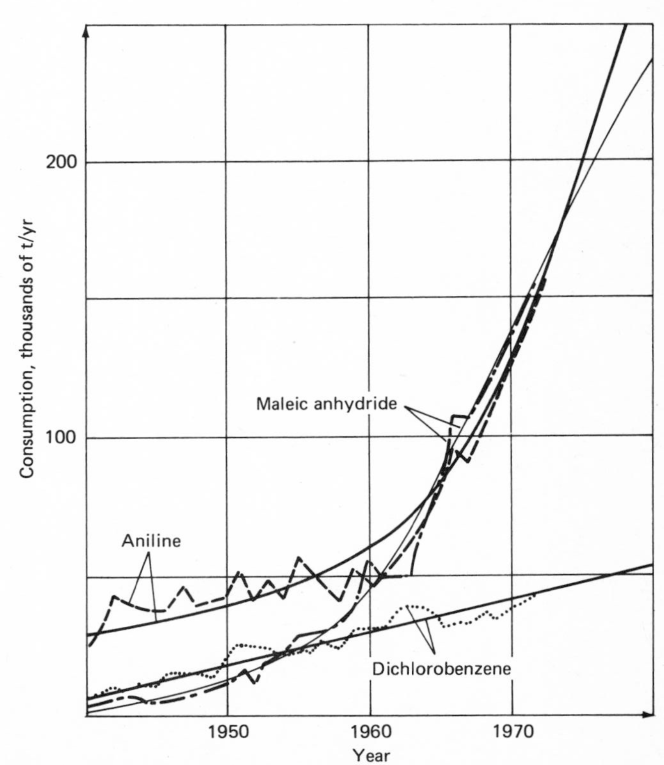

Fig. E1.2 United States consumption of benzene for the production of maleic anhydride, aniline, and dichlorobenzene.

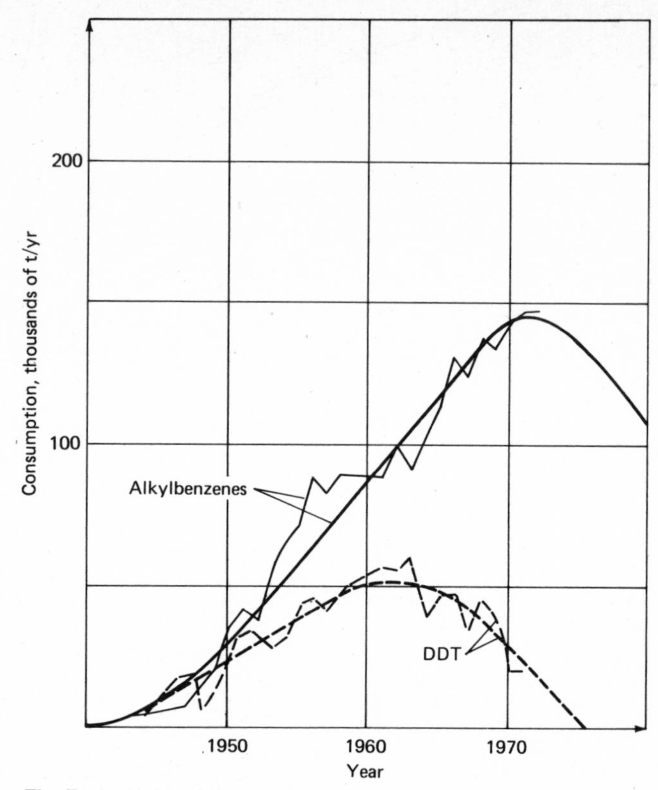

Fig. E1.3 United States consumption of benzene for the production of alkylbenzenes and DDT.

rate [Eqs. (1.3)–(1.7)]. Applying this technique to the case of Western Europe,

- The average annual growth from 1960 to 1973 = 14%.
- The average annual growth from 1974 to 1975 = 11.5%.
- The average annual growth of the GNP = 4%.

Calculating k Eq. (1.3) can be rearranged to

$$-kt = \ln \frac{r - r_\infty}{r_0 - r_\infty}$$

If time t is 2 years for 1974–1975, the rate of actual growth, r, is 11.5%, the rate of growth during the exponential period, r_0, is 14%, and the limiting rate of growth, r_∞, is taken as the growth of the GNP, or 4%. Thus,

$$-2k = \ln \frac{0.115 - 0.04}{0.14 - 0.04} = \ln 0.75 = -0.2877$$

TABLE E1.2 Benzene Consumed for Chemical Production, % of Total

	United States			Western Europe			Japan		
Consuming Chemical	1973	1975	1980	1973	1975	1980	1973	1975	1980
Alkylbenzene	2.5	2.0	1.0	5.0	4.0	3.0	3.5	3.0	2.5
Maleic anhydride	3.5	3.5	3.0	4.0	3.5	3.0	1.0	1.0	1.0
Aniline	3.5	4.0	4.0	5.5	6.0	6.0	1.5	2.0	2.5
Cumene	15.5	15.0	15.0	19.5	17.5	14.0	15.5	14.5	15.5
Cyclohexane	18.5	17.0	14.5	17.0	16.0	14.0	25.0	25.0	27.5
Ethylbenzene	51.0	53.5	58.0	45.0	49.0	57.5	51.0	52.5	49.5
Phenol (besides cumene)	1.5	1.5	1.5	0.5	0.5	0.5	0.5	—	—
Miscellaneous	4.0	3.5	3.0	3.5	3.5	2.0	2.0	2.0	1.5
TOTAL, %	100.0	100.0	100.0	100.0	100.0	100.0	100.0	100.0	100.0
Total demand, 1,000 t/yr	4,710	5,450	7,750	4,070	5,000	8,700	1,920	2,500	3,900
Average annual growth rate, %		7.2	7.2		12.1	11.6		11.2	9.4

and $$k = 0.1438$$

Then applying Eq. (1.7) to calculate the demand D in 1980 from the demand D_0 in 1975,

$$D = D_0 \exp\left[0.04 + \frac{0.14 - 0.04}{5(0.1438)}\left(1 - e^{-0.1438(5)}\right)\right]5$$

Thus, if D_0 for Western Europe is 5 million tons per year in 1975 (see Table E1.2), the projected demand D for 1980 is 8.7 million tons per year.

Similarly, in the case of the United States,

$$r_0 = 8.7 \qquad r = 7.2 \qquad r_\infty = 4$$

$$\frac{r - r_\infty}{r_0 - r_\infty} = 0.68085$$

$$\ln\left[\frac{r - r_\infty}{r_0 - r_\infty}\right] = -0.3844$$

$$k = 0.1922 \qquad \frac{r_0 - r_\infty}{k} = 0.2445$$

$$t = 2 \text{ and } 5$$

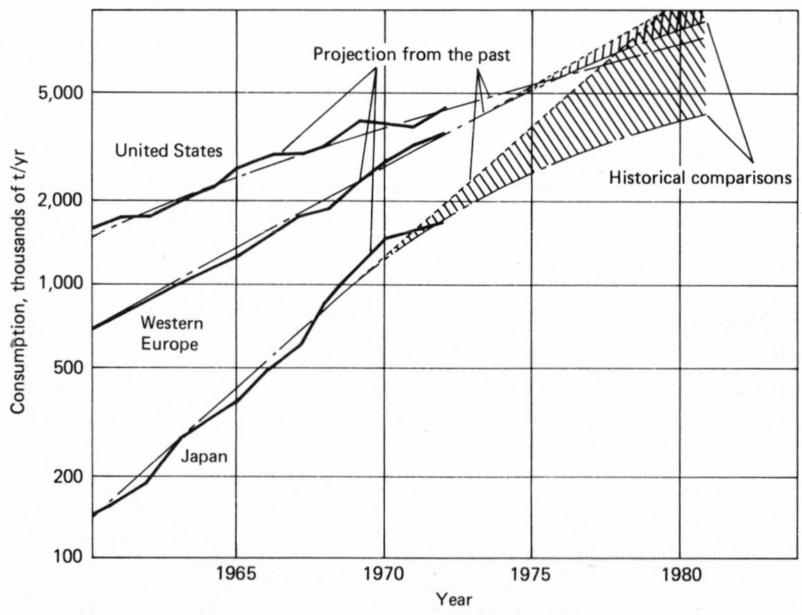

Fig. E1.4 Consumption of benzene in United States, Western Europe, and Japan: forecast.

so that if $D_0 = 5.45$ million, then $D = 7.741$ million or about 7.75 million, for a production in 1980 of about 7.75 million tons per year.

This same method ($r = r_o$) cannot be used for Japan, where the growth of benzene consumption has been very rapid and is in the exponential period. It is necessary to assume that the demand will evolve toward the the poorest rate of growth in a manner analogous to the evolution seen in the United States and Western Europe (historical comparisons, Sec. 1.3.3). Thus for 1980, the demand in Japan is projected at around 4 million tons per year (Fig. E1.4).

E1.1.3 THE BALANCE BETWEEN PRODUCTION AND CAPACITY

The study summarized in Table E1.3 permits an examination of the overall operating ratio in 1973 and its projection to 1975 for the United States, Western Europe, and Japan. The ratio between production and available capacity is first of all examined by means of data on reported plant capacities and announced projects. This ratio, expressed as a percentage, is the average operating ratio of the units. From the point of view of such an examination, Table E1.3 shows

The importance of steam pyrolysis as a source of benzene, especially in Western Europe and Japan.

The installation of new units for hydrodealkylation of pyrolysis gasoline.

An increase in the average operating ratio, which attains, especially in Japan, values higher than 90%, indicating a shortage in production capacity. Over a long period of time, an operating ratio of 80% generally corresponds to a stabilized relation of supply to demand.

Next, production should be related to the type of consumption, whether internal consumption or external sales. The balance of exports versus imports for each region (Table E1.3) shows that there is most often a deficit (numbers in parantheses). From a worldwide view, imports are necessary for the United States and Western Europe. Deficits appear in these areas to the amounts of 110,000 and 130,000 tons/yr, respectively, in 1973, and should reach 340,000 and 320,000 tons/yr, respectively, in 1975. Although Japan exports 120,000 tons/yr of benzene in 1973, it should need to import about 170,000 tons/yr in 1975, and continue importing from then on.

Finally, Table E1.3 introduces a general balance (bottom line) to show the deviation between results of the calculations and the real internal consumption. Thus, in the United States in 1973, this deviation is 40,000 tons/yr.

TABLE E1.3 Production-to-Capacity Ratio for Benzene

a. United States

Source	1973 Cap.	1973 Prod.	1973 %	1975 Cap.	1975 Prod.	1975 %
Catalytic reforming, direct	3,243	2,550	78.5	3,245	2,770	85.5
Pyrolysis gasoline, direct	582	485	83.0	582	520	89.5
Hydrodealkylation						
of pyrolysis gasoline compounds	349	320	91.5	349	320	91.5
of catalytic reforming compounds	1,583	1,050	66.5	1,583	1,220	77.0
Total from petroleum	5,757	4,405	76.5	5,757	4,830	84.0
Total from coal	332	275	83.0	332	290	87.5
Total	6,089	4,680	77.0	6,089	5,120	84.0
Balance exported (or imported)		(70)			(340)	
Internal consumption		4,710			5,460	
Balance		40			0	

b. Western Europe

Source	1973 Cap.	1973 Prod.	1973 %	1975 Cap.	1975 Prod.	1975 %
Catalytic reforming, direct	1,216	1,015	83.5	1,335	1,125	84.5
Pyrolysis gasoline, direct	1,688	1,250	74.0	2,080	1,560	75.0
Hydrodealkylation	625	480	77.0	835	710	85.0
of pyrolysis gasoline compounds	800	645	80.5	800	700	87.5
of catalytic reforming compounds						
Total from petroleum	4,329	3,390	78.5	5,050	4,095	81.0
Total from coal	892	550	61.5	892	600	67.5
Total	5,221	3,940	75.5	5,942	4,695	79.0
Balance exported or (imported)		(130)			(320)	
Internal consumptions		4,070			5,015	
Balance		0			0	

c. Japan

Source	1973 Cap.	1973 Prod.	1973 %	1975 Cap.	1975 Prod.	1975 %
Catalytic reforming, direct	289	215	74.5	289	270	93.5
Pyrolysis gasoline, direct	1,188	960	81.0	1,188	1,130	95.0
Hydrodealkylation						
of pyrolysis gasoline compounds						
of catalytic reforming compounds						
Total from petroleum	2,097	1,685	80.5	2,097	1,960	93.5
Total from coal	554	355	64.0	554	370	67.0
Total	2,651	2,040	78.0	2,651	2,330	88.0
Balance exported or (imported)		120			(170)	
Internal consumptions		1,920			2,500	
Balance		0			0	

Cap.: capacity, 1,000 t/yr.
Prod.: production, 1,000 t/yr.
%: production-to-capacity ratio × 100.

E1.1.4 ESTIMATING NEW BENZENE CAPACITY FOR 1980

This study is based on the following assumptions:

- Starting from 1975, no allowance is made for an increase in benzene production from reforming and hydrodealkylation, for which capacities are arbitrarily assumed constant.
- Primary benzene from pyrolysis gasoline is estimated from the increase in capacity of steam pyrolysis for ethylene production, based on data from outside this study.
- The large demand for benzene suggests a slight increase in coal-tar benzene produced from coking.

A total is then made of the production coming from reforming and hydrodealkylation units existing in 1973 and of the estimated production from new steam pyrolysis and coal-tar capacity for the period of 1973–1980. The difference is calculated between this total and the internal demand (whose growth is calculated from the predicted growth of benzene derivatives in Table E1.2), and that difference indicates the new capacities of reforming and hydrodealkylation plants to be built, aside from foreign trade (Table E1.4). Since from 1975 on, operating ratios should reach their limits, new demand must be met by new installations and not an increase in production from existing units.

E1.2 CONCLUSION

The following growth in capacities are forecast between 1975 and 1980 for reforming and associated units, as well as hydrodealkylation:

United States	1,430,000 t/yr
Western Europe	2,730,000 t/yr
Japan	270,000 t/yr

These calculations cannot take into account economic crises and socioeconomic accidents. Thus the preceding calculation, by adopting 1975 instead of 1973 as the year of reference, leads to lower results, because production failed to grow and even declined during 1974–1975.

TABLE E.I.4 Forecast of New Benzene Capacity,* 1,000 t/yr

a. United States

Benzene Source	Capacity	Consumption		
	1973	1973	1975	1980
Catalytic reforming	3,243	2,550	2,770	2,770
Pyrolysis gasoline	582	485	520	1,690
Hydrodealkylation	1,932	1,370	1,540	1,540
TOTAL	5,757	4,405	4,830	6,000
Total from coal	332	275	290	320
TOTAL	6,089	4,680	5,120	6,320
Balance exported or (imported)		(110)	(340)	
New capacity				1,430
Internal consumption		4,710	5,460	7,750

b. Western Europe

Benzene Source	Capacity	Consumption		
	1973	1973	1975	1980
Catalytic reforming	1,216	1,015	1,125	1,125
Pyrolysis gasoline	1,688	1,250	1,560	2,710
Hydrodealkylation	1,425	1,125	1,410	1,410
TOTAL	4,329	3,390	4,095	5,245
Total from coal	892	550	600	725
TOTAL	5,221	3,940	4,695	5,970
Balance exported or (imported)		(130)	(320)	
New capacity				2,730
Internal consumption		4,070	5,015	8,700

c. Japan

Benzene Source	Capacity	Consumption		
	1973	1973	1975	1980
Catalytic reforming	289	215	270	270
Pyrolysis gasoline	1,188	960	1,130	2,300
Hydrodealkylation	620	510	560	560
TOTAL	2,097	1,685	1,960	3,130
Total from coal	554	355	370	500
TOTAL	2,651	2,040	2,330	3,630
Balance exported or (imported)		120	(170)	
New capacity				270
Internal consumption		1,920	2,500	3,900

*These forecasts allow for the growth in capacity of steam pyrolysis for ethylene, as well as for benzene from coal, but not for growth in foreign trade.

Chapter 2

Elements of Economic Calculation

Profitability calculations require a comparison between revenues and costs for the manufacturing project. Revenues come from sale of the manufactured products and can be forecast from a study that has established the probable market volume and an estimated price. The costs are identified as *fixed,* as *labor,* or as *variable.*

Fixed costs accrue from the existing installation, whether or not it is in production. They represent compensation for the invested capital, expressed as depreciation, as well as the cost of maintaining the plant and insuring it. They depend on the size of the investment.

Labor identifies the compensation paid as salaries and wages to the people who operate and maintain the plant, as well as their supervisors. Although operating labor may be curtailed, if a plant is shut down, labor requirements are generally independent of the rate of production, so that labor costs are a function of the investment.

Variable costs comprise those expenses incurred in production, i.e., the costs for raw materials, utilities like fuel and water, consumed chemicals and catalysts, and so forth. They vary directly with the amount of production.

2.1 THE INVESTMENT

The amount of capital investment corresponds to the total money spent to carry an industrial project from preliminary technical and economic studies to actual start-up of the plant. The source of these funds depends somewhat on the size of the project relative to the size of the investing company. They can be appropriated from the company's working capital or they can be borrowed. Usually, investment funds are drawn from a combination of working capital, fixed assets, and loans, according to the company's financial policies of debt liquidation and investment.

2.1.1 BATTERY-LIMITS INVESTMENT

The battery-limits investment amounts to the cost of the manufacturing installation, i.e., of the necessary equipment, preparation of the site and construction, with the cost of engineering studies sometimes also included (although this latter is better kept separate). Usually, the *battery limits* represents a geographic frontier over which are imported

Raw materials

Utilities (electricity, water, fuels, refrigeration, etc.)

Required chemicals, catalysts, and solvents

Also, the battery limits represents the frontier over which are exported

Manufactured products

By-products, including fuel gas, tar, and various residues

Utilities to treatment for recycle

Usually, the only storage included in the battery limits is surge capacity for intermediate products; storage capacity for raw materials and finished products is included in the *off-site* investment.

2.1.2 OFF-SITES

Off-sites investment covers the cost of

Units for the production and distribution of utilities (steam boilers, cooling towers, etc.)

Roads

General offices, repair shops, garage, warehouse, laboratory, cafeteria, etc.

Waste-water treating facilities and sewage system

Generally, off-sites can be described as all those things, necessary to the functioning of a production unit, that can be shared with other production units in a large plant. Thus off-site storage units make it possible to have on hand sufficient reserves of raw materials, feedstock, catalysts, etc., to avoid any interruption in supply, as well as to store the finished product until it can be delivered to other units or to the customers. That part of the general services and storage units earmarked for a particular unit depends on the internal organization of the operating company. For new plants, it is easy to divide the off-site investment among the several units according to a prescribed rule, such as according to the turnover of finished products or according to the relative investments for the units.

When a new unit is to be located within an existing industrial site, it is necessary to take into account the existing off-site equipment and its capacity relative to the possibility of adding new equipment or increasing capacities of the existing off-site units.

Exact calculations can be made for specific projects, but when a process unit is destined for an unknown site, a precise calculation of the off-sites is not possible. Since the routine utilities (electricity, steam at different pressures, cooling water, etc.) are the most important off-site facilities, it is customary to charge for such utilities at their cost. On this basis, it is reasonable to exclude both the utilities and their distribution systems. On the other hand, if unusual utilities are needed, these are normally included in the battery-limits investment. If, for example, low temperatures are needed, the refrigeration unit is included, either in the battery limits or in a general installation tied directly to the process unit under study.

2.1.3 ENGINEERING FEES

When a specialized engineering company undertakes the construction of a plant, that company plays a role analogous to an architect; it enters at a preliminary stage of the project by establishing a proposal, which is compared to proposals of other engineering companies. When one company is chosen on the basis of such proposals, it agrees to accept a degree of responsibility for a number of services: design calculations, detailed specifications for the equipment, layout of equipment in the plant, soliciting bids from specialized equipment suppliers, coordination and control of construction, and start-up of the unit. One engineering company may even subcontract a part of the studies and work to other engineering companies.

Engineering fees represent the payment for these services. They can vary according to the amount and the quality of the work required for a project, as well as according to the nature and capacity of the planned installations. They are not proportional to the capacity of the units. For a first approximation, they can be estimated according to the relative investment in the various types of facility, whether process unit, utility, storage, or general service.

2.1.4 SPARE PARTS

Any new installation should include a supply of spare parts, in order to be prepared for any mechanical problems that might arise. These spare parts can represent considerable investment when a high rate of corrosion or wear is foreseen or when the breakdown of a vital piece of equipment could cause a shutdown, with disastrous results on the production and consequently on the profits of the project.

The problem is not acute in the industrialized countries, and the investment for spares is generally small and even negligible. However, rotating equipment, such as pumps and small compressors, is often furnished with an installed spare, and this is equivalent to including an investment for spare parts within the battery-limits investment.

2.1.5 ROYALTIES

The company that has done the research and pilot-plant development and has carried a process or product through to commercial manufacture holds certain rights that it can sell to an eventual user. There are two types of payment or royalty for such rights: *paid-up* and *running.*

2.1.5.1 Paid-up Royalties

A lump-sum royalty can be paid to the owner of a process, generally depending on the size and installed capacity of the manufacturing unit for which the rights are granted. The fee can vary according to the technology from a few dollars per installed ton of capacity for refinery processes to hundreds of dollars per installed ton of capacity for special products. Payments may be scheduled so as to be smaller as plant capacity is larger.

The actual payment is not generally made in one lump sum, but in parts determined by the contract (for example, 25% at the time of signing the contract, another 25% during construction of the plant, and the remaining 50% after the acceptance tests of the plant).

2.1.5.2 Running Royalties

Sometimes royalties are paid as annuities for the duration of the depreciation of the plant or as long as the patents protecting the invention are in effect. Most often, these running royalties are related to the actual production and not to the installed capacity; and the payments are thus dependent on variations in the plant's production and subject to the hazards of the market.

As a first approximation, the following rule holds: a paid-up royalty is equivalent to the running royalty accumulated over 10 years.

Intermediary royalty arrangements include initial cash down payment followed by a continuing payment over a specified period and according to a calculation procedure defined by the contract.

2.1.6 PROCESS DATA BOOK

During the period when engineering studies are being carried out to identify the best of the available technology, potential licensors will usually furnish technical and economic data for use in the analysis. When a given technology has been selected, however, and agreements made, the licensor of the selected process must give the licensee a document containing enough information about the process to allow the selected engineering company to undertake its own studies and work. Transmittal of this document, called the *process data book*, usually entails a payment fixed by contract.

Also, the licensor provides another document called the *operating manual*, which complements the process data book, but which is normally used at the time of start-up. Given to the engineering company, the operating manual permits verification of certain calculations. The cost of producing this manual is part of the payment tied to the delivery of the process data book.

2.1.7 LAND COSTS

The purchase of land on which to build a plant is included in the investment. Land costs depend on the selected site and thus vary considerably from one project to another. Local advantages may in some cases make a site desirable, whereas constraints can in other cases dissuade development. Without knowing the details, it is impossible to include these figures in the estimating calculations. However, one can differentiate between the following:

1. Projects intended for an industrialized site, where land has been purchased for previous constructions.
2. Projects intended for a new site (*grass roots*), where it is necessary to take into account numerous local considerations in addition to the acquisition and preparation of the land.

2.1.8 INITIAL CHARGE OF CATALYSTS, SOLVENTS, AND CHEMICALS

After construction is complete, there still remain costs for catalysts, adsorbents, solvents, etc., that must be placed in the new plant before it can be brought to production. These products make up the initial charge. When they are used up within the depreciated life of the plant, their cost can be taken as part of the operating cost and drawn from working capital. When the rate of consumption is so low that the initial charge lasts as long as the depreciated life of the plant, the cost of the initial charge should be taken as part of the investment.

Certain precautions are necessary to an analysis of the expense of initial charges. Thus a complete spare charge of a catalyst or adsorbent might be purchased and kept on hand where a possibility for rapid deterioration is foreseen. Also, part of the cost of the initial charge of material may be recovered as that charge is replaced, and the recovered value can be considered as income derived from the feedstock, i.e., from the working capital. This latter situation often arises with precious-metal catalysts that are of considerable value, even when spent.

2.1.9 INTEREST DURING CONSTRUCTION

A certain period of time must elapse between the decision to construct a plant and the start-up of that plant. For a simple unit, this construction delay should not exceed 12 to 15 months, but larger plants employing complicated processes require at least two years.

During this period it is necessary to release a large part of the investment, in order to finance feasibility studies, design studies, civil engineering work, orders of bulk construction materials, equipment purchases, installation and erection of equipment, and so forth. The problem becomes that of supplying a progressively increasing amount of money without obtaining any revenue. This money can come either from a combination of direct loans and the company's own resources or solely from the company's own resources. If borrowed, use of the money must be paid for in the form of interest, from the moment of the loan; if the company's own funds or plant funds are immobilized, it is equally necessary to pay for use of the involved sums on the basis of a fictive interest rate that reflects the financial policy of the company and its management methods. (For discussion, see Sec. 2.2.2.2.)

These interests are said to be *intercalated* and depend on the current prime lending rate of interest. They are sometimes neglected in making evaluations, because the sum is not exact, compared to the investment, which is known.

2.1.10 START-UP EXPENSES

In the course of starting up, the plant is put into order, specifications of the final products are verified, the normal theoretical capacity is shown to be feasible, required performances are achieved for various systems, such as catalysts, solvents, and adsorbents, and control of the production is shown to be according to the operating manual.

Technical responsibility for the start-up is generally in the hands of the process licensor, with assistance by personnel from the company who will normally operate the unit, as well as responsible people from the engineering company that has studied, designed, and built the plant.

The expenses for these services are carried by the operating company, which pays the licensor and the engineering company for their assistance. In addition, various adjustments or modifications carried out during the start-up period prevent the unit from immediately producing product to the required specifications, so that neither the consumed raw materials, chemicals, and utilities nor the labor costs for plant personnel nor the costs of services are compensated by the sale of finished products. Accordingly, all these costs are designated as *start-up expenses.*

The time necessary to get a unit working depends on its complexity, and the corresponding expenditures vary from relatively modest (for starting up a simple unit built on a well organized existing site) to considerable sums (for complicated installations, such as a steam pyrolysis unit and its satellite operations of hydrogenation, butadiene extraction, etc.). Thus start-up expenses are often estimated as months of operating costs beyond the capital costs, i.e., in consumption of raw materials, chemicals, utilities, and labor. In the event that the finished product has value or can be recycled when not up to specification, the costs of raw material may be excluded from the estimated start-up costs.

2.1.11 WORKING CAPITAL

Money set aside as working capital is not the same kind of investment as the battery limits, off-site, engineering fees, spare parts, land costs, start-up costs, and other investments treated in this chapter. Working capital corresponds to a temporary tying up of funds necessary for operating the installation, and contrary to the other types of investment, calls for an eventual return that is close to its initial value, without taking into account depreciation or possible changes in currency values during the period that these funds are tied up.

For economic calculations, working capital, f, is considered as a nondepreciable cost on year 0 and is included with the investments. It can be recovered at any given time after n years, which corresponds to the life of the unit. If discounting is not taken into consideration, it retains its initial value; if discounting is applied, it reaches a present worth of

$$\text{Present worth} = \frac{f}{(1 + i)^n}$$

where i is the rate of interest compounded annually.

Working capital consists of a current asset comprising

- The value of the normal supply of raw materials, chemicals, catalysts, solvents, etc.
- The value of the products contained in the processing units
- The value of the normal stored contingent of finished products
- The expenses that would be incurred in continuing to (1) supply product and receive raw materials according to contract and (2) pay plant personnel

normally employed in operating and maintaining the installation, in the event of a temporary shutdown

The value of spare parts, packaging materials, etc.

The cash requirements for bridging delays in payment for sale of products

In practice, the required working capital can be estimated in various ways, one convenient solution consisting of the monthly operating cost minus the financing charges.

Working capital may also include a collection of materials, ranging from precious-metal catalysts to carbon black, that can be considered as temporarily stored products. This is particularly true for platinum and other noble-metal catalysts for catalytic reforming or xylene isomerization; when the catalyst is deactivated the precious metal can be recovered by means of certain treatments, allowing for some loss. This item, which can form a large part of the working capital, differs from current assets that depend on incurred risks, since it corresponds to an amount of money that can be precisely determined on the basis of purely technical considerations.

2.2 FIXED COSTS

These are the annual charges accruing from the investment for the plant. They are called *fixed* because they are constant for a given capacity, no matter what the production volume. In case of a long shutdown, for example, the fixed costs are not eliminated but continue to appear on the financial balance sheet.

These costs differ according to policy, between companies and even between different plants of the same company. They include the following principal items: depreciation, interest, maintenance, taxes, insurance, and overhead.

2.2.1 DEPRECIATION

2.2.1.1 Purpose

Depreciation is an accepted means for recovering invested capital, not including working capital. Its justification is based on progressive obsolescence of the investment, which arrives the following three ways:

1. During operation, progressive deterioration and wear of materials of construction and operating equipment bring about a lower production efficiency, which is reflected in rising operating and maintenance costs and more and more frequent nonscheduled shutdowns. Thus the value of the plant as a production tool decreases.

2. Competing new plants employing improved processes place older units in difficulty from the point of view of profitability or quality of their products. Also, larger plants made possible by technical improvements and expanded markets can take advantage of economies of increased scale and similarly force older plants of modest size into a less profitable frame of operation.

3. New products better adapted to the demand may invade the market and cause the products of older plants to face sales difficulties. This is particularly true of finished products, such as rubber, plastics, synthetic fibers, and detergents.

Among these three types of obsolescence, only the first, wear and deterioration, can be estimated with any precision; the two others, obsolescence of processes and products, are difficult to foresee. However the combined effect in reducing the value of plants leads industrial companies to depreciate the invested capital, i.e., to recover the investment during the plant's effective period of activity. The time allowance for depreciation is thus tied to the evaluation of the risks incurred and varies according to the type of production. The notion has been accepted legally, and taxable profits are calculated after deduction of the depreciation.

2.2.1.2 Calculations

Of all the elements that make up an investment, only the working capital is not depreciable. To calculate depreciation, it is necessary to consider (1) the salvage value and (2) the life of the plant.

2.2.1.3 Salvage Value

The equipment in a plant can be converted to other uses or sold; and its value at the time of conversion or sale depends on the cost of replaced equipment or the selling price. This value can be determined during the period of depreciation or after the conversion or sale takes place.

Accounting methods that reproduce the actual cash movements at the times they take place must calculate the salvage value only at the year of resale, or if no resale takes place, at the end of the depreciation period. If equipment is sold, such accounts should show any gain in dollar value and the associated taxes; if the equipment is not sold, a time limitation for the cash flows should be established.

However, accounting methods are used in a much more rudimentary manner in process evaluation. The purpose is usually that of studying comparable technology in theoretical installations; and the primary concern is to define identical conditions for the different processes, particularly, with respect to time. Thus it might be appropriate for purposes of calculation to estmate the plant's value at an arbitrary date. To simplify the calculation, this value could be taken as the fraction of the capital not yet depreciated—or even zero, in the event the equipment is totally depreciated.

2.2.1.4 Life of the Plant

As long as a unit is economically viable and meeting costs of maintenance and repair, as well as market demands, it can continue to function. Thus it is conceivable that the life of a plant can be longer than the period of depreciation, with or without supplementary investment, changes, or additions in equipment. On the other hand, the plant may be shut down even before the depreciation is complete.

In actual practice, the period considered for an investment project is tied to the actual life of the equipment and not to the life of the scheduled depreciation. In evaluating a potential investment project, however, the calculations may be simplified by assuming that the actual life of the plant coincides with the scheduled period of depreciation and the salvage value determined only at the end of that period.

There are several methods of calculating depreciation, including straight line, declining balance, sum-of-years digits, and sinking fund.

2.2.1.4a Straight-Line Depreciation

If I is the initial value of a depreciable item and n the duration in years of its depreciated life, the annual depreciation will be constant and equal to

$$A = \frac{I}{n}$$

According to accepted simplifying assumptions, the salvage value of the unit I_r is taken as zero at the close of the depreciation period, so that the rate of depreciation, a_l, is given by the equation:

$$a_l = \frac{1}{n}$$

and is constant for all the years, allowing for complete depreciation of the unit by the nth year.

2.2.1.4b Declining-Balance Depreciation

With this method, depreciation over each year is taken as a fixed fraction of the value of the unit at the beginning of that year. If a_d is this fixed fraction, the depreciation at the first year is $a_d I$, and the value of the equipment is $I(1 - a_d)$ at the end of the first year, $I(1 - a_d)^2$ at the end of the second year, and $I(1 - a_d)^n$ at the end of the nth year. If the salvage value corresponds to the investment value not yet depreciated, this relation shows salvage value I_r as

$$I_r = I(1 - a_d)^n$$

And the rate of depreciation predicated on a salvage value is given as

$$a_d = 1 - \left(\frac{I_r}{I}\right)^{1/n}$$

Under these conditions, the annual depreciation A_p for the year p is

$$A_p = (I - I_r)(1 - a_d)^{p-1}a_d$$

where p is always smaller than n.

Calculated in this manner, depreciation is faster at the beginning of the life of a project and is not complete at the year n (see Fig. 2.1). This method has two advantages over the straight-line method. First, it allows for recovery of a greater fraction of the investment during the early years of a project and for thus reducing risks tied to rapid obsolescence; and second, relative to the straight-line method, it diminishes gross income during the early life of the project to thus reduce taxes in a manner favoring reinvesting the recovered capital. However, tax rules are often applied to limit the depreciation rate a_d. Thus until 1954 in the United States, values of a_d were limited to those that would give no more than 150% of the linear depreciation. After 1954, the allowed values were increased to those that would give double of that of linear depreciation, so that the method came to be called *double declining balance.*

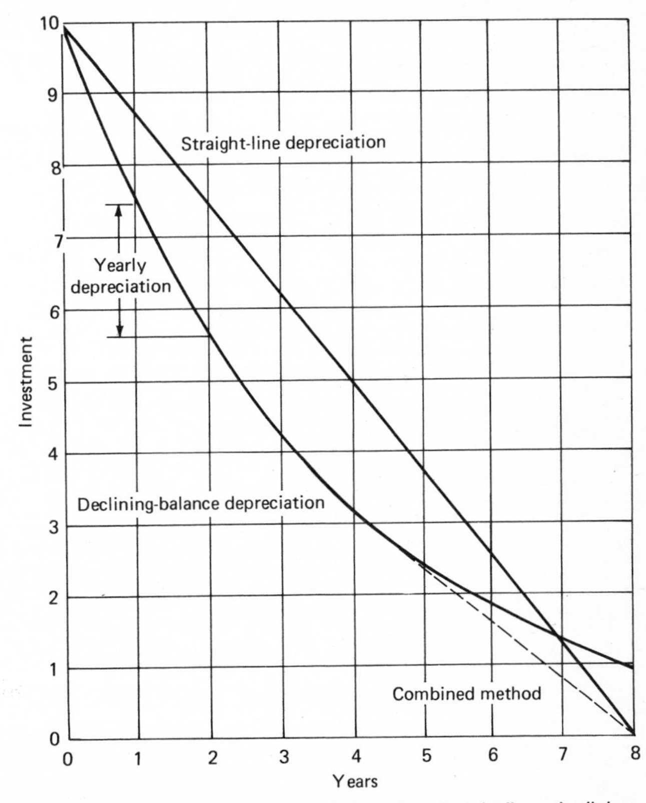

Fig. 2.1 Comparison of depreciations by straight-line, declining-balance, and combined methods.

Practices vary from one country to another, and even from one type of equipment to another, so that analyses of foreign investments should take local rules into account. In France, for example, Article 37 of the law of December 28, 1959, and a decree of May 9, 1960, foresees a declining-balance system according to the life of the plant:

Less than 3 years $a_d = 1a_l$
From 3 to 4 years $a_d - 1.5a_l$
From 5 to 6 years $a_d = 2a_l$
More than 6 years $a_d = 2.5a_l$

When a rate greater than linear depreciation is adopted for a_d, the salvage value can never be zero. This complication can be eliminated by assuming a fixed factor during the first years of a project and ending up with a linear depreciation, i.e., the *combined method.*

Legally, such a combined method can be applied, providing the rate of straight-line depreciation that follows the declining balance does not exceed the rate which would be calculated for straight-line depreciation over the whole life of the project. Thus, if the combined method is to be used for an 8-year project, it is possible to assume 5 years of declining-balance depreciation at a fixed factor of $a_d = 2(a_l) = 2(\frac{1}{8}) = 0.250$, and then to follow this with straight-line depreciation during which the rate a_l equals $(I_r)_5/3$ or $I(1 - a_d)^n/3 = I\ (1 - 0.250)^5/3 = I\ (0.0791)$. (See Fig. 2.1.)

The above methods are legal in France and other foreign countries. Still other methods, such as the sum-of-years digits and sinking fund, can be used in the United States.

2.2.1.4c Sum-of-Years-Digits Depreciation

This method is related to the constant rate method through a calculation technique. The annual depreciation in the year p is

$$A_p = \frac{n - p + 1}{1 + 2 + \cdots p + \cdots n}(I - I_r) = \frac{2(n - p + 1)}{n(n + 1)}(I - I_r)$$

Since this method offers higher rates of depreciation during the early life of the project, it is used frequently in the United States.

2.2.1.4d Sinking-Fund Depreciation

This method consists of operating at a constant rate, while discounting the reimbursed sums. The annual depreciation, constant and discounted, is then

$$A = \frac{i}{(1 + i)^{\mathrm{n}} - 1}(I - I_r)$$

where i is the rate of discounting used (see Sec. 2.2.2).

At the end of the period of depreciation, the total of the sum recovered and discounted should be equal to the depreciable capital. Since the sinking-fund method leads to a less rapid recovery of the investment than linear depreciation, it increases the initial risk of the investor and is rarely used.

Not only do the rules of authorized depreciation vary from one country to another, they also differ according to the kind of installation (see Table 2.1).

2.2.2 INTEREST

Financing for a project is represented by a collection of sums equal to the total investment. These sums can either be borrowed or come from funds already at the disposal of the investor.

In practice the distinction between borrowed capital and available capital is often delicate. A company with a certain standing in the industry may assure financing from its own resources, which consist as much of its own capital as of loans of various kinds and origins. The company treasury books all capital outlays and capital requirements accountable to the operation of existing plants, and in the final analysis provides the sums and advice relating to policies on new construction and new development. Under these conditions, it is impossible to be precise about the financing for one specific project among numerous others envisaged by the company; instead, the total sum required is assumed to be borrowed from the company's general treasury.

When a project involves a smaller company or a company formed especially to manage the installation, the sources of capital are differentiated more easily. In a general way, three types of financing may be distinguished: actual loans obtained directly from accepted loan organizations; working capital or funds

TABLE 2.1 Typical Depreciation Periods for Different Types of Plants

Plant Type	Depreciation Period, yr
Petrochemical units	6–8
Petroleum refineries	8–10
Utility-generating plants	12
General installations	12
Mineral processing plants	15
Industrial structures	20
Commercial or residential structures	20–50
Furnishings	10
Tools and equipment	5–10
Automobiles and rolling machinery	4–5

recognized as such; and capital provided by the general treasury as an allocation or fictitious loan.

Whatever the means of obtaining it, such capital requires compensation. In the case of actual loans, compensation is made through the payment of interest; in the case of working capital and general treasury funds, compensation is accounted for through the notion of discounting.

2.2.2.1 Borrowed Capital

Borrowed money is repaid by means of interest, which can be simple, compound, or average.

2.2.2.1a Simple Interest

Simple interest is a constant payment for the use of a unit of capital over a unit of time, the year. At the end of n years, the sum to be reimbursed, S_n, for the loan of capital, S_0, at the simple-interest rate i is given by

$$S_n = S_0(1 + i)n$$

When the loan is for less than a year, simple interest is proportional to the number of days of the loan.

2.2.2.1b Compound Interest

Simple interest yields the same amount of money, whether paid partially at the end of each time period or totally at the close of the loan. When it is paid at the end of the loan, the lender experiences a loss, because interest-money paid at the end of intervening time periods could meanwhile have been used otherwise. Compound interest is used to mitigate such losses; it assumes that all interest not returned at the end of each payment period acquires its own interest as an integral part of the loaned capital for the remainder of the loan period.

Under these conditions, the sum S_n due at the end of n unit periods for a loan of S_0 is given by $S_n = S_0(1 + i)^n$, where i represents the rate of interest.

Generally, the year is chosen as the unit time for expressing compound interest rates, but shorter time periods, such as a month or even a day, are often required for short-term loans. Since it is necessary to compare the rates given for such different time bases, the year is adopted as the reference unit of time. This leads to one of the following:

A rate in terms of an annual interest compounded for the appropriate fraction of the year (i_r),

A rate in terms of a nominal annual interest calculated as the number of periods in a year times their interest rate (i_a). This is the usual method of comparison.

Whether compounded annually, monthly, or daily, compound interest is the most employed by industry. In certain situations, infinitessimal compounding periods are used, leading to continuously compounded interest. Under these conditions, an instantaneous interest, i_a, yields

$$S_n(\text{years}) = S_0 \exp i_a n$$

where S_n is the capital at the end of n years, and S_0 is the original capital loan. The equation for the corresponding annual compound interest i_r is given by

$$i_r = \exp i_a - 1$$

2.2.2.1c *Average Interest*

When straight-line depreciation is used for the calculations, it is sometimes useful to apply a straight-line method for calculating interests. This leads to the introduction of an average annual rate of interest, i_{ave}, as given by

$$i_{\text{ave}} = i_r \frac{n + 1}{2n}$$

where i_r = compound interest
n = number of years in the depreciation period
and where i_{ave} is applied to the original investment S_0.

Both working capital and construction loans involve interest rates that can be converted to annual interest. Since the time periods are relatively short, the practiced interest rates are higher than those for loans on the investment.

2.2.2.2 Discounting

The notion of *discounting* is called on to define the minimum cost relating to the total capital investment for the various projects of a company. When working up studies it is not possible to determine which part of the investment should be self-generated and which part borrowed, since the available funds are not specifically allocated to a particular project. If in a particular case it is possible to segregate capital borrowed for the overall investment and establish a unit cost for that capital in relation to the going interest rate, discounting is used to define the similar cost in relation to the company's own funds. It is within the framework of a profitability analysis (Sec. 2.6) that this concept is applicable under a variety of forms. When a business ties up its resources to finance a project, its capital is not working and must be compensated for on account of

- Inflation
- The interest that might be earned if the funds were used for outside investment

The cost of borrowed funds generated by the company's treasurer in conjunction with the company's overall cash flow requirements

Income from projects that might either show a quicker payoff or offer more diversification

Consequently, an investment cannot be regarded simply as disposing of a certain sum to be recovered in n years of depreciation. Instead, the capital that has been committed to a project should be compensated for at a certain annual rate called the *discounting rate,* i.e., receiving or paying 1 dollar today or $(1 + i)^n$ dollars in n years, where i is the discounting rate.

A future sum of money, S, available at year n, thus has the same value as a present sum, S_0, available at the year 0, according to the following relation:

$$S_0 = \frac{S_n}{(1 + i)^n}$$

S_0 is the present worth of the sum S_n recovered in the year n; and $1/(1 + i)^n$ is the discounting coefficient of the year n, as it relates to the year 0. The equation resembles that for the rate of compounded interest.

The discounting rate i is analogous to a rate of interest. It differs from interest in that it varies from company to company, being generally defined by the average cost of the company's financing and the balance between available capital and needs for completing projects. In France, where many companies are controlled by a government planning agency, the rate of discounting is determined by the *Commissariat au Plan,* and under the *Plan VI,* it is equal to 10% in current francs. In the absence of any other criteria, this official French discounting rate can be used in various profitability calculations.

Also, in the case of company funds available at "no cost," discounting should be used at least to assure that money will be recovered at a value equivalent to its initial value.

2.2.2.3 Remarks

Income tax rules allow for the interest on loans to be deducted from taxable profits. Thus in Secs. 2.4 and 2.5 interest is included in fixed costs and thereby in the operating cost, so that interest payments enter into the deduction of profits before taxes as part of the operating costs.

On the other hand, it is impossible to operate in this manner when the capital comes out of company funds for which the origin is unknown. Only on the level of general accounting can the interest on loans be deducted from the gross profits, and the rate of discounting must take this into account when used for a particular project.

The values of $1/(1 + i)^n$ that allow for discounting a sum are given in Table 4.12 for various values of i and n.

2.2.3 MAINTENANCE

Maintenance can be divided as follows:

1. Routine maintenance, such as periodical inspection of various parts of the installation
2. Planned modifications made during scheduled shutdown
3. Urgent repairs required by unforeseen events

The costs of maintenance include

Costs of labor required by the operators to keep the equipment functioning well, including specialists to take care of complicated or delicate equipment —all attached to the general services of the plant

Costs for equipment that has been charged to the investment but has to be replaced before the end of the life of the unit (most often a percentage of the investment that varies according to the technology)

Also, spare parts, which have been included as part of the investment, are necessary for the maintenance of the unit.

Annual costs of maintenance in the refining and petrochemical industry amount to 3–4% of the battery-limits investment if the process involves only noncorrosive substances and the equipment is conventional. Such costs go up markedly for plants handling corrosive substances and using special equipment—to about 10% per year of the investment. For example, plants handling solid products like minerals experience relatively high maintenance costs on the order of 10% of the investment per year, whereas maintenance for conventional distillation or extraction units is 2–3% per year.

From a purely logical point of view an estimate of maintenance costs as a percentage of investment does not appear satisfactory, since those costs can diminish quite a lot in such cases as a long shutdown.

2.2.4 TAXES AND INSURANCE

These include

Miscellaneous taxes, such as real estate taxes and patent fees.

Miscellaneous insurance covering not only the investment in equipment, structures, initial charge of catalyst, etc., but also the stored feedstocks and products. These insurance costs have a periodic character, but are found to be practically identical from year to year.

Precise estimates require the aid of specialists in taxes and insurance. Such costs vary not only with type and size of the unit, but also according to the site, the region, and the immediate location of the installation. Certain local advantages can be awarded to enterprises by the authorities of a community or region, or even nation, in view of favoring certain types of industrial development that might create new jobs. The usual practice is to take a fixed percentage of the investment (usually around 2% per year) as annual costs for taxes and insurance.

2.2.5 OVERHEAD

The costs that can be charged to the nonproductive elements of a plant (workshops, personnel facilities, administration, etc.) or of a company (office management, contracts, licenses, patents, accounting, management, etc.) must obviously be supported by the productive elements. These costs can be itemized variously, depending on the enterprise; they can be estimated from the investment, often at about 1% of the investment per year.

2.3 VARIABLE COSTS

Because they are proportional to the actual production of a plant, rather than to its nominal capacity, some costs vary throughout the life of a project. Such costs are generally expressed as annual costs but can also be reported as cost per ton of raw material or product. When expressed in terms of raw material or product, the so-called "variable costs" tend to become constant for any given capacity.

2.3.1 PRODUCTS AND RAW MATERIALS

There are

Costs calculated from the material balance of the unit (raw materials, products, coproducts, and by-products).

Costs resulting from the use of chemicals during separation operations (solvents) or final purification (acids or alkalis) or from the use of catalysts.

2.3.1.1 The Material Balance

Generally, this calculation is done by recording the consumption of raw materials per 1,000 units of product. By-products (other materials often produced in small quantities) are conveniently priced independently of production capacity; handled in this way, they become a credit against the costs of raw materials.

However, in some operations such as the steam pyrolysis of naphtha several coproducts are the object of a market study to establish a price-quantity relationship, and the price of those coproducts depends on the quantity manufactured. In such cases, the calculation is often made in terms of an *average ton* of product. With steam pyrolysis of naphtha, the average ton would include ethylene, propylene, butadiene, benzene, and toluene. If the economic study has independently established a relation between the prices of the various constituents of such an average ton, the cost of raw material can be proportioned among the various coproducts.

It becomes apparent that knowledge of a realistic price is indispensable to economic analysis. If the company making the study is also the producing company, the prices of products are generally known. Others can make use of "posted prices" and "spot" prices published weekly in journals such as *European Chemical News, Chemical Marketing Reporter, Chimie Actualités,* and *L'Usine Nouvelle.* Prices are also published monthly in journals such as *Information Chimie, Petroleum Press Service,* and *Chimie et Industrie.*

Posted prices do not usually correspond to the prices at which products are actually bought and sold, coming from either published quotations or costs cataloged in import-export statistics. For this reason, the posted prices are generally higher than the average actual price during periods of economic stability, especially for products coming from units with low capacity or for products having an essentially captive use.

Spot prices correspond to the sales of shipments in limited tonnage; they characterize an up-to-the-minute situation of supply and demand and are differentiated from the posted prices that represent the average price for a determined period.

Normally, all buying and selling of chemical products takes place through a contract, which often accords the advantage of a refund to the buyer, as against posted prices. This refund takes into account the amount of tonnage involved and the duration of the contract, which may include other clauses such as a reciprocity of furnishing raw materials, or in a period of economic instability, formulas for price adjustment. The exact terms of these contracts are generally confidential, so that interpretation of the information transmitted by specialized technicoeconomic publications becomes a delicate task.

There do exist certain systematic studies that have the enormous advantage of reflecting a more accurate price, since they result from inquiries carried out on the producing companies, and since they are concerned with the near totality of tonnages involved. Although these studies have the inconvenience of applying only to certain countries and of appearing only after a 1- to 2-year delay, the inconvenience is justified by the volume and complexity of information handled. Examples are reports by the United States Tariff Commission, and the *Office Central des Statistiques de la CEE* for countries of the European Economic Community, and the *Direction Générale des Douanes* for France.

When a product is captive, it is often difficult to establish the actual price for this product, even within a company. Generally, only an approximation is possible, through an artificial calculation that takes into account the financial structure of the company and certain of its rules. In this way, an *average contract price* is determined, based on products made directly from the material in question.

In pricing the by-products, a potentially inferior quality should be taken into account, as well as the possibility that the by-product may be sold in a saturated market. It may be necessary to use refunds to get rid of a by-product, or to realize only a partial sale at a given period, with the balance of the by-product priced as fuel, if it is such that it can be burned. Sometimes supplementary costs for disposing of by-products must be anticipated.

2.3.1.2 Chemicals and Catalysts

The term *chemicals* is here used for materials that are occasionally used in small quantities for secondary treatment: neutralization, hydrolysis, saponification, washing, purifying, etc. Their costs represent not only the consumption of chemicals, due to the various secondary operations of treating, but also the losses of solvents and solids, etc., due to evaporation, entrainment, degradation, erosion, and so forth.

The consumption of catalysts calls for a distinction between homogeneous and heterogeneous catalysts. In homogeneous catalysis the catalyst must often undergo a physical or chemical treatment to restore its power after being separated and before being recycled to the principal reactor. The costs of reactants and chemical or mechanical losses during the course of this treatment must be figured in the economic calculation. When a homogeneous catalyst is destroyed, so that it is no longer economical to recover its constituents, a new catalyst is used; and the catalyst consumption is directly proportional to the tonnage of the manufactured product.

In heterogeneous catalysis the cost of the catalyst is fixed according to its life, a characteristic that is generally established and guaranteed by the licensor of the process. At the end of its life, the catalyst must be replaced; i.e., a new charge must be bought and put into place. The corresponding costs are prorated from the actual production during the catalyst's lifetime. If, for example, the life of the catalyst is two years, the cost of a half charge is calculated annually under consumption.

In these costs, it is necessary to take into account the salvage value of the spent catalytic system. Indeed certain manufacturers repurchase the used charge in order to recover certain constituents. For example, one can try to reuse the molybdenum in iron molybdate, the catalyst employed for the oxidation of methanol to formaldehyde. The salvage value of such used catalysts will be as high as the compounds are valuable.

Noble-metal catalysts impregnated on a ceramic support are handled as though the corresponding quantities of precious metal were rented. That portion of their costs accountable to the initial charge falls into two categories:

1. Costs related to the support, which make up part of the investment
2. Costs related to the precious metal, which result from a temporary immobilization of funds similar to working capital

When the spent catalyst needs to be replaced, it is usually processed by the catalyst manufacturer, who recovers the precious metals. Under these conditions, costs spread over the life span of the replacement charge will comprise the purchase of a new support, the impregnation of that support, the catalyst promoter, and the losses of precious metals, i.e., the difference between the quantity impregnated and the quantity recovered.

2.3.2 UTILITIES

The term *utilities* indicates nonprocess commodities used for supplying process needs, as well as for plant maintenance, i.e., electricity, steam, plant air, cooling water, process water refrigeration, plant water, etc. Since consumption of the most important utilities (steam, electricity, cooling water) is a function of the operating capacity, utility costs are made proportional to production. Whether for estimating or for planning studies, the costs kept for utilities take into account the following:

Manufacture
Distribution
Recycling reusable utilities, such as condensate, warm cooling-tower water, and so forth
Additives such as cooling-water treating chemicals and boiler feedwater chemicals, makeup water, etc.

These costs are thus divided among the following:

Investment costs for equipment to produce, distribute, and recycle the utilities
Variable costs for supplies of utility commodities
Costs for operating and maintaining the equipment

A certain profit should also be figured on the capital invested in the equipment.

This way of operating has the advantage of according a certain autonomy to the units that produce the utilities and allows for the centralization and rationalization of the energy and other needs of the different installations in an industrial complex. Each element in the ensemble can bill its consumption of the utilities in the same way that it does the purchases of raw materials or chemical products.

Since each unit has its own function and a variable operation, however, it will need to vary its consumption of utilities from the central network. Thus, for example, there can be demand for an increase or reduction in saturated steam supply, or a need to obtain steam at a higher pressure. Investment costs arising from such needs should be attributed wholly to the unit imposing the variation without modification of the initial costs of the utilities.

In the particular cost of supplying refrigeration at a given level, the necessary investment is most often assigned to the consumer, taking into account only the required energy and renewal of the refrigeration fluid. The same method is used when the consumption of a utility is so large that it cannot be supplied from a central installation serving other units, for example, a unit for liquifying natural gas.

The price of utilities varies according to the site and ease of production. Northeastern United States, Scandinavia, and Alpine regions are known for furnishing electricity at low prices, while certain countries of Asia and Africa often compensate for the lack of water by the low cost of fuel. Thus each project is associated with a well-determined price for utilities. However, utilities do tend to stay within certain limits that allow for averaged values when other data are lacking.

2.3.3 MISCELLANEOUS VARIABLE COSTS

Other costs that can be included with the variable costs are

Running royalties that are proportional to the operating capacity of a unit

Costs for catalyst regeneration, such as burning out coke from catalysts (which requires supplies of oxygen, etc.)

Costs of preparing those homogeneous catalysts that must be synthesized only at the moment of being used

2.4 LABOR

The costs of labor are sometimes integrated with the variable costs, but it is preferable to treat labor separately, since these costs are not proportional to production but follow a certain relation to the design capacity. When production lines become too complex because of their size and are doubled, labor costs grow from one to a higher discrete level.

The term *labor* usually includes only the personnel on shift, i.e., those in charge of the operation and maintenance of the equipment; it does not include the costs of work done outside the battery limits, such as the work of general and exceptional maintenance or work in the repair shops, control shops, laboratories, warehouses, etc.

In certain methods of economic calculation, a breakdown of personnel is made according to its responsibility, whether manufacturing, maintenance, control, etc. In this book, the costs of maintenance are counted wholly in the fixed costs, as a percentage of the investment, and the costs of control are included in the overhead costs.

A work shift lasts 8 hours. A continuous workday of 24 hours thus calls for three shifts of workers. When weekly and annual holidays, sick leaves, social benefits, etc., are taken into account, four to five shifts of workers are necessary to assure continuous operation for a year.

The technical press quotes numerous examples of work-force needs or of the number of workers in a shift, according to the type of installation under consideration. Also, certain short-cut methods determine the number of operator hours for a specific new industrial project according to relations such as

$$\frac{\text{Number of operator hours}}{\text{Ton of product}} = t\,\frac{\text{number of steps in the process}}{(\text{capacity in tons per day})^{0.76}}$$

where t is a coefficient equal to 23 for discontinuous operations; 17 for continuous operation at an installation with average instrumentation; 10 for continuous operation at an installation with extensive instrumentation.

Another formula approximates labor as proportional to the 0.2–0.25 power of the capacity. However, such information should be used with caution because increasing automation tends to reduce the need for personnel. Each company usually has a particular labor policy that fixes the real number of employees according to the sophistication of the installation.

Salaries also depend on company policy, the site of the project, etc. Usually the work of supervisors and engineers is not charged to specific units by the hour but extends over several operating units at a time. Thus it is conventional to take a fixed percentage of the total operator's pay for supervision, unless the equipment is too complex. In the exceptional case, where a shift must remain under the direction of a supervisor all the time, this percentage is taken on the total of the salaries for the shift. It is generally on the order of 20–25%.

2.5 OPERATING COST AND EXPLOITATION COSTS

Fixed costs, plus variable costs and labor make up the operating cost, i.e., all of the costs that are entailed in completing the manufacturing project, including those for recovering the invested capital while paying off interest.

Since the fixed costs and labor are generally expressed in dollars per year, it is convenient to add to them the variable costs, which are tied directly to the plant's output, expressed in dollars per ton, for determining costs for the quantity of product actually obtained in 1 year. It is necessary to take into account two parameters: First, actual annual production can be noticeably different from the nominal capacity forecast for the installation in the engineering calculations. Most often it will be lower because the market is developing progressively. In exceptional cases, it can have a higher value if the overdesign factors used in sizing the equipment permits this. A rate of feed, or use rate, that represents the relation of actual-to-capacity production is introduced.

Second, the calculation for determining the nominal capacity of an installation takes into account a reasonable number of hours or days of shutdown in the course of a year, in order to reserve time for maintenance, replacing catalysts, etc. Thus the designated tonnage of product is figured at less than a year's operation. A coefficient called *stream factor* is used to represent the theoretical duration of operation as a percentage of 365 days. This stream factor is most often in number of actual hours or days of production in a year and is usually estimated at about 8,000 hours per year or 333 days per year.

Besides dollars per ton, the operating cost can also be given in dollars per year, a convenient expression for accounting studies or profitability calculations. By dividing the annual costs by the actual or forecast annual feed or product, the costs can be expressed in terms of either raw material or product, as desired.

These relations are used, particularly by the technicoeconomic publications, to express cost of transformation, production cost, manufacturing cost, etc. However, there are possibilities for confusion, since all these different terms cover the same content.

It is useful to distinguish operating cost from the exploitation cost, which takes into account only the expenses of running the unit without depreciation and interest. In the balance of this book, we will use only the terms *operating costs,* or when these are reported in tons of product, *manufacturing cost,* and *costs of exploitation* as described above. In certain cases the economic calculation should also include various additional expenses, such as financing costs to facilitate promotion and commercialization of the products, to cover research and development taken up for the study of the project, to adapt an unusual or remote plant site, etc.

2.6 PROFITABILITY STUDY FOR A PROJECT

Numerous techniques allow evaluating the economics of a project, but many of these find limited use because of the lack of information, a lack that has two causes:

1. Certain profitability calculations require a large number of precise facts. In a long-term market study, for example, raw materials and products must be forecast in terms of quantity, quality, and price.

2. Economic evaluations must proceed, at least at first, without information on outlets, supply, site conditions, financing possibilities, etc. Thus a distinction should be made between (a) profitability studies that are essentially an easily used, simple tool and (b) the analysis of an actual project through an exhaustive investigation of all parameters.

2.6.1 ECONOMIC CRITERIA

Among the various profitability criteria, some take into account the exploitation income (see Sec. 2.5); others do not. Those that do are preferable, although operating costs and manufacturing costs alone offer a means for preliminary comparisons.

Seeing that the gross annual income (difference between the revenue from sale of products and the operating cost) serves as a basis for taxes, these taxes should enter into calculations for the profitability of a project. Tax rates vary according to the location, the fiscal system in effect, and the method of financing; and their applications are more or less complex. By way of example, in the United States the rate schedule is as follows:

- For profits amounting to less than $25,000, a tax of 22% is applied.
- For profits amounting to more than $25,000, a surtax of 26% is added to the tax of 22%.
- If the profits are used as additional capital, one tax only of 25% is applicable; and if the profits are used as an investment, a reduction of 7% is allowed.
- In addition, a supplemental surtax of 10% is added.

As a first approximation, the following average rates of taxation may be assumed for the various countries:

Country	Percent
United States	52
Canada	41
West Germany	51
France	50
Italy	35
United Kingdom	53.75
Japan	50

When it is not possible to determine the actual tax, a calculation is either made from the gross annual income or an arbitrary tax of 50% is assumed. Such calculations are useful for comparing similar projects in the same area.

The profitability criteria most often employed in project analysis are as follows:

- The time for reimbursement, called *net recovery time.* The term *payout time* corresponds to this but actually represents a simplified form of this criterion.
- Discounted cash flow (DCF) or net present value.
- Rate of return, or internal return.

Other criteria are also employed, but will not be developed at length in this chapter, with the exception (up to a point) of the method for equivalent-maximum-investment period (EMIP). There will also be brief accounts of turnover ratio, relative capital gain, and the return on investment (ROI).

Most of these parameters are general economic notions and not of immediate interest for project analysis. Taking into account the precision of the calculations, the discounted cash flow often appears to be too exact, since it theoretically requires forecasts of supplies and outlets. The EMIP method is of marginal interest, since it effectively requires knowing the production schedule and thus the forecasts on products.

In applying profitability criteria, the term *break-even point* often appears. This can have various meanings according to the specific use. Therefore, it is important to recall the initial meaning: When a chart is constructed (Fig. 2.2) in which revenue and operating costs are both plotted against the production rate, the two plots will intersect at the break-even point, which thus identified is the rate of use for which the direct receipts and costs are in equilibrium. The break-even point for a given type of plant comprises an independent economic criterion for considering the profitability of a project.

2.6.2 PAYOUT TIME

The *payout time* (POT) for recovering invested capital is defined as the period over which the total receipts minus all deductions including taxes are equal to the entire investment of depreciable capital necessary for purchasing, constructing, and starting up (etc.) the unit. From a strictly rational point of view, this criterion should carry with it the idea of discounting; and it will be studied in this sense in the following sections. For the present, however, we offer a simplified version currently employed in the United States and widely used by plant designers and operators to balance increased costs of equipment against improvements in efficiency. Discounting such as is practiced by economists does not appear in explicit fashion in such studies.

Thus the criterion of payout time has the following hypotheses:

1. Prices and quantities of raw materials, by-products, reactants, catalysts, utilities, labor, and manufactured products are all constant over time, so that revenues are identical from year to year.
2. The capital is borrowed and therefore recoverable with the help of an interest rate.
3. Profits are subject to deductions in advance.
4. Depreciation is linear and salvage value neglegible.

Under these conditions, if

B = gross annual profit before taxes,
a = rate of tax on the gross profit,
I = investment corresponding to the depreciable capital,
A = annual depreciation constant,

the payout time, in years, is given by

$$\frac{I}{B(1 - a) + A}$$

The payout time thus appears as the time during which the investment I is recovered in the form of actual profit (taxes deducted) and allows for depreciation. The term

$$B(1 - a) + A$$

represents the flow of liquid assets and is generally known as *cash flow.* When the taxes amount to 50% of gross profits, the expression is

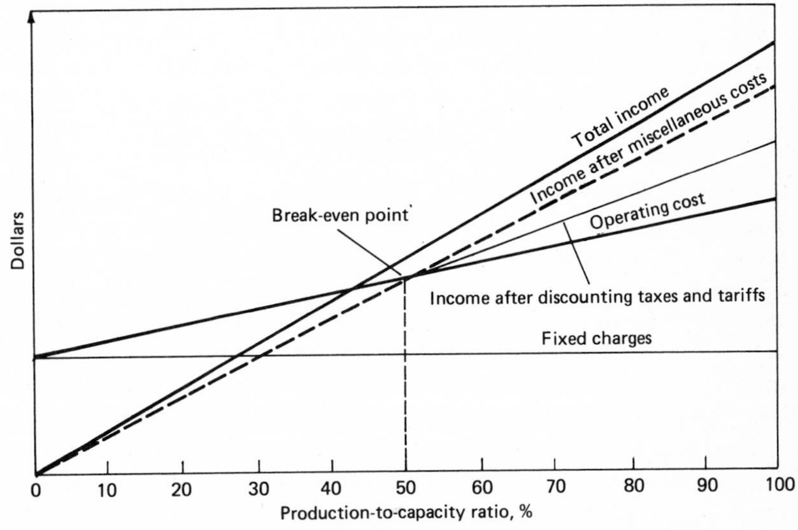

Fig. 2.2 Definition of the break-even point.

$$\text{Payout time} = \frac{I}{0.5B + A} \tag{2.1}$$

Equation (2.1) shows that payout time, whose inverse is sometimes taken to be the return on investment, is actually less than the period of depreciation if the project is profitable. Consequently, the shorter the payout time, the better the economics of the unit.

Convention and the need to simplify calculations both put I as the investment for battery-limits and off-sites material only. With a complicated petrochemical plant, a payout time of 5 years is generally considered adequate; refinery processes and certain organic chemical plants may require faster returns. Also, when I represents the depreciable capital in Eq. (2.1), a payout time of 7 years is often considered acceptable for petrochemicals.

As previously noted, evaluation in terms of payout time assumes knowledge of the selling price of the finished product. When those data are not known, a payout time of 5 years can be assumed and a required average selling price back-calculated. Thus if R designates the relationship of investment to payout time (I/P) and C the operating cost, the average required selling price of product V is given by the expression:

$$V = \frac{R - A}{1 - a} + C$$

Or in the case of a tax rate amounting to 50% of gross profit,

$$V = 2(R - A) + C$$

Although the value of V thus obtained gives the average value of manufactured products required to recover the invested capital in a certain time, it is usually considered the minimum selling price when the payout time is set at five years. This preliminary method thus allows for a rapid analysis of the economic situation for a project. It is well adapted for preliminary estimates, but it also has all the disadvantages that go with the use of payout time as a criterion (see Sec. 2.6.3.1).

One practical limitation of payout time lies in the assumption of borrowed capital involving interest. In a number of cases, particularly in the refining and petrochemical industries, the life of a unit is relatively brief, and the initial investment is recovered rapidly. Because of this, it is often costly or even impossible to obtain the necessary loans. Financing must come from the company's own assets, which benefit from more favorable rates. Therefore, the precise origin of the money cannot be determined, and from a strictly rational point of view it ought to be discounted. This contradiction may be avoided by assuming that the capital is all borrowed from the company's assets at an assumed rate of interest. But this is in reality discounting, and one must be careful to use a rate of interest equal to the rate of discounting that would normally apply.

2.6.3 DISCOUNTING

Evaluation methods based on discounting differ from those based on payout time in the following ways:

- They take into account the receipts and expenses of each year, corresponding to a production program and probable price variations that are the results of a prospective study and shown in a schedule.
- They assume that the necessary capital is furnished through the general accounting of the company making the study. Consequently, the type of financing does not enter, or in cases where the percentage borrowed is known, it is possible to define an *opportunity cost.*
- They permit calculations either before or after taxes.

2.6.3.1 Determining Discounted Cash Flow, or Net Present Value*

The variations of the annual revenue as a function of time is calculated. This requires identification of those conditions most likely to be met during the study period.

First, a given system of taxation is assumed, and the effect of tax variations is examined only as a note. Then, since the project generally covers several years, the effects of inflation may be considered. The calculation can thus be conducted in two ways, variable currency and fixed currency. For variable currency the cash flow is evaluated in dollars for the year under consideration. For fixed currency a year of reference is chosen, and all transactions carried out during the life of the project are accounted for in dollars at that year of reference. It is possible to change from a discounting rate for constant dollars, i, to a discounting rate for variable dollars i', by means of the relation $i' \simeq i + d$, with d being the rate of monetary depreciation.

In estimating, it is often simpler to think in terms of constant dollars, since inflation is unpredictable over a long time. Because of this, it is necessary to be precise about two essential points:

1. The date (year) in which the study is made, since this date serves as reference in the course of the calculations. Generally, a year 0, taken as the time when the unit begins to produce, is the reference.
2. The duration (years) of the project. Usually, this duration period, n, is equal to the depreciation period.

* *Translators' Note:* It is assumed the reader will recognize that the term *discounted* must modify some monetary item such as sales value, feedstock costs, wages, and so forth. The thrust of this text is to assure that the application of discounting to the various segments of an economic calculation is consistent.

2.6.3.1a Financing by a Company

Usually, financing for a project is assured by the company itself, and calculations are preferably done in terms of a constant value of currency, using the reference year 0 as the year of start-up. The revenues and expenses that result from exploiting the unit are then determined, i.e., the cash flow. For each year of exploitation, p, the cash flow $(CF)_p$ is written in the following form:

$$(CF)_p = V_p - D_p - I_p - (V_p - D_p - A_p)a$$

or

$$(CF)_p = (V_p - D_p - A_p)(1 - a) + A_p - I_p$$

where a = tax rate on gross profits
V_p = revenues for the year p
D_p = costs of exploitation for the year p
A_p = depreciation for the year p
I_p = total investment for the year p

The operating cost can be introduced:

$$C_p = D_p + A_p + F_p$$

where F_p is the cost of financing. (In the present case $F_p = 0$, since the calculation is made without taking into account the part played by borrowed capital.) The cash flow then becomes

$$(CF)_p = (V_p - C_p)(1 - a) + A_p - I_p \tag{2.2}$$

with $F_p = 0$.

Using the discounting rate i to discount this expression of actual revenue for the year p gives

$$\frac{(CF)_p}{(1 + i)^p} = \frac{(V_p - C_p)(1 - a) + A_p - I_p}{(1 + i)^p} \tag{2.3}$$

with $F_p = 0$.

From this equation (2.3), the following consequences can be deduced according to where the value of p falls between 0 and n:

- For the year $p = 0$, receipts and expenses are nil, depreciation has not yet gone into effect, and I_0 is equal to the initial funds placed; i.e., the total investment I_0 is equal to the depreciable capital I plus the working capital f, or $I_0 = I + f$, so that $(CF)_0 = -(I + f)$.
- For any subsequent year p, the additional investment I_p is generally nil, unless supplementary investments are made, in which case it would be necessary for A_p to take into account not only the initial but also the additional funds.
- For the year $p = n$, the recoverable capital is represented by $-I_n$, which comprises both the working capital f and the salvage value I_r. In process evaluation, where it is usually a matter of comparing imaginary projects without recourse to accounting, working capital f is constant and

salvage value I_r is nil or equal to the depreciable capital for the year n. Thus the accounting problem of dealing with increased prices is avoided. It follows that

$$\frac{(CF)_n}{(1+i)^n} = \frac{(V_n - C_n)(1-a) + A_n + (I_r + f)}{(1+i)^n}$$

with $F_n = 0$.

2.6.3.1b The Apparent Cost of Self-financing

In certain cases it is possible to distinguish that part of an investment financed by loans from the part made up out of the company's own funds. In such cases, a question arises over the best use of the company funds. Equations (2.2) and (2.3) are still viable, only with the financing charge F_p not equal to 0. The finance charges represent interest for the year p on the borrowed capital, either initially or for a supplementary investment.

For the year $p = 0$, I_0 should represent the actual funds placed by the company. If I is the initial depreciable capital consisting of both loans, E, and the company's own funds, P, i.e., if $I = E + P$, then the expression for the initial funds placed, I_0, is

$$I_0 = I + f - E = P + f$$

with f equal to the working capital, which can be taken as equally drawn from loans and the company funds. The financing charges F_p should then be corrected.

No matter what the year is, the term for additional investment, I_p, is generally no longer nil but equal to the reimbursement R_p of the borrowed capital for that year, so that

$$\sum_{p=1}^{p=n} R_p = E \qquad (R_0 = 0)$$

If there are no supplementary investments in equipment, Eq. (2.3) becomes

$$\frac{(CF)_p}{(1+i)^p} = \frac{(V_p - C_p)(1-a) + A_p - R_p}{(1+i)^p}$$

with $F_p \neq 0$.

For the year $p = n$, the recoverable capital $-I_n$ now includes not only the salvage value and working capital $(I_r + f)$ but also the reimbursement for the year n. It follows that

$$I_n = R_n - I_r - f$$

A similar but roundabout method of taking into account a company's financing structure while carrying on the profitability study consists of modifying the rate of discounting. Thus if the initial capital is made up of 20% of the company's own funds, for which the rate of discounting is 15%, and the remaining

80% is made up of loans discounted at 5%, the resulting rate of discounting would be taken as equal to 7%.

2.6.3.1c *Profitability Calculations That Omit Taxes*

When the profitability calculations do not include advance deductions for taxes, the term for tax rate on gross profits, *a,* becomes 0 in the expression (2.3). In the most common case, when financing is handled within the company, the equation becomes

$$\frac{(CF)_p}{(1+i)^p} = \frac{V_p - C_p + A_p - I_p}{(1+i)^p}$$

with $F_p = 0$.

Furthermore, since

$$C_p = D_p + A_p + F_p$$

the above equation can be written as

$$\frac{(CF)_p}{(1+i)^p} = \frac{V_p - D_p - I_p}{(1+i)^p}$$

Where the financing structure is known, so that the cost of financing F_p is not assumed 0, this leads to

$$\frac{(CF)_p}{(1+i)^p} = \frac{V_p - D_p - F_p - I_p}{(1+i)^p}$$

In this equation, I_p takes into account payments on the principal.

2.6.3.2 Criteria of Profitability When Discounting

Equation (2.3) serves to identify the various criteria of profitability when discounted as

Cumulative net present value
Rate of return
Payout time

2.6.3.2a *Cumulative Net Present Value*

The cumulative net present value is the sum total of the discounted cash flows for the project, from its inception to its completion. When *B* is the cumulative net present value,

$$B = \sum_{p=0}^{p=n} \frac{(CF)_p}{(1+i)^p}$$

Discounting is made for the year 0, with year *n* normally the last year of operation. In process evaluation, *n* is often assumed to be the end of the period of depreciation. Mathematically, the cumulative net present value B thus becomes

$$B = \sum_{p=0}^{p=n} \frac{(V_p - C_p)(1 - a) + A_p - I_p}{(1 + i)^p}$$

with $F_p = 0$.

Generally, the total investment for the year p, I_p, will have the following values, depending on the year:

- Initially, $I_o = I + f$.
- On the year p, $I_p = 0$.
- On the year n, $I_n = -(I_r + f)$.

When these values are substituted into the equation for B, we obtain

$$B = -(I + f) + \sum_{p=1}^{p=n} \frac{(V_p - C_p)(1 - a) + A_p}{(1 + i)^p} + \frac{I_r + f}{(1 + i)^n}$$

From this, we have the equation

$$B = -\left[I - \frac{I_r}{(1 + i)^n} + f\left(1 - \frac{1}{(1 + i)^n}\right)\right] + \sum_{p=1}^{p=n} \frac{(V_p - C_p)(1 - a) + A_p}{(1 + i)^p} \tag{2.4}$$

When the type of financing, and thus the financing charges F_p are known, Eq. (2.4) becomes

$$B = -\left[P - \frac{I_r}{(1 + i)^n} + f\left(1 - \frac{1}{(1 - i)^n}\right)\right] + \sum_{p=1}^{p=n} \frac{(V_p - C_p)(1 - a) + A_p - R_p}{(1 + i)^p}$$

When taxation is not taken into account, $a = 0$, and Eq. (2.4) becomes

$$B = -\left[I - \frac{I_r}{(1 + i)^n} + f\left(1 - \frac{1}{(1 + i)^n}\right)\right] + \sum_{p=1}^{p=n} \frac{V_p - D_p}{(1 + i)^p}$$

When receipts and expenses, as well as depreciation, are the same each year and $F_p = 0$, the relation

$$\sum_{p=1}^{p=n} \frac{(V_p - C_p)(1 - a) + A_p}{(1 + i)^p}$$

can be put into the following form:

$$[(V_p - C_p)(1 - a) + A_p] \sum_{p=1}^{p=n} \frac{1}{(1 + i)^p}$$

It should be noted, with respect to this last equation, receipts and expenses, on the one hand, and depreciation, on the other, do not behave the same whether the value of the currency is variable or fixed. Linear depreciation leads to an identical value each year only with fixed currency. To assume that the depreciation is the same each year is tantamount to assuming a fixed currency. However, to fix the value of receipts and expenses with a variable currency is the same as forecasting a loss of activity for the plant during the life of the project.

The value of the expression

$$\sum_{p=1}^{p=n} \frac{1}{(1 + i)^p}$$

which is equivalent to

$$\frac{(1 + i)^n - 1}{i\,(1 + i)^n}$$

is solved for various values of i and n in Table 4.12. When the method of financing is allowed for, and the financing charges not assumed 0, this simplified expression and the table can only be used when both the interest charges and the payment on the principal are the same from year to year—a situation that can be simulated through the notion of *average rate of interest* described in Sec. 2.2.2.1*c*.

When $F_p = 0$ and the salvage value is nil at the end of n years, Eq. (2.4) can be written

$$B = -\left[I + f\left(1 - \frac{1}{(1 + i)^n}\right)\right] + [(V_p - C_p)(1 - a) + A_p] \times \frac{(1 + i)^n - 1}{i(1 + i)^n}$$

This is the form generally used for profitability calculations.

The discounted-cash-flow (DCF) method allows for examining a project over long enough period of time for economic parameters to be considered in the calculations. When the problem is that of trying to decide whether to go through with a certain project, this method indicates that the project should be taken on if the discounted profit is positive, i.e., if the discounted receipts are greater than the discounted expenses. When the problem is that of trying to select among several possibilities, the project whose cumulative net present value is the highest would be chosen.

2.6.3.2b Rate of Return with Discounting

This method comes directly out of the preceeding one, since the rate of return, i_r, for a project is equal to the value of that discounting rate i which will reduce to zero the cumulative net present value for n years.

One should look for a value of i_r such that

$$\sum_{p=0}^{p=n} \frac{(CF)_p}{(1 + i_r)^p} = 0$$

[This means that B, as well as its equivalent in Eq. (2.4) must become zero.] In order to do this, a curve is made giving the cumulative net present value as a function of the rate of discounting. Tables 4.12a and 4.12b permit the necessary calculations. The rate of return, i_r, is that discounting rate i at which the curve crosses the axis of zero cumulative net present value (Fig. 2.3).

In order for a study to be judged profitable by a company, the discounted rate of return should be higher than that company's discounting rate, which means that the cumulative net present value is positive when the company's rate of discounting is applied. Conversely, if the company's discounting rate gives a negative cumulative net present value, the project should be put aside.

The utility of the notion of *rate of return* is that of an economic criterion which

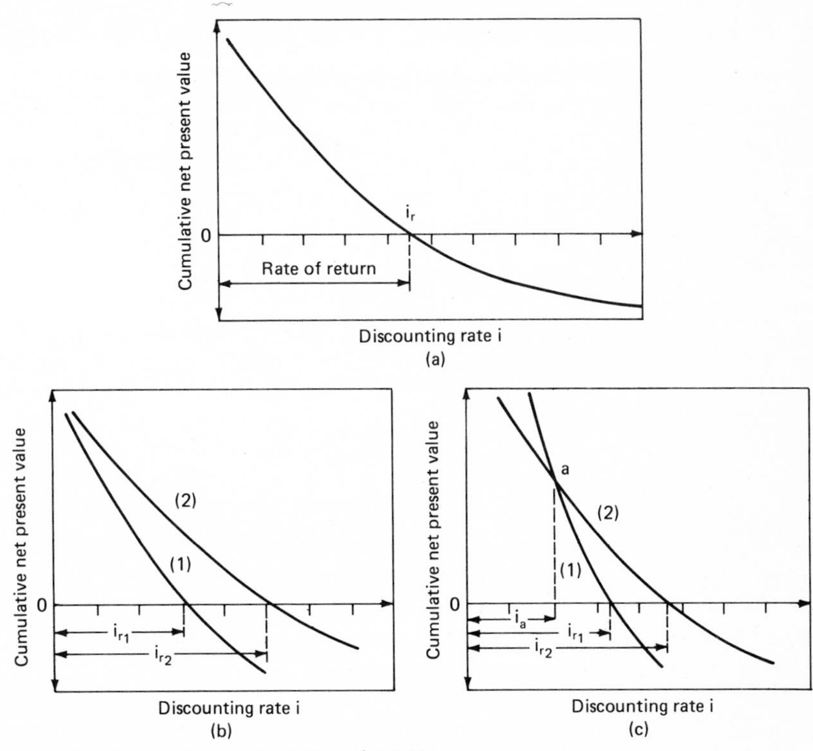

Fig. 2.3 Curves determining rates of return.

is independent of the discounting rate while revealing more. However, certain limitations should be placed on the use of this criterion; although the project with the highest rate of return is theoretically the best, certain precautions should be taken, and the economic comparison should not rest solely on this calculation. In particular, one should examine the relative slope of the curves (Fig. 2.3) as well as their relative position.

Slope of the curves Actually, the best project often has the slope with the smallest angle (curve 2 in Fig. 2.3*b*), since such a project will be more sensitive to variations in revenue and the hypotheses defined from the start by the market study.

Position of the curves When the curves cut each other (Fig. 2.3*c*), the relative profitability for the projects can only be established as a function of the discounting rate corresponding to this intersection. Actually, the project with the highest rate of return is not necessarily the one with the highest cumulative net present value. Project 2 in Fig. 2.3, whose rate of return, i_{r2}, is greater than that of project 1, has the greater cumulative net present value only if the company's discounting rate is higher than the point of intersection of the curves, i.e., higher than i_a. In effect, Eq. (2.4) and those equations derived from it give increasing values of i_r as the investment, I, decreases. It follows that the project with the smallest initial investment has an advantage—a conclusion that can contradict the criterion of rate of return.

The rate of return for the improved revenue due to a modification to an existing installation can be determined analogously, by adapting Eq. (2.4) to include only the additional investment necessary for the modification.

2.6.3.2c Payout Time with Discounting

The *payout time* for a project is the period whose termination date, k, marks the beginning of the period when the cumulative net present value becomes positive. Date k is thus determined such that

$$\sum_{p=0}^{p=k} \frac{(CF)_p}{(1+i)^p} > 0$$

This date is the moment when income from the project has paid off the initial capital at a rate equal to the operating company's rate of discounting. The *recovery time*, i.e., payout time, is the time lapse that separates this date from the beginning date of the economic calculations. Mathematically, it is necessary to determine k, taken between 0 and year n, such that (with financing charges, F_p equal to 0):

$$\sum_{p=0}^{p=k} \frac{(V_p - C_p)(1-a) + A_p - I_p}{(1+i)^p} = 0$$

In the general case, financing is handled by the company, so that with k less than n and F_p equal to 0,

$$-(I+f)+\sum_{p=1}^{p=k}\frac{(V_p-C_p)(1-a)+A_p}{(1+i)^p}=0 \tag{2.5}$$

Actually, the salvage value I_r and the working capital f are recovered only in year n at the end of the project.

Under specific situations, the method of financing is defined, and the mathematical expression becomes

$$-(P+f)+\sum_{p=1}^{p=k}\frac{(V_p-C_p)(1-a)+A_p-R_p}{(1+i)^p}=0 \tag{2.6}$$

with $F_p \neq 0$.

Taxes are not taken into account, so that $a = 0$, and

$$-(I+f)+\sum_{p=1}^{p=k}\frac{V_p-D_p}{(1+i)^p}=0$$

If receipts, costs, and depreciation are all assumed constant and financing is handled by the company, Eq. (2.5) becomes

$$-(I+f)+[(V_p-C_p)(1-a)+A_p]\frac{(1+i)^k-1}{i(1+i)^k}=0$$

In general, when comparing two projects, the one with the shorter payout time is the best. However, the comparison can also be made by examining profitability over longer periods. Calculating a payout time neglects all the elements affecting the years after date k, and leads to results different from those obtained by classing projects in the order of decreasing cumulative net present value, which takes into account the receipts and costs throughout the depreciable life of the plant. Thus when two projects are compared (Fig. 2.4), the receipts from one (project 1, Fig. 2.4*b*) are higher at the end of n years, although the payout time for that project is longer and thus makes it appear less profitable.

Such a situation rarely occurs when comparing two projects of the same type, but does occur more frequently with comparisons between installations of a different nature, such as between a refinery unit and a petrochemical plant. Consequently, the criterion of payout time is most useful for comparing competitive technologies in a given set of manufacturing conditions.

If one had perfect knowledge of the future, a calculation of discounted payout time would be superfluous. Discounting considers the investments whose returns rest on data to appear in the far future; it permits a choice between projects whose recovery dates are compatible with the risks connected to completing the project. Therefore, it permits evaluating the risks incurred at the moment of decision, especially when competition is tough; which is why it is of great interest to first judge the profitability of a study

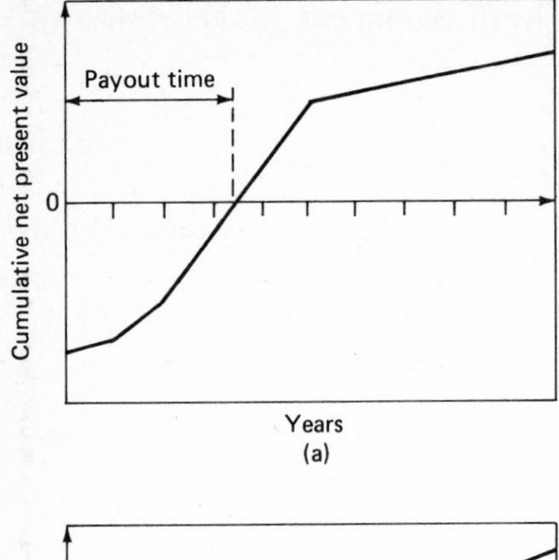

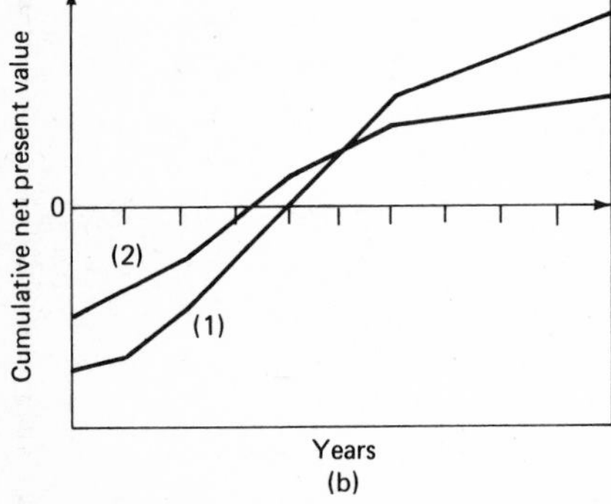

Fig. 2.4 Discounted payout time.

on the basis of this criterion, particularly in its simplified form: discounted payout time.

2.6.4 THE OTHER METHODS

Various other profitability calculations employ more or less complex ideas in an attempt to correct the faults of the preceding methods through an emphasis on one or another aspect of the calculations. However, such complex methods often assume a precision in the data that is illusory. Some examples are turnover ratio, return on investment (ROI), and booked rate of return, which are all derived from pure accounting. Other methods, such as equivalent-maximum-investment period (EMIP), fall closer to the preoccupations of cost estimating.

We will discuss turnover ratio, capital growth, return on investment, and the equivalent-maximum-investment period.

2.6.4.1 Turnover Ratio

Turnover ratio describes the relation of total annual revenue to invested capital. It affords a means for classifying projects of different natures, with conceptually different processing schemes and different investments. It is a measure of the recovery rate of appropriated money.

2.6.4.2 Capital Growth

Capital growth describes the relation between the cumulative net present value of a project and its investment cost. If n is the life of the unit, and the symbols for the discounting equations employed, the capital growth, expressed as a percentage, is given by

$$\frac{\displaystyle\sum_{p=0}^{p=n} \frac{(CF)_p}{(1+i)^p}}{I} \cdot 100$$

The profitability of a project is as great as its rate of capital growth is high. However, when the problem is that of determining the optimum size of an installation, this criterion is not generally applicable. In fact, there is a solution only when the curve of cumulative net present value versus the investment takes on an S shape. In addition, the highest rate of capital growth corresponds to a lower investment than that indicated by the maximum cumulative net present value; it indicates a smaller installation (Fig. 2.5).

2.6.4.3 Return on Investment

This criterion of profitability is used frequently in the United States and England; and it has passed into "franglais" under the form "return." The method basically consists of calculating the ratio of the annual return (receipts V minus

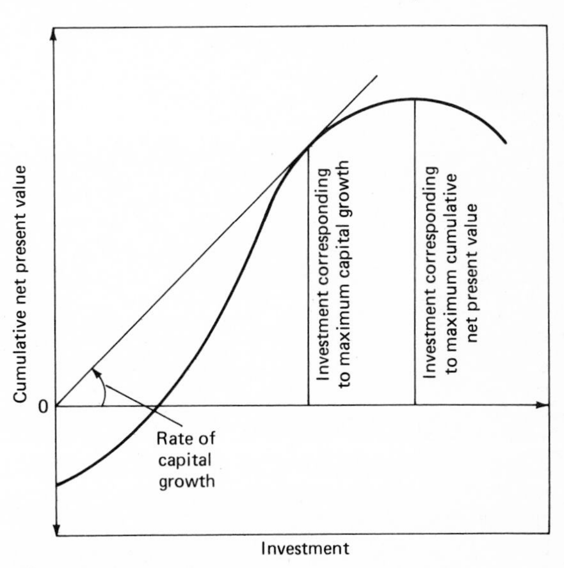

Fig. 2.5 Rate of capital growth.

costs D) to the depreciable capital I. Using the mathematical symbols established for the discounting calculations, the return on investment (%) is

$$\frac{V - D}{I} 100$$

This amounts to judging profitability before taxes; consequently, depreciation and finance charges are not taken into consideration. Often, a return on investment of a given percent is added to the operating cost to determine a selling price through the reverse calculations, as has been done with payout time. A return on investment of 30%, for example, corresponds to a payout time of 5 years.

The return on investment should not be confused with rate of return, for which the attained percentage level is generally much lower.

Another definition of return on investment takes into account taxation and thus depreciation and the finance charges. In this case the return on investment is the reciprocal of payout time, so that if we let $C = (D + A + F)$, the return on investment by this definition is

$$\frac{(V - C)(1 - a) + A}{I} 100$$

2.6.4.4 Equivalent-Maximum-Investment Period (EMIP) and Its Complement, the Interest Recovery Period (IRP)

With the EMIP method, a graph is first established showing the cumulative net present value versus time, when time 0 is taken as the moment of decision to build the plant. The curve thus establishes different phases of the project's progress (Fig. 2.6) as

1. Preliminary studies, during which expenses are incurred with purchase of a license, engineering calculations, calling for bids, etc.

2. Construction, during which the necessary funds become available.

3. Start-up and acceptance tests.

4. Coming on stream, which can be done gradually and can present various irregularities, according to the development of the markets for the manufactured products.

In the absence of precise data, these phases can be approximated by straight lines (*OA, AB, BC,* and *CD-DE* in Fig. 2.6*a*). Also, the calculations are simplified, if the unit is assumed to be operated at full capacity.

The point where the curve intersects the time axis is called the *break-even point*—a definition that differs from the break-even point described earlier (see last two paragraphs of Sec. 2.6.1).

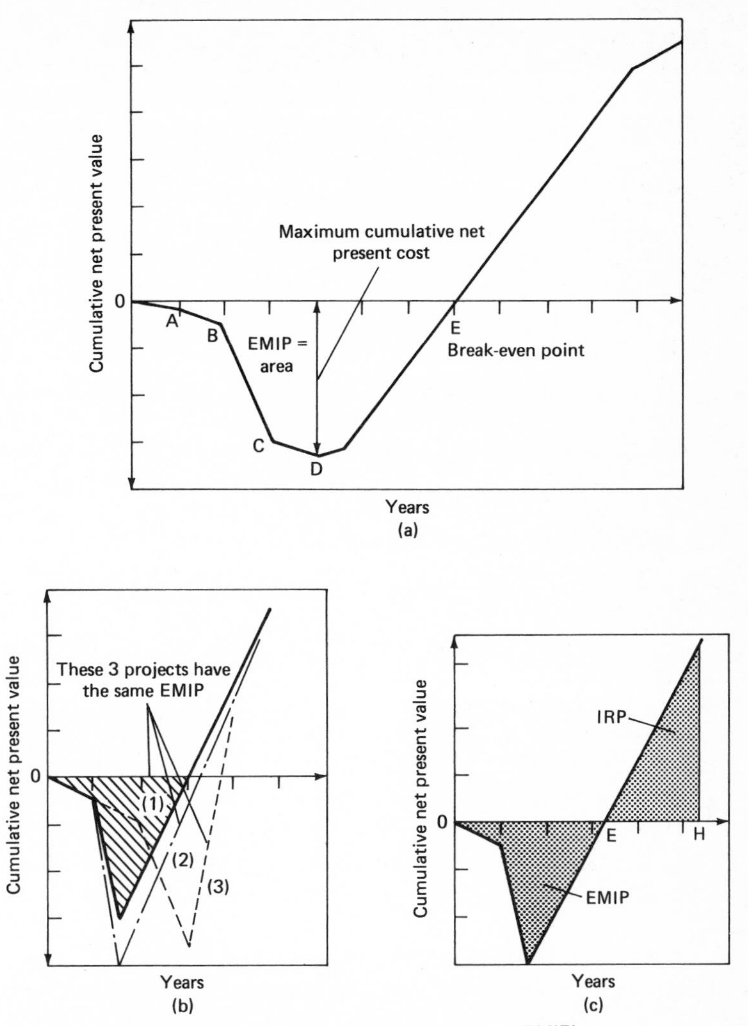

Fig. 2.6 The equivalent-maximum-investment period (EMIP).

Under these conditions, the EMIP becomes the relationship of (A) the area bounded by the time axis and the cumulative-net-present-value curve to the break-even point, expressed as dollars per year and represented by polygon *OABCDEO* in Fig. 2.6*a*, to (B) the maximum cumulative net present cost, expressed in dollars and represented by the distance from point D to the time axis in Fig. 2.6. The EMIP therefore has the dimension of time, and it is equivalent to the length of one side of a rectangle whose other side is the maximum cost and whose area is that given above. The curve to the break-even point both characterizes the project and allows for defining the EMIP. This curve can be constructed by dividing the maximum accumulated costs into the accumulated revenues (i.e., divide the accumulated revenues by the distance between point *D* and the time axis in Fig. 2.6). This method of describing the polygon *OABCDEO* gives the EMIP directly.

A higher profit from a project is indicated by a lower value of EMIP, which is thus similar to recovery time, except that it includes everything from the inception of the study. When comparing several projects, the procedure of dividing accumulated revenues by accumulated costs does not distort results, since it is specific for each case. Therefore, the EMIP method offers a relatively convenient way of comparing several versions of the same study (Fig. 2.6*b*).

To avoid the limitations of a reduced time span (an inconvenience of this criterion), the average slope of the revenue curve can be extended beyond the break-even point, so as to simulate the method of cumulative net present value. To do this, some authors define an *interest recovery period* (IRP) as the period between the break-even point and the time when the accumulated revenues (equal to the area beneath the extended revenue curve) is equal to the EMIP (Fig. 2.6*c*). Since the IRP thus defined is inversely proportional to the average slope of the revenue curve, a shorter IRP time indicates a more profitable project. This calculation can be convenient for taking market variations into account.

Example 2

Calculating the Profits from Separating C_8 Compounds from Their Mixture

Ethylbenzene, *p*-xylene and *o*-xylene are separated by superfractionation and adsorption from a distillate fraction of C_8 aromatics obtained through catalytic reforming.

E2.1 STATEMENT OF THE PROBLEM

In a sample flow scheme for treating the aromatics from catalytic reforming, the following operations are successively performed (Fig. E2.1):

1. C_9^+ aromatics are removed by distillation.

2. Ethylbenzene is recovered by superfractionation.

3. *p*-Xylene with *m*-xylene is distilled out of the xylene mixture, and 99.9% *p*-xylene recovered by adsorption.

4. *m*-Xylene is distilled out of the residual C_9 aromatics left in the mixture.

Isomerization of the *m*-xylene and the C_8 aromatics is not considered. The problem is to know if there may not be a better process scheme: for example, first adsorbing the *p*-xylene or first separating the *o*-xylene.

Certain aspects of this comparison will be immediately apparent to experienced process design engineers. For example, all the schemes (Figs. E2.1 to E2.3) boil xylenes one more time than would be required if the C_9 fraction

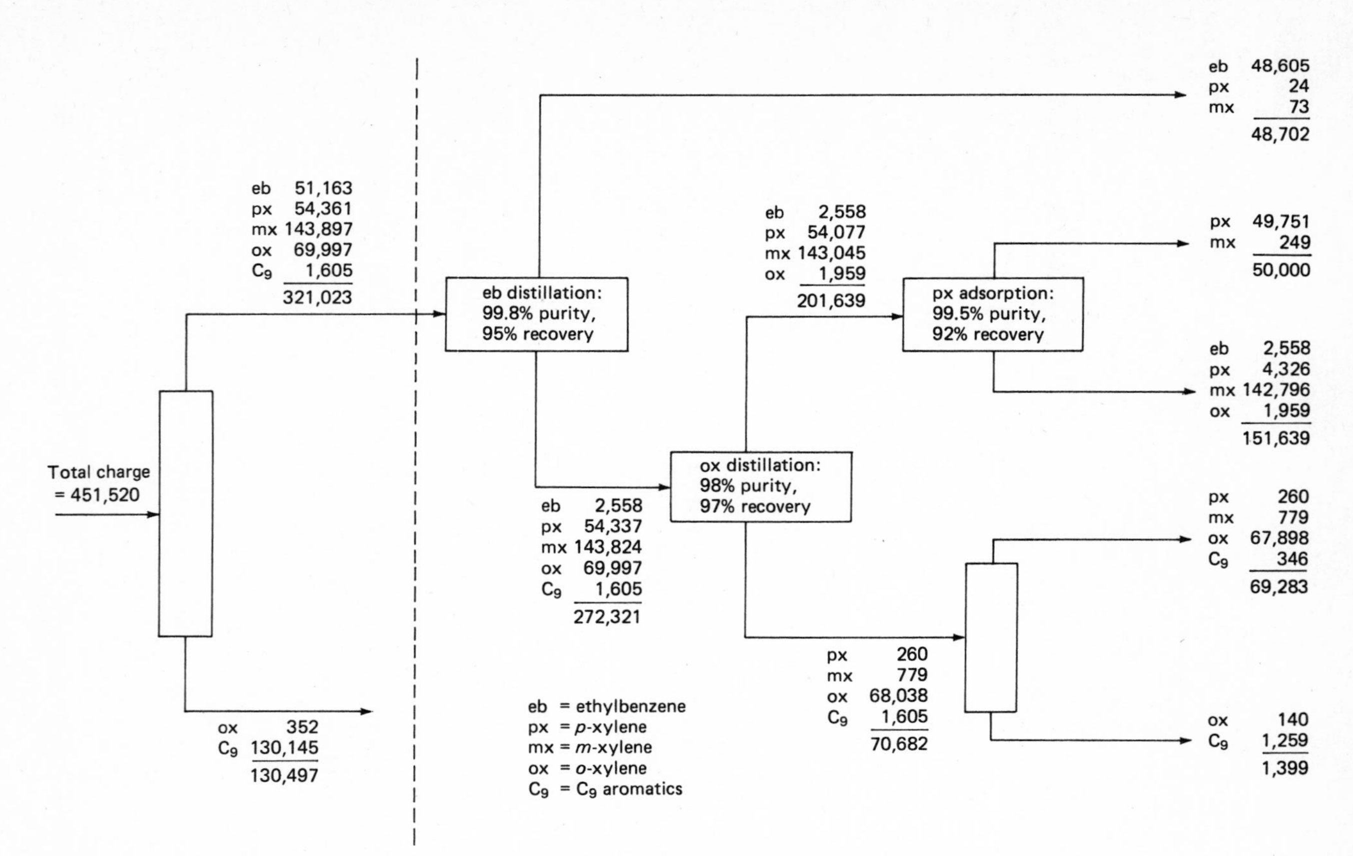

Fig. E2.1 Flow scheme for separating C_8 aromatics (flows are in metric tons).

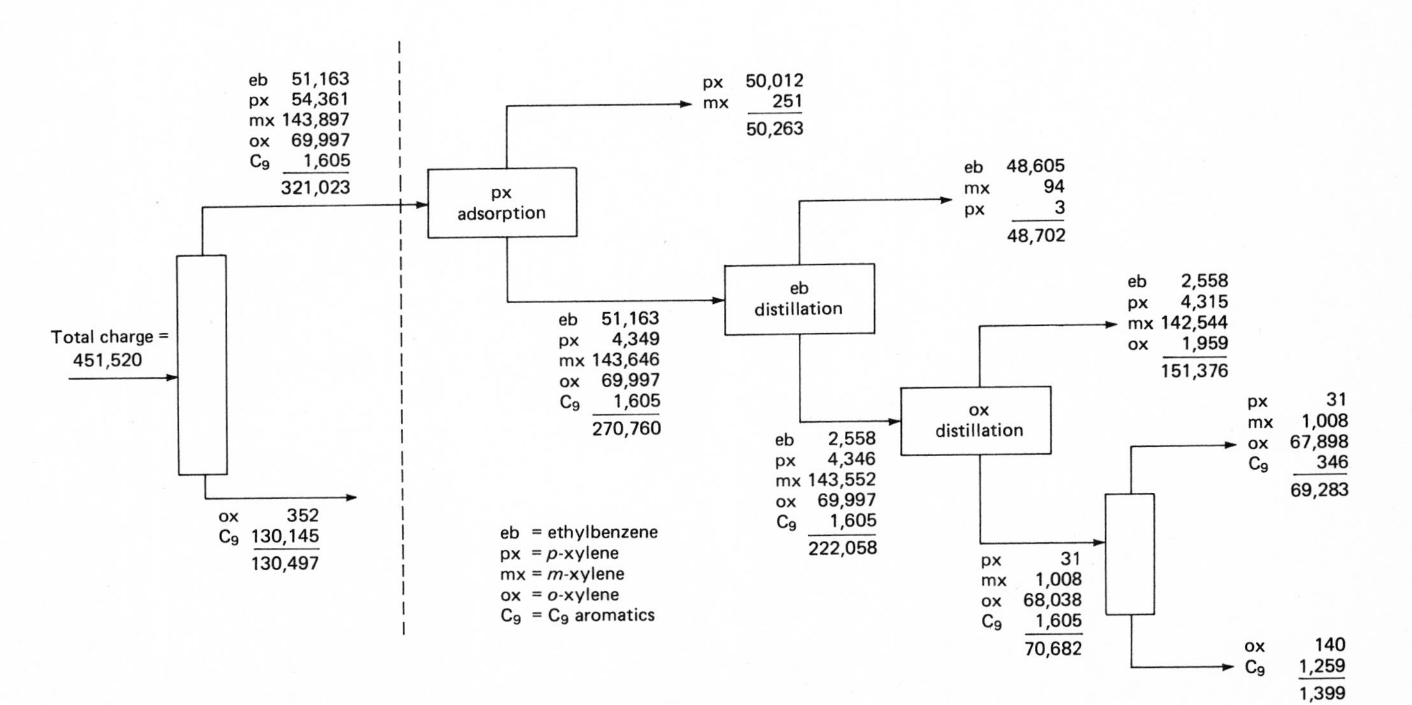

Fig. E2.2 Flow scheme for separating C_8 aromatics (flows are in metric tons).

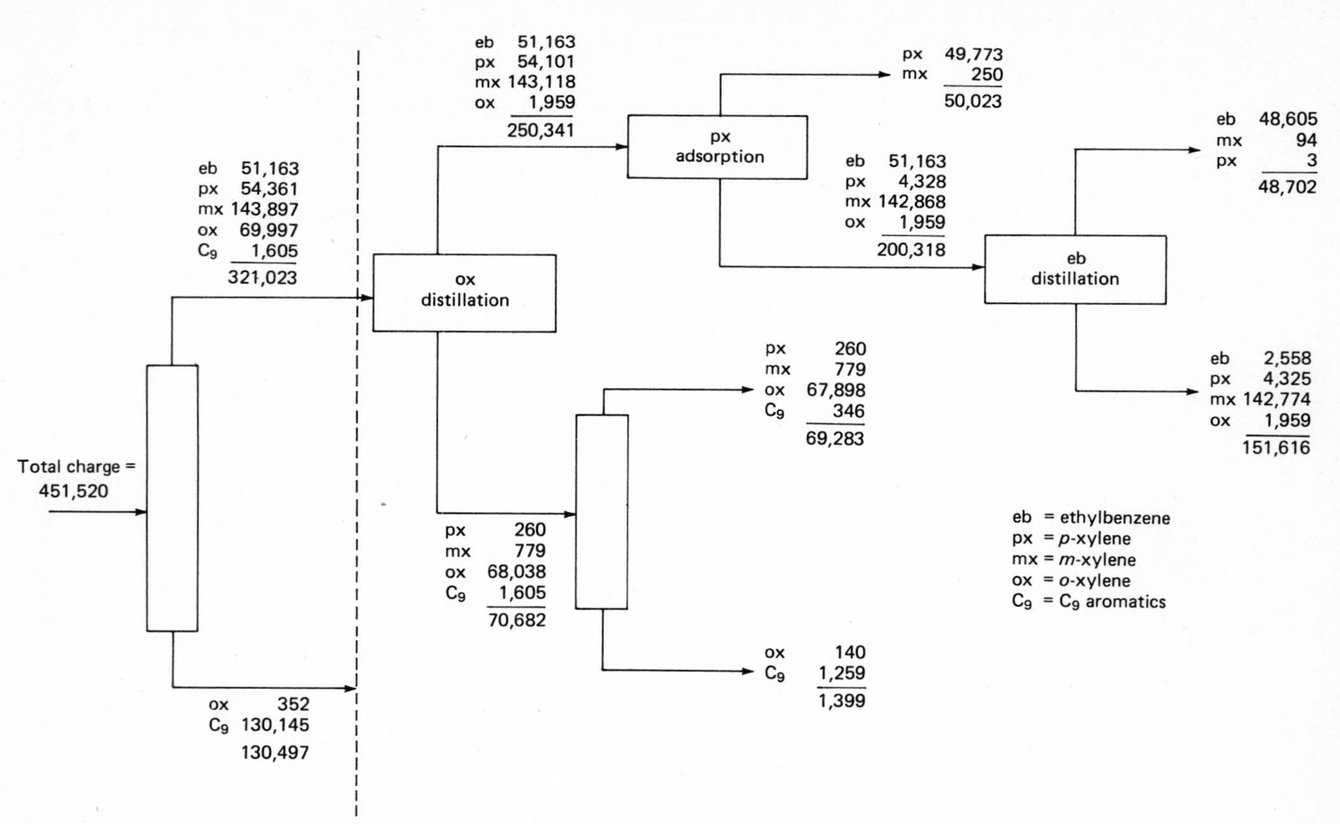

Fig. E2.3 Flow scheme for separating C_8 aromatics (flows are in metric tons).

were removed as a single residual distillation cut at the end of the process. However, the purpose of this study is to show the accumulated effects of all such considerations on the economics. Material balances are shown in the figures. Specific sizing procedures and estimating data (see Apps. 1 through 11) can be used to calculate costs for the specific operations of distillation and adsorption shown in the schemes. When the numbers have been assembled (Table E2.1), it is a matter of comparing the following three criteria: (1) payout time, not discounted, (2) cumulative net present value, and (3) rate of return, always assuming that

- Profits are after taxation (rate $a = 0.50$).
- Financing is handled by the company.
- The salvage value of the units is 0 at the end of the depreciation period.
- The receipts and costs are the same from year to year.
- The minimum acceptable discounting rate is 10%.

E2.2 SOLUTION OF THE PROBLEM

The solution of the problem is found in the results of calculating payout time, cumulative net present value, and rate of return.

E2.2.1 CALCULATING PAYOUT TIMES

From Sec. 2.6.2., the payout time (POT) is given by

$$POT = \frac{I}{B(1 - a) + A}$$

where $B = V - C$.

Using the data of Table E2.1, payout times for the three process schemes can be calculated as shown in Table E2.2. According to these results, the most economic flow scheme is that illustrated by Fig. E2.3, which also corresponds to the scheme with the lowest initial investment.

E2.2.2 Calculating the Cumulative Net Present Value

From Sec. 2.6.3.2a, the applicable equation is

$$B = -(I + f) + \sum_{p=1}^{p=n} \frac{(V_p - C_p)(1 - a) + A_p}{(1 + i)^p} + \frac{I_r + f}{(1 + i)^n}$$

with $F_p = 0$.

The simplifying assumptions give $I_r = 0$, with V_p, C_p, and A_p constant; which means that the subscript p loses its significance. As shown previously (Sec. 2.6.3.1a) the operating cost C_p is given by

$$C_p = D_p + A_p + F_p$$

and with the cost of financing F_p equal to 0, because financing is handled by the company, we get

$$B = -I - f + [(V - D - A)(1 - a) + A] \sum_{p=1}^{p=n} \frac{1}{(1+i)^p} + \frac{f}{(1+i)^n}$$

If symbols α and β are substituted for the more complex terms, as

$$\alpha = \frac{1}{(1+i)^n} \qquad \beta = \sum_{p=1}^{p=n} \frac{1}{(1+i)^p} = \frac{(1+i)^n - 1}{i(1+i)^n}$$

$$B = -[I + f(1 - \alpha)] + [(V - C)(1 - a) + A]\beta$$

and assuming that $a = 0.5$, $n = 8$, and $i = 0.10$, calculations will yield the results shown in Table E2.3. As did Table E2.2, this table indicates that the process illustrated by Fig. E2.3 offers the best economics—not a surprising conclusion, since the assumption of constant revenues and costs causes the slope of the straight line of accumulated discounted cash flow versus time to

TABLE E2.1 Economic Data* for Sample Problem

	E2.1	E2.2	E2.3
Battery-limits investment	65.5	65.5	60.0
Total investment I	120.3	122.3	112.6
Working capital f	15.3	15.5	15.3
Operating costs C_p	122.57	125.04	121.08
Exploitation costs D_p	97.73	99.80	97.75
Straight-line depreciation for 8 years A_p	15.04	15.29	14.08
Receipts, V_p	133.25	133.37	133.26

*Expressed as millions of French francs at 5.00 francs per dollar.

TABLE E2.2 Payout Time* for Sample Problem

	E.2.1	E.2.2	E.2.3
Receipts V	133.25	133.37	133.26
Operating cost C	122.57	125.04	121.08
Gross profits $V - C$	10.68	8.33	12.18
Net profits $(V - C)(1 - a)$	5.34	4.16	6.09
Depreciation A	15.04	15.29	14.08
Cash flow $(V - C)(1 - a) + A$	20.38	19.45	20.17
Depreciable capital I	120.3	122.3	112.6
Payout time, yr	5.9	6.3	5.6

*Expressed as millions of French francs at 5.00 francs per dollar.

TABLE E2.3 Cumulative Net Present Values* for Sample Problem

	Flow Scheme		
	Fig. E2.1	Fig. E2.2	Fig. E2.3
Depreciable capital I	120.3	122.3	112.6
Working capital f	15.3	15.5	15.3
Project life in years n	8	8	8
Discounting factor from Table 4.12a, $\alpha = 1/(1 + i)^n$	0.4665	0.4665	0.4665
1. TOTAL DISCOUNTED INVESTMENT $I + f(1 - \alpha)$	128.5	130.6	120.8
Annual receipts V	133.25	133.37	133.26
Exploitation costs D	97.73	99.80	97.75
Depreciation A	15.04	15.29	14.08
Financing costs F	0.00	0.00	0.00
Operating cost C	112.77	115.09	111.83
Net profits at a tax rate $= 0.5(V - C)(1 - a)$	10.24	9.14	10.72
Depreciation A	15.04	15.29	14.08
Average annual revenue	25.28	24.43	24.80
Cumulative discounting factor from Table 4.12b $\beta = [(1 + i)^n - 1]/i(1 + i)^n$	5.335	5.335	5.335
2. CUMULATIVE NET PRESENT REVENUES	134.9	130.3	132.3
Cumulative net present value **2 − 1**	6.4	−0.3	11.5

*In millions of French francs at 5.00 francs per dollar.

stay the same before and after its intersection with the time axis (see definition of payout time, Sec. 2.6.2).

E2.2.3 CALCULATING THE RATE OF RETURN

Using Tables 4.12a and 4.12b, the variations in cumulative net present value can be established as a function of the rate of discounting, as shown in Table E2.4. Calculations at the several discounting rates are identical to those shown in Sec. E2.2 for the calculation at a discounting rate of 10%. When the several discounting rates are plotted against cumulative net present value, the curves of Fig. E2.4 result; the intersections of these curves with zero cumulative net present value yield the following rates of return:

Flow scheme E2.1: 11.2%
Flow scheme E2.2: 9.9%
Flow scheme E2.3: 12.3%

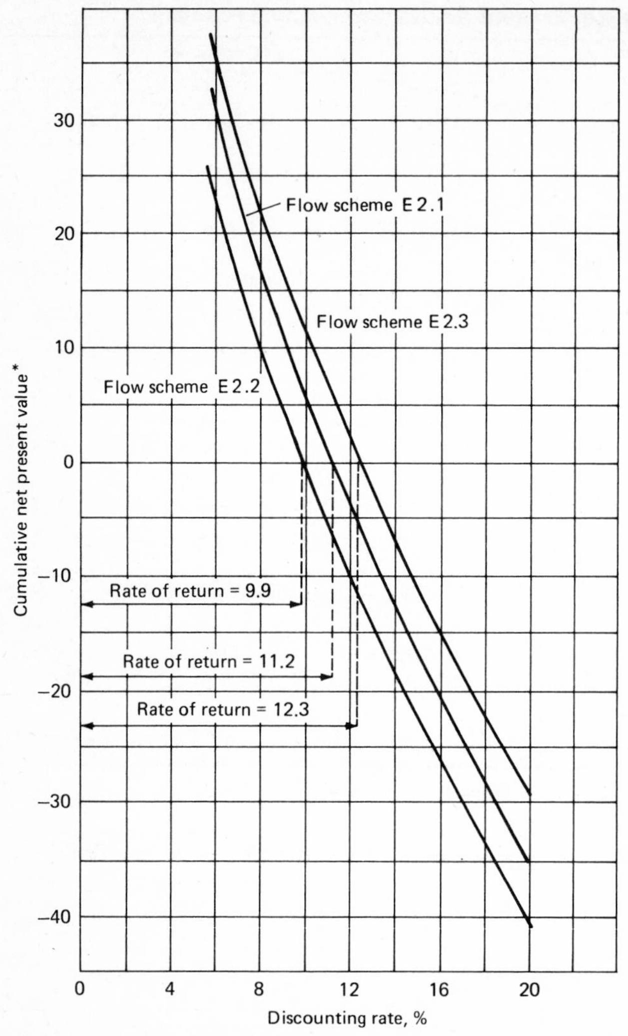

*In millions of French francs @ 5.00 francs per dollar

Fig. E2.4 Identifying rates of return for sample problem. (Cumulative net present value is shown in millions of French francs at 5.00 francs per dollar.)

This comparison confirms the superior economics of the flow scheme in diagram E2.3. Since the return for the flow scheme in diagram E2.2 is less than the company's rate of discounting, that process should be considered unacceptable. The calculations also indicate that the economics of the other two flow schemes are not particularly favorable.

TABLE E2.4 Cumulative Net Present Value at Different Discounting Rates

Discounting rate i, %	6	8	10	12	15	20
$\alpha = 1/(1+i)^n$ for $n = 8$	0.6274	0.5403	0.4665	0.4039	0.3269	0.2326
$\beta = [(1+i)^n - 1]/i(1+i)^n$ for $n = 8$	6.210	5.747	5.335	4.968	4.487	9.837
Flow sheet no.	(1) = Total investment*					
E2.1	126.0	127.3	128.5	129.5	130.6	132.0
E2.2	128.1	129.4	130.6	131.5	132.5	134.2
E.2.3	118.3	119.6	120.8	121.7	122.9	124.3
Flow sheet no.	(2) = Cumulative net present revenue*					
E.2.1	157.0	145.3	134.9	125.6	113.4	97.0
E.2.2	151.7	140.4	130.3	121.4	109.6	93.7
E.2.3	154.0	142.5	132.3	123.2	111.3	95.2
Flow sheet no.	(2) − (1) = Cumulative net present value*					
E.2.1	31.0	18.0	6.4	−3.8	−17.2	−35.0
E.2.2	23.6	11.0	−0.3	−10.1	−23.1	−40.5
E.2.3	35.7	22.9	11.5	1.5	−11.6	−29.1

*In millions of French francs at 5.00 francs per dollar.

Chapter 3

Investment Costs

Investment costs, such as have been defined in Sec. 2.1, are determined in engineering companies by an "estimating department," whose function is to effectively prepare a "bill" upon which to base a contract for engineering and constructing the plant. Such an estimate frequently requires guaranteed prices from equipment suppliers, careful accounting of bulk materials, and thousands of hours of engineering time. It is not necessary to achieve an equal precision in evaluating a project; a relative investment obtained by less precise but faster and less expensive methods will suffice.

Therefore, we propose to avoid the precise data of an estimating department and to analyze the different rapid estimating techniques so as to achieve enough precision to make a decision or at least to justify the expense of a more extensive cost study.

We will employ simplified calculation methods derived from those used in engineering offices, but the resulting precision should not be compared to that which would be achieved by an estimating department. Instead of obtaining quotations from equipment vendors, for example, we use available data and correlations. On the other hand, we try below to describe the characteristics of the best estimating methods so as avoid any supplementary errors that could add to the uncertainty of the technical information.

Since project analysis often calls for comparing an existing plant with one that is assumed or in the state of research and development, the uncertainty can be reduced by estimating the investment for each one of the two on the same basis. When technical data for the existing plant are imprecise or incomplete, its investment can be calculated by general methods, but the uncertainty accumulates.

3.1 THE STRUCTURE OF INVESTMENTS

The various types of investment that have been described in Sec. 2.1 all contain an essential element upon which the calculation of other items depend: the equipment. Production units consist mostly of equipment—auxiliary units and buildings to a lesser degree. For this reason, the investments analyzed here are classified as to the importance of the contained equipment, as either (1) battery limits and on-plot or (2) off-sites and off-plot.

Sometimes published data do not make a distinction between these two categories. Some authors, such as Lang, Bauman, and others, would associate the elements of the production units with contingent services, thus indicating equipment of another class that cannot be disassociated from the battery-limits ensemble.

For these reasons, it is necessary to be skeptical of any general information relative to a broad investment and to try to be as precise as possible about the type of units involved. Most often the information given out by engineering offices deals only with the battery-limits units. Along with this practice, there is a tendency among authors to publish the amount of the battery-limits investment as a characteristic of the technology. In other instances—a new installation, for example—a total operating budget, called *grass-roots* investment might be mentioned. With only such information, it is very difficult to find the structure of the investments.

3.1.1 THE BATTERY-LIMITS INVESTMENT

Most often, this investment comprises

1. The cost of the primary equipment, including
   - **a.** Towers and tanks
   - **b.** Storage tanks (for intermediate products)
   - **c.** Reactors
   - **d.** Heat exchangers, reboilers, condensers, and evaporators
   - **e.** Furnaces and fired heaters
   - **f.** Pumps and drivers

   - **g.** Miscellaneous equipment, such as filters, centrifuges, ejectors, dryers, etc.
   - **h.** Instrumentation (if there is automatic control)
2. The cost of secondary equipment, including
   - **a.** Platforms and structures
   - **b.** Piping and valves
   - **c.** Thermal insulation
   - **d.** Instrumentation
   - **e.** Electrical systems
   - **f.** Buildings, including the control room
   - **g.** Painting
3. The cost of civil engineering and erection, including
   - **a.** Site preparation
   - **b.** Foundations
   - **c.** Installation of the equipment
   - **d.** Railroad loops and sidings
   - **e.** Road work
4. The indirect construction and shipping costs, including
   - **a.** Renting and installing special equipment such as cranes
   - **b.** Temporary buildings
   - **c.** Taxes, insurance, and similar costs for the construction site
   - **d.** Transportation of the equipment
   - **e.** Contingencies to take care of extra costs related to unforeseen delays due to strikes, bad weather, technical changes affecting construction, and so forth

When the evaluation is made for a whole complex, including a group of related production units, the total investment is the sum of the specific investments related to each of the production units considered alone. Figure 3.1 shows the simplified composition of a typical investment.

3.1.2 THE OFF-SITES

The cost of the off-site facilities depends largely on the location of the plant. If a new unit is to be built at an existing plant, the available utilities and storage facilities should be taken into consideration, as well as the effects of any concurrent construction or expansion projects.

It is often difficult to determine the off-sites investment for projects not yet completely defined. However, a convenient solution is to assume the off-sites

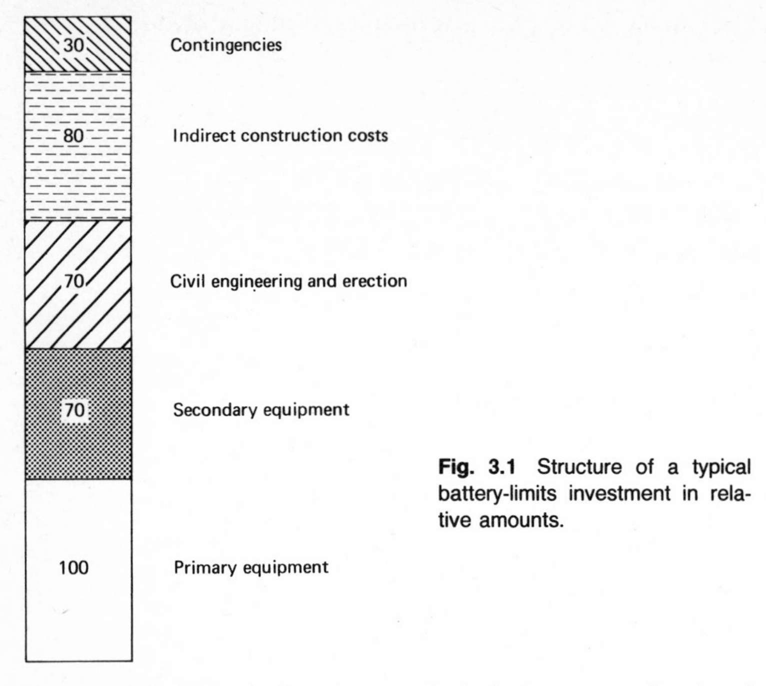

Fig. 3.1 Structure of a typical battery-limits investment in relative amounts.

is a fixed percentage of the battery-limits investment, allowing, in certain cases, for special utilities or storage facilities, which require a separate calculation. When the investment is for a completely new plant or complex, the estimation of off-sites is generally easier. Existing units do not have to be taken into account and utility and storage needs, as well as other auxiliary buildings, are completely accountable to the project under consideration. The calculation can be made either by comparing existing plants or by using curves of average costs.

3.2 THE ACCURACY OF ESTIMATING METHODS

As already stated, project analysis employs simplified cost-estimating methods with prices based on average values. Although the purpose is to compare the options available in several projects and not to fix the value of any one of them, the limits of the techniques should be known, in order to avoid the temptation of making the project comparison a tool for absolute judgment.

W. T. Nichols is one of the first to have tried to relate the accuracy of an estimate to the method of estimating, i.e., to the details of the calculations. His study is summarized in a plot that gives the highest limit of probable error on investments as a function of the available information and the volume of data. His conclusions, a little dated now (1951), have been reviewed and brought up to date by J. W. Hackney; and the improved version has been adopted by the American Association of Cost Engineers. The principal elements of these conclusions are shown in Table 3.1, which indicates the effects of various types of information and studies on the accuracy of an estimate. The cost of an estimate increases fivefold, as the accuracy is increased from ± 30 to $\pm 10\%$.

3.3 KEEPING COST DATA CURRENT

Monetary inflation, as well as increases in wages cause a continuous increase in the cost of process plants. The consequences for project analysis are that

All data related to investment costs should have reference dates.

A comparison of two projects completed at different dates requires that the investment costs be transposed from one date to the other.

3.3.1 COST INDEXES

Cost indexes are used to transpose cost data from one date to another; they take into account the average change in prices of the various elements of investment. Thus if A_1 is the cost index for the year 1 and A_2 that for year 2, the ratio of the investments for the same unit between years 1 and 2 will be equal to the ratio of the indexes, or

$$\frac{I_1}{I_2} = \frac{A_1}{A_2}$$

3.3.2 THE DIFFERENT COST INDEXES

Indexes are calculated from current data for characteristic elements of the economy, such as sheet steel, equipment, and wages. The sensitivity of any given index to one or another types of plants depends on the weight given the different characteristic elements. Accordingly, different indexes are more or less suited to different aspects of the economy; and there are a great many indexes found in technical publications. Many professional groups have them; frequently an engineering company will develop and maintain a private index for its particular needs. Also, different coefficients can be used as well, for each type of equipment, such as towers, compressors, furnaces, etc.

TABLE 3.1 Estimating Accuracy Afforded by Design Work

		Type of Estimate for Which the Design Work is Required % Uncertainty				
		Detailed ±5%	Current ±10%	Preliminary ±20%	Planning ±30%	Order of Magnitude
	Type of Design Work					
Site	Location	x	x	x	x	
	General Description	x	x	x		
	Soil study	x	x	x		
	Location and size of roads, ditches, fences	x	x	x		
	General plot plan and topographical map	x	x			
	Plot plan with equipment locations	x				
Fluid circulation	Flow scheme				x	
	Preliminary drawings			x		
	Engineering drawings	x	x			
Equipment	Preliminary sizing and material specifications			x	x	
	Engineering specifications	x	x			
	Individual specification sheets	x	x			
	General arrangement:					
	Preliminary		x	x		
	Final	x				
Buildings and structures	Approximate dimensions; type of construction			x	x	
	Foundation layout		x	x		
	Construction plot plan	x	x	x		
	Preliminary drawings of structures			x		
	Plan and elevation drawings	x	x			
	Detailed drawings	x				
Utilities	Approximate requirements				x	
	Preliminary heat balances			x		
	Preliminary flow sheets			x		
	Final heat balances	x	x			

		Detailed ±5%	Current ±10%	Preliminary ±20%	Planning ±30%	Order of Magnitude
	Final flow sheets	x	x			
	Drawings of details	x				
Piping	Preliminary drawings			x	x	
	Engineering flow diagrams	x	x			
	Plan and elevation drawings	x				
Insulation	Preliminary specifications			x		
	Preliminary list of equipment and piping to insulate		x			
	Final specifications	x	x			
	Detailed drawings	x				
Instruments	Preliminary list			x		
	Final list	x	x			
	Detailed drawings	x				
Electrical systems	Preliminary list of motors and approximate horsepower			x	x	
	Final motor list and horsepowers	x	x			
	Substations with horsepower specifications	x	x	x		
	Distribution specifications	x				
	Preliminary lighting specifications			x		
	Preliminary specifications of circuits and feeders		x			
	Distribution diagrams	x	x			
	Detailed drawings	x				
Labor	Engineering and drafting	x	x	x	x	
	Construction crafts	x				
	Supervision	x				
Processing scheme	(product; capacity; location; utilities; buildings and off-sites; feedstock and product storage)					x

3.3.2.1 French Cost Indexes

Although there appears to be no regularly published French cost index in the chemical field, R. Boulitrop (in *Techniques de l'Ingénieur*) has proposed a general industrial index for which factors are published monthly in the *Bulletin Officiel du Service des Prix* (*Institut National de la Statistique et des Études Économiques*) or in *Usine Nouvelle.* The factors employed in this index are

The weighted total average wages S in the mechanical, electrical, and refractory products industries.

The average price of 4-mm-thick basic open-hearth sheet steel (Martin steel).

The same two journals also publish indexes for relations such as

$$A_2 = A_1\left(0.10 + 0.35\,\frac{F_2}{F_1} + 0.05\,\frac{C_2}{C_1} + 0.50\,\frac{S_2}{S_1}\right)$$

where F = ferrous metals index
C = nonferrous metals index
S = salary index

Also, more exact formulations have been developed for specific types of equipment, for example,

For equipment made of ordinary steel:

$$P = P_0\left(0.10 + 0.30\,\frac{T_{ma}}{T_{ma_0}} + 0.60\,\frac{S}{S_0}\right)$$

For equipment made of stainless steel:

$$P_i = P_{i_0}\left(0.10 + 0.55\,\frac{A_{idf}}{A_{idf_0}} + 0.35\,\frac{S}{S_0}\right)$$

For equipment installations:

$$P' = P'_0\left(0.10 + 0.90\,\frac{S}{S_0}\right)$$

In these equations, P, P_i, and P' are the current index values; subscript 0 indicates the reference values, and

T_{ma} = index based on the average cost of a sheet of basic open-hearth steel with a thickness between 2 and 5 mm (basis = 100 on Jan. 1, 1960)

A_{idf} = index based on the average cost of $\frac{18}{8}$ stainless steel (basis = 100 on Jan. 1, 1960)

S = index based on the average cost of labor in the mechanical and electrical industries (basis = 100 on Jan. 1, 1973)

The evolution of costs in France is indicated by monthly values of these indexes, relative to the reference, as shown in Table 3.2.

3.3.2.2 U.S. Cost Indexes

Several indexes are published regularly in the United States. The best known are the following.

3.3.2.2a The Engineering News Record *(ENR) Construction Cost Index*

Appearing weekly in *Engineering News Record* magazine and bimonthly in *Chemical Engineering* magazine, this index includes the average numbers obtained for 20 different cities in the United States; it takes into account the costs of assumed quantities of steel shapes, structural timber, and cement, as well as 200 hours of labor. The base is 100 in 1913.

The index fails to take into consideration improvements in technology, because of the way it is made up. When applied to the chemical industry, it varies too much from actual experience.

3.3.2.2b The Marshall and Swift Index (M&S) (previously the Marshall and Stevens Equipment Cost Index)

Resulting from the periodic determination of detailed costs of equipment, as well as costs for the corresponding plants, there are two indexes prepared by Marshall and Stevens, Inc.:

1. "All industry" index comprising the arithmetic average for 47 types of industrial, commercial, or construction equipment.

2. "Process and related industries" index comprising the weighted average of the cost index of units in 12 industries, including petroleum, rubber, cement, glass, paints, paper, clay, electrical equipment, mines, crushing and grinding, refrigeration, and steam producing.

These indexes (basis = 100 in 1926) are published bimonthly in *Chemical Engineering* magazine.

3.3.2.2c The Nelson Refinery Construction Cost Index

W. L. Nelson has introduced various composite indexes that are weighted averages of equipment, plant costs, material prices, and labor costs. The Nelson refinery inflation construction cost index (more commonly called the *Nelson index*) and the Nelson true refinery construction cost index relate to the productivity obtained in constructing complete refineries. Others such as the Nelson refinery operating cost index, which is the weighted average costs of fuels, labor, investment, and chemical products, relate to operating costs.

TABLE 3.2 Cost Trends for Steel and Labor in France for the First Half of the 1970s

Cost Index Published	A_{idf}	T_{ma}	S
1970			
January	150	142	68.4
February	150	142	69.4
March	150	143	70.3
April	154	145	71.3
May	154	145	72.2
June	154	145	72.5
July	154	145	72.8
August	154	145	n.p.*
September	158	145	73.4
October	158	133	74.5
November	158	133	75.0
December	158	133	n.p.
1971			
January	158	133	77.2
February	161	133	78.1
March	161	143	78.8
April	161	143	80.0
May	161	143	80.6
June	161	143	80.9
July	161	143	81.6
August	161	143	81.6
September	161	143	82.5
October	161	143	n.p.
November	161	138	n.p.
December	161	134	n.p.
1972			
January	161	128	87.0
February	161	128	87.7
March	161	128	89.2
April	161	131	90.2
May	161	131	90.8
June	161	131	91.8
July	161	138	92.4
August	161	138	92.4
September	168	138	93.4
October	171	141	95.3
November	171	141	95.9
December	171	141	n.p.
1973			
January	171	141	100.0
February	171	141	100.7
March	171	141	101.6
April	171	151	103.2
May	171	152	103.9
June	171	158	104.8
July	171	158	106.4

TABLE 3.2 Cost Trends for Steel and Labor in France for the First Half of the 1970s *(Continued)*

Cost Index Published	A_{idf}	T_{ma}	S
1973 *(continued)*			
August	171	168	106.4
September	171	168	107.8
October	171	168	110.6
November	171	172	111.5
December	171	178	112.5
1974			
January	171	182	117.1
February	179	198	118.5
March	179	198	120.5
April	179	211	122.8
May	179	230	124.4
June	192	230	125.4
July	192	237	127.8
August	192	243	127.8
September	192	243	129.6
October	195	243	131.2
November	195	243	133.2
December	195	238	134.9
1975			
January	195	233	142.1
February	202	220	143.2
March	202	205	145.5
April	202	206	147.2
May	202	206	148.3
June	199	206	149.7
July	199	206	154.8
August	199	206	154.8
September	199	198	156.6
October	194	198	158.1
November	194	198	159.4
December	194	198	160.2

*n.p. = index not published.

Of these, the one most used is the inflation index, which can be compared directly with other current total averages developed by different authors. It is built on three principal components: equipment (12%), materials (ferrous metals: 20%, nonmetallic materials of construction: 8%), specialized labor (39%), and nonspecialized labor (21%). The evolutionary trends, as well as the composite inflation index that comes from these, are shown in Table 3.3. The equipment index is of itself the result of a weighted average of costs of (1) pumps and compressors, (2) electrical equipment (motors, transformers, welding equipment, lighting, etc.), (3) drivers other than electric, (4) instruments, and (5) heat exchangers.

The Nelson composite indexes, particularly the inflation index, are pub-

TABLE 3.3 Cost Trends for Equipment, Materials, and Labor for Refinery Construction:
The Nelson Refinery (Inflation) Index

Year	Miscellaneous Equipment	Materials	Labor	Nelson Index
1946	100.0	100.0	100.0	100.0
1947	114.2	122.4	113.5	117.0
1948	122.1	139.3	128.0	132.5
1949	121.6	143.6	137.1	139.7
1950	126.2	149.5	144.0	146.2
1951	145.0	164.0	152.5	157.2
1952	153.1	164.3	163.1	163.6
1953	158.8	172.4	174.2	173.5
1954	160.7	174.9	183.3	179.8
1955	161.5	176.1	189.6	184.2
1956	180.5	190.4	198.2	195.3
1957	192.1	201.9	208.6	205.9
1958	192.4	204.2	220.4	213.9
1959	196.1	207.8	231.6	222.1
1960	200.2	207.8	241.9	228.3
1961	199.5	207.7	249.4	232.7
1962	198.7	205.9	258.8	237.6
1963	201.4	206.3	268.4	243.6
1964	206.8	209.6	280.5	252.1
1965	211.6	212.0	294.4	261.4
1966	220.9	216.2	310.9	273.0
1967	226.1	219.7	331.3	286.7
1968	228.8	224.1	357.4	304.1
1969	239.3	234.9	391.8	329.0
1970	254.3	250.5	441.1	364.9
1971	268.7	265.2	499.9	406.0
1972	278.0	277.8	545.6	438.5
1973	291.4	292.3	585.2	468.0
1974	361.8	371.3	623.6	522.7
1975				
January	414.8	417.0	652.5	558.3
July	415.2	416.7	686.1	578.3

lished in the first issue of each month of the weekly *Oil and Gas Journal.* A detailed recapitulation of the partial indexes is given in the first number of January, April, July, and October. This deals with the cost of utilities, of mineral and organic chemical products, of labor, of equipment, and of materials—each of which carries a reference indicating the source of the data. Table 3.4 shows the trend of the cost indexes for types of equipment widely used in the refining and petrochemical industries.

The inflation indexes and the true construction cost index, as well as the components, labor, equipment, and materials, all use a basis of 100 for the year 1946; the others use 100 for 1956.

3.3.2.2d The Chemical Engineering *(CE) Plant Cost Index*

Of more recent creation, this index is published in each bimonthly issue of *Chemical Engineering* magazine. The basis is 100 in 1957–1959 (Table 3.5). It is the weighted average of four principal components: equipment, rotating machinery and structures (61%), erection (22%), buildings (7%), and engineering (10%). The first component is itself made up of the following components: assembled equipment (37%); pumps and compressors (7%); other machines (14%); piping, valves, and fittings (20%); instrumentation and controllers (7%); electrical equipment (5%); structures, insulation, and painting (10%).

Figure 3.2*a* affords a comparison of these four U.S. indexes, all transposed to a common basis of 100 in 1958. This comparison shows that the CE cost index and the M&S equipment cost index undergo practically the same variations. They are, by their makeup, the most representative of the chemical industry in general.

The ENR index undergoes a quasi-exponential evolution that is not in agreement with the actual observed progression of costs for plants. This may be due to the manner of weighing the components. An excessive weight is given to labor and wooden structures, while the effect of steel usage, by comparison, remains modest.

TABLE 3.4 Cost Trends for Different Types of Equipment: Component Indexes for the Nelson Index

Year	Pressure Vessels	Pumps and Compressors	Heat Exchangers	Electric Motors	Furnaces	Instruments
1957	176.3	206.7	203.6	181.2	115	187.4
1958	175.1	214.7	181.2	186.4	111	194.9
1959	177.2	226.5	178.9	186.4	109	201.0
1960	181.0	228.3	194.6	182.2	98.0	202.5
1961	177.3	228.8	188.1	172.8	95.0	207.5
1962	181.5	222.5	183.6	166.4	90.4	214.8
1963	183.1	224.3	189.1	165.1	91.1	224.5
1964	184.6	231.5	206.1	161.0	91.5	225.3
1965	186.0	245.6	218.4	159.8	90.2	221.5
1966	189.9	260.5	235.2	163.7	91.8	229.7
1967	192.7	272.2	237.9	171.0	93.0	232.3
1968	198.5	284.5	223.4	175.3	95.4	239.1
1969	207.8	298.6	235.8	184.7	100.8	252.8
1970	224.3	313.1	253.8	199.9	108.6	278.8
1971	237.0	330.6	268.0	206.4	118.0	306.5
1972	246.4	337.5	274.3	211.0	125.3	328.4
1973	262.5	346.9	313.7	219.8	130.0	338.0
1974	318.3	416.3	501.0	250.4	157.5	376.5
1975						
February	359.0	501.5	539.4	299.1	177.6	413.1
August	356.7	513.8	547.6	304.8	178.7	423.2

TABLE 3.5 Cost Trends by Selected Indexes in Europe and America

Year	United States *Chemical Engineering* (average)	United Kingdom *Process Engineering* (January)	West Germany *Chemische Industrie*	The Netherlands WEBCI (average)
1950	73.9	41.7	53.5	—
1951	80.4	48.8	63.7	—
1952	81.3	53.5	72.7	—
1953	84.7	52.6	74.3	—
1954	86.1	54.1	74.3	—
1955	88.3	59.4	77.0	—
1956	93.9	62.5	80.7	—
1957	98.5	64.5	84.5	—
1958	99.7	65.6	88.2	—
1959	101.8	67.5	88.8	71
1960	102.0	68.2	91.4	75
1961	101.5	69.7	95.5	79
1962	102.0	70.9	100.0	81
1963	102.4	72.0	101.1	84
1964	103.3	74.3	104.6	89
1965	104.2	77.0	108.7	93
1966	107.2	81.6	111.1	100
1967	109.7	83.0	107.8	100
1968	113.7	87.9	102.8	104
1969	119.0	92.7	110.1	105.1
1970	125.7	100.0	125.7	119.4
1971	132.3	109.4	135.5	129.2
1972	137.2	118.9	140.0	137.0
1973	144.1	128.0	145.5	144.0
1974	165.4	146.0	156.6	170.0 (Oct.)
1975	179.4 (Jan.) 182.1 (July)	190.0	163.8 (Jan.) 167.6 (July)	—
1976	—	236.0	—	—

The Nelson refinery (inflation) construction cost index holds an intermediate position between the ENR and the CE indexes. Thus, as its name indicates, it reflects changes registered in the petroleum industry, most particularly in refining.

3.3.2.3 Western European Indexes

A number of indexes, published in Great Britain, West Germany, and the Netherlands, have recently been developed to reflect local cost trends in Western Europe.

3.3.2.3a British Indexes

The *plant construction cost index* can be mentioned here. This index is calculated as follows:

$$P = 0.46M + 0.09E + 0.18C + \frac{0.075N(I^0)^2}{V}$$

where M, E, C = price indexes of equipment and energy used by the mechanical, electrical, and construction industries, respectively

I^0 = index for the average wage in the British construction industry

V = index for the cost of investments in the food, tobacco, refining, chemical, painting, and ink-making industries

N = number of employees in the construction industry

The necessary information is obtained from the British Department of Trade and Industry and Department of Employment and Productivity. This index is published monthly in *Process Engineering* magazine. Previously established with a basis of 100 in 1965, the index has been modified and adapted to a basis of 100 in 1970 (Table 3.5). *Process Engineering* also publishes the *Productivity Index* for manufacturing.

Also, the magazine *The Cost Engineer* publishes monthly the composite cost indexes for four types of petrochemical units. These differ by (1) the average weight given the principle equipment and secondary equipment, and (2) the weight given to complementary items such as erection. The component indexes that make up the composite are obtained from the Board of Trade and the Ministry of Labor. The basis is 100 in 1958.

3.3.2.3b West German Indexes

Chemische Industrie publishes monthly a cost index specifically for chemical plants, which is made up of the prices of the following components of installed plants: nonmoving equipment (26.7%), rotating machines (13.3%), piping (16%), steel structures (5%), electrical equipment (7%), instrumentation (7%), industrial buildings (12%), and indirect construction costs (13%). The evolution of these component indexes is determined from documents published by Statistichen Bundesamtes zur Verfügung. The basis is 100 in 1962.

3.3.2.3c Dutch Indexes

The Dutch organization WEBCI has constructed a cost index weighted for chemical plants as follows:

- Year of reference: 1966 (basis 100)
- Average composition: primary equipment (33%); piping (8%); electrical systems (3%); instrumentation (8%); civil engineering and steel structures (6%); construction, engineering, and general (42%)
- Component indexes published by Centraal Bureau Voor de Statistick

Figure 3.2*b* compares the European indexes with the U.S. CE Cost Index on a common basis of 100 in 1962.

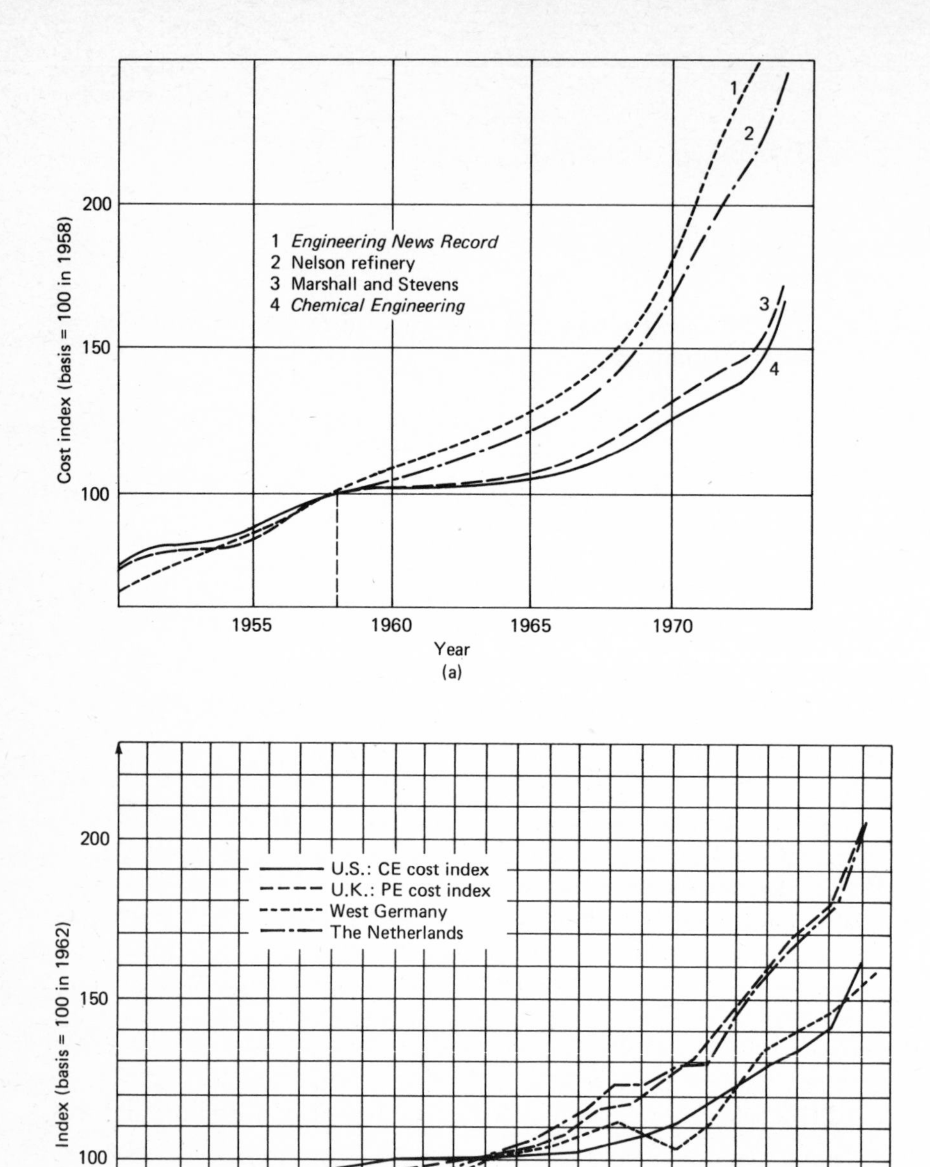

Fig. 3.2 Comparative cost trends shown by (*a*) United States indexes and by (*b*) United States and European indexes.

3.4 EFFECTS OF SITE LOCATION

The effect on costs of locating in one country or another often invites comparison. Similar plants of the same size are likely to require different investments if built at different sites, because of local variations in the costs for labor, supervision, and materials; of different standards for equipment, construction, and safety; and of different practices within engineering companies.

Most importantly, site location affects the location of two activities: equipment fabrication and construction. In heavily industrialized countries, the equipment is often fabricated in the same area as the plant is constructed; in developing countries, however, the two activities must be treated separately, since equipment either made of high-alloy steel or depending on a particular technology will generally be imported along with specialized personnel to install it. In the latter case it is necessary to include the cost of shipping the imported equipment plus the expense of bringing in the qualified personnel. Conversely, the advantages of low-cost local unskilled labor should be taken into account.

Unfortunately, information on the influence of the site location is relatively rare, and reliable correlations have not been possible.

In 1963, W. L. Nelson related the cost of plants built in the United States and abroad to local wage practices. He showed, in a general way at that time, that inexpensive local labor (South America, Southeast Asia, Near East, Spain, etc.) went with high construction costs. The exceptions were Japan and Belgium, which had free-enterprise governments and highly qualified but poorly paid personnel, as well as Sweden and Spain, which had a large number of government-owned companies and a labor force with high wages (in Sweden) and low wages (in Spain).

In 1969, R. J. Johnson published the relative costs (basis: U.S. Gulf Coast) of equipment and plants for the principal industrialized countries (Tables 3.6 and 3.7). The following average relative coefficients from various up-to-date data can be added to these tables:

Country	Coefficient
Australia	1.20
India	1.35
Sweden	1.09
Spain	1.00

The relative values given by R. J. Johnson for erected plants (battery-limits investments) lead to lower costs than those furnished by W. L. Nelson. These differences come as much from inaccuracy of the methods and the available information as from the change in conditions between 1963 and 1969—condi-

TABLE 3.6 Relative Costs of Process Equipment at Different Locations

	United States*	United Kingdom	Netherlands	Belgium	West Germany	France	Italy	Japan
Furnaces	1.00	0.95	0.90	0.90	0.90	1.00	1.00	0.90
Tanks	1.00	1.00	1.10	1.00	0.98	0.95	0.85	0.80
Heat exchangers	1.00	1.10	1.00	1.00	0.90	1.10	0.85	1.15
Pumps	1.00	0.90	0.90	0.85	0.85	0.82	0.80	0.80
Compressors	1.00	0.85	0.85	0.85	0.75	0.80	0.80	1.00
Piping	1.00	0.91	1.00	0.95	0.91	0.88	0.87	0.90
Structures	1.00	0.90	0.95	0.95	0.90	0.90	0.85	0.85
Instruments	1.00	1.05	1.05	1.05	1.10	1.05	1.00	1.00
Insulation	1.00	0.95	1.00	1.00	0.90	0.90	0.90	0.95
Electricity	1.00	0.95	0.95	0.93	0.93	0.90	0.88	0.85
AVERAGE	1.00	0.96	0.97	0.96	0.91	0.92	0.89	0.91

*Gulf Coast, 1969.

TABLE 3.7 Relative Cost of Process Plants at Different Locations

	United States*	United Kingdom	Netherlands	Belgium	West Germany	France	Italy	Japan
Materials and equipment	1.00	0.96	0.97	0.96	0.91	0.92	0.89	0.91
Land	1.00	0.95	0.95	0.95	0.90	0.90	0.85	0.80
Engineering	1.00	0.75	0.80	0.90	0.80	0.90	0.80	0.70
AVERAGE	1.00	0.91	0.92	0.94	0.88	0.91	0.86	0.83

*Gulf Coast, 1969.

tions that were particularly dependent on the variations in rate of exchange for money and changes in the economies of the different countries.

In 1973, J. Cran made a systematic study of costs of plants in various industrialized countries, using Great Britain as a reference. For a certain number of refining and petrochemical plants, on which he had data for identical production capacities, he determined the representative average costs from distribution curves and then the corresponding location indexes. The results are overall relative coefficients by country. For example, a factor of 1.2–1.3 is obtained for the United States relative to Great Britain.

More recently, Ph. Terris has proposed a breakdown of the costs of erected plants (Table 3.8) in order to compare investments in France with those in developing countries. To do this, he made three distinct categories of countries according to the level of development, particularly according to the aptitude in civil engineering work and erection.

Type I: Countries already having industrial experience, especially in petrochemicals, for which outside help consists of supervision and supporting personnel

Type II: Countries with local labor having good experience in civil engineering and public works and with modern equipment in those fields

Type III: Countries where the labor is poorly qualified and where the equipment for civil engineering and erection is nonexistent

None of these countries have heavy-metal or mechanical industries; and equipment is for the most part imported.

In the course of becoming industrialized, those countries that belong to the less favored categories progressively approach those countries that are better equipped. The coefficients that relate to them should thus be improved from time to time.

Finally, in a book published in 1974, K. M. Guthrie presents coefficients for location that take into account mid-1970 equipment costs, labor costs, and indirect costs (Table 3.9).

TABLE 3.8 Cost of Refineries and Petrochemical Plants According to Industrialization

	France	Type I	Type II	Type III
Primary equipment	38	42	40	42
Secondary equipment	19	21	20	21
Erection (including supervision)	26	31	44	54
Civil engineering (including supervision)	17	18	19	30
TOTAL	100	112	123	147

3.5 AVAILABLE ESTIMATING METHODS

This section discusses the principal estimating methods to be found in the literature; it does not attempt to present a logical system capable of being used without ambiguity. (That is done in the second part of this book.) Consequently, nomenclatures for the calculation methods reviewed are those of the authors; and there is some conflict of meanings for the same term between the different methods.

Three distinct approaches—prorating known investments, factoring major equipment, or factoring materials characteristics—are to be found among the various techniques for calculating the basic investment of a plant, i.e., the investment for battery limits plus off-sites. The methods based on known investments allow for determining the investment for new units by extrapolation. The methods based on primary equipment apply constant multiplying factors in order to arrive at the total investment. The methods based on materials arrive at the total investment by means of multiplying factors that depend on the nature of the materials (size, quality, operating conditions, etc.). The calculations using multiplying factors are more complex but more exact than those that depend on extrapolation.

3.5.1 EXTRAPOLATING KNOWN INVESTMENTS

In these methods it is assumed that investments are known for at least one plant of reference, either for the entire plant or for a given part of it. Included are exponential extrapolation, factoring operating units (R. D. Hill), factoring functional units (F. C. Zevnik and R. L. Buchanan), factoring products and

TABLE 3.9 Relative Costs of Equipment, Labor, and Indirect Costs at Different Locations*

Country	Coefficient	Country	Coefficient
Canada	1.05	Australia	1.07–1.11
United States	1.00	West Germany	0.92
Mexico	1.08	Belguim	0.96
Puerto Rico	1.04	Denmark	0.86
Trinidad	1.08	Spain	0.76
Argentina	1.06	France	0.98
Bolivia	0.98	Northern	
Brazil	1.10	Ireland	0.87
Chile	1.08	Southern	
Peru	0.86	Ireland	0.79
Uruguay	1.06	Italy	0.96
Venezuela	0.92	United Kingdom	0.89

*In mid-1970

functional units (E. A. Stallworthy), factoring product characteristics and equipment (G. T. Wilson), and factoring delivered equipment costs (D. H. Allen and R. C. Page).

3.5.1.1 Exponential Factors: The Six-tenths Factor

This method rests on the fact that the costs of two units using the same process and having different production capacities are related by

$$\frac{I_1}{I_2} = \left(\frac{C_1}{C_2} \right)^f \tag{3.1}$$

where I_1 and I_2 are the investments for the two units, C_1 and C_2 their capacities, and f is the factor (improperly called *factor,* because it is an exponent).

The use of this exponent was first of all conceived for extrapolating costs of various categories of equipment. Its use for estimating process units was not discovered until later.

Before applying the above expression, it is important to verify that the unit under question does not represent a variant flow scheme for the given process. Sometimes a change in capacity calls for multiplying or dividing the number of production lines, which introduces discontinuities in the law of extrapolation and calls for defining the range of capacities to which the coefficient applies.

The exponent f is frequently called the *six-tenths factor,* because experience with many plants shows that the exponent falls between 0.5 and 0.7. When the investment is known for the same process design at two different capacities, the exponent can be determined as the slope of a line connecting the two cost-capacity points on a log-log scale. This method can be used with assurance as long as it is certain that the plants are the same, with the same number of pieces of equipment.

Various publications furnish information on exponential factors for the principal types of equipment or for process units. Most of these are listed in the bibliography. Thus, from the extrapolation factor f and data from a plant of reference, it is possible to rapidly calculate the investment for a plant with different capacities; and the factor is very useful as long as one recognizes its limitations, which come from

Not having a precise value for the exponent

Not knowing the range of capacities at which the exponent applies

Not being able to determine the exact investment for the plant of reference

Various authors have suggested methods for reducing these limitations. Thus the equipment can be classified according to type (furnaces, pumps, reactors, exchangers, etc.) then each type given an individual extrapolation exponent and the investments for all the types of equipment added. By comparing this calculated investment with the reference investment, Eq. (3.1) can be used to determine the characteristic exponent for the plant.

3.5.1.2 Estimates Based on Plant Characteristics

A number of rapid methods give less precise estimates based only on the general process flow without details or even a material balance. On the other hand, these methods usually require careful attention to the different stages of the process, especially the purification stages, in order to establish as complete a flow scheme as possible and not overlook any unit operations. Various authors have tried to improve the precision of these methods while holding to the initial purpose of working with a minimum of data.

3.5.1.2a R. D. Hill's Method

This is applied to plants operating at moderate pressure where only fluids circulate. Knowing the process flow scheme allows for calculating the number of operating units, which are classed as follows:

1. Mild steel equipment, including reactors, evaporators, blowers, and cyclones, counts for one unit.
2. Stainless steel equipment, including furnaces, centrifuges, compressors, and refrigeration cycles, counts for two units.
3. Each feed stream or effluent stream needing an intermediate supply counts for one unit.
4. Each solid or gaseous effluent counts for two units.

The cost assigned to each of these units is $30,000 for a production capacity of 10 million pounds per year (4,500 metric tons per year) and a Marshall-and-Stevens cost index of 185 (1954). The total cost of the plant is obtained by multiplying the base cost by the calculated number of units. An overall extrapolation exponent of 0.6 is used for different capacities, and a factor of psi/100 is applied for higher pressures. A number of coefficients, such as those suggested by C. H. Chilton for piping, instruments, buildings, auxiliaries, engineering, and miscellaneous (see Sec. 3.5.3), are used to estimate the grass roots investment.

3.5.1.2b The Method by F. C. Zevnik and R. L. Buchanan

Applying to chemical plants handling fluids, this method takes into account the following characteristics:

- The capacity of the unit
- The number of *function units,* which are defined as that equipment necessary for a given unit operation such as distillation, stripping, compression, etc.
- A complexity factor, which is determined by taking the extreme operating conditions of temperature (factor F_T) and pressure (factor F_p) and the materials required, whether common steel, special steels, alloys, etc. (factor F_a), and then calculating the complexity factor (CF) as

$$(CF) = 2.10^{(F_T + F_p + F_a)}$$

with

$$F_T = \begin{cases} 1.765 \times 10^{-4}T - 0.053 & \text{where T} > 300°\text{K} \\ -2 \times 10^{-3}T + 0.6 & \text{where T} < 300°\text{K} \end{cases}$$

$$F_p = 3 \times 10^{-4}(\log_{10} \frac{P}{P_{\text{atm.}}} - 1)$$

$$F_a = \begin{cases} 0 & \text{for cast iron, mild steel, and wood} \\ 0.1 & \text{for aluminum, copper, stainless type 400AISI} \\ 0.2 & \text{for stainless type 300, monel, nickel, and Inconel} \\ 0.3 & \text{for Hastelloy} \\ 0.4 & \text{for precious metals} \end{cases}$$

The cost of a functional unit for the process under consideration (unit cost) depends on the production capacity and the complexity factor (Fig. 3.3). The battery-limits investment is obtained by multiplying the number of functional units by the unit cost. A total grass-roots investment is obtained by means of a constant factor. If a cost index is needed, the *Engineering News Record* index is used. The final result is

$$I_g = \frac{N(\text{CPFU})1.33(\text{ENR})}{300}$$

where I_g = overall investment
N = number of functional units
(CPFU) = cost of a functional unit
(ENR) = ENR cost index on the date of estimating

The advantages of this method lie in the speed with which an estimate can be obtained in the absence of precise information. However, it does not lead to results that can be used for less complex processes. (The authors have used it successfuly for the production of cyanhydric acid, ammonia, acetylene, and phthalic anhydride.) If the flow scheme is simple (i.e., with only one synthesis unit that achieves practically complete conversion with high selectivity and easy purification), the estimated investments are as much as 100% higher than reality (for example, the hydrogenation of benzene to make cyclohexane).

Actually, the value of this method depends on two things:

1. The correct estimate of a number of functional units whose common cost is directly proportional to the total investment. Such an estimate is very delicate and calls for much attention and experience. Actually, data on processes are often incomplete as far as preparation of the feedstock, finishing the products, catalyst handling and start-up are concerned.

2. A convenient factor of complexity, in particular the coefficient F_a which depends on the type of material used. This assumes that there has been a careful study of the corrosion possibilities. If such a study cannot be done (for example, by comparing the process with similar situations), the accuracy of the method is affected, and it becomes considerably less attractive.

A major drawback of this method consists of taking a uniform base cost for the functional units. Because of this, the method does not allow for considering the actual circulation of the fluids and thus the cost of each of the functional units taken singly. To a lesser degree, the use of only one factor of complexity representing extreme conditions represents a weakness. Several authors, among them E. A. Stallworthy, G. T. Wilson, and D. H. Allen, have tried to find various improvements.

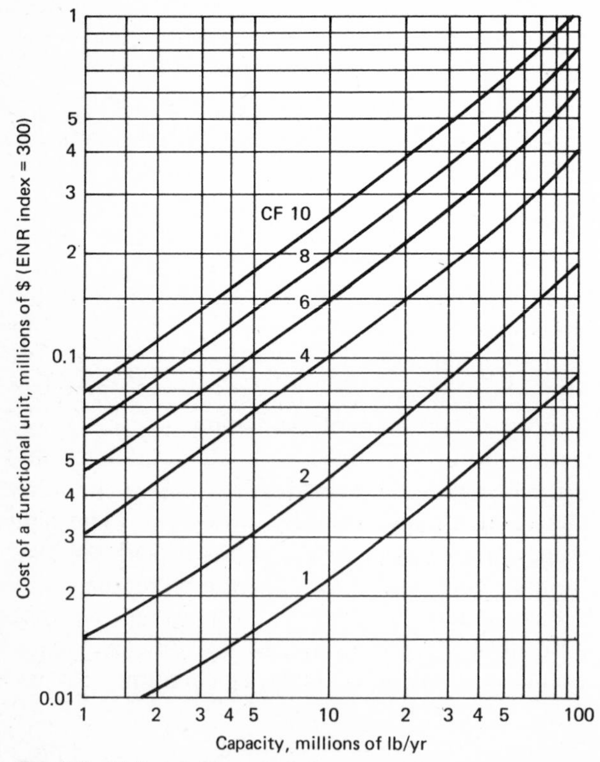

Fig. 3.3 Unit costs for calculating battery-limits investments.

3.5.1.2c E. A. Stallworthy's Method

The influence of effluents other than the primary product is included. The battery-limits investment in 1967 is given by

$$I = \frac{0.0075}{A} \sum_{i=1}^{i=s} (NF_M F_P F_T R)_i$$

where I = investment in British pounds
S = total number of both primary and secondary products
R = ratio of the rate of flow of product i to the rate of flow of the primary product
N = number of functional units used by product i
F_M = a factor specific to the materials of construction required for product i
F_p = a factor related to the operating pressure required for product i
F_T = a factor related to the operating temperature required for product i

The coefficient A, expressed in British pounds of 1967, is a function of the manufacturing capacity in terms of primary product. If V is this manufacturing capacity, metric tons per year, A is given as

$$A = 0.62 \times 10^{-5} (V)^{-0.65}$$

The values of factors F_M, F_p, and F_T are found by weighing the operating conditions and materials of construction.

In addition to the difficulties of characterizing the functional units and determining their number, this method suffers from

The definition of coefficient A, which is calculated from the capacity of the primary product alone

The calculation for ratio R, which calls for knowing the material balance around the plant

3.5.1.2d G. T. Wilson's Method

If the factors for product i are assumed to be constant in the general equation given by E. A. Stallworthy, the equation takes the form $I = k(V)^{0.6-0.7}$, in which the constant k is independent of V. C. A. Miller has shown (Sec. 3.5.4.4) that coefficient k diminishes as the size of the plant increases. The method of G. T. Wilson borrows both from the work by E. A. Stallworthy and the adaption from the Lang factor (Sec. 3.5.2) made by C. A. Miller. It allows for determining the battery-limits investment, expressed in 1971 British pounds, from the following relation:

$$I = fN(\text{AUC})F_M F_P F_T$$

where f = investment factor obtained from Fig. 3.4 as a function of AUC and depending on the nature of the products, whether fluids, solids, or mixed

N = number of pieces of primary equipment, excepting pumps, required for a unit operation in the plant

AUC = average unit cost of the primary equipment, expressed as 1971 British pounds, as a function of the average rate of flow, V, in metric tons per year, through the equipment, and given by

$$\text{AUC} = 21V^{0.675}$$

F_M = factor for the type of material used (Various values of F_M are presented in Table 3.10.)

F_p = correction factor for pressure, expressed in psia and obtained from Fig. 3.5*a*

F_T = correction factor for temperature, expressed in degrees Celsius and obtained from Fig. 3.5*b*

When operating conditions vary markedly from one primary piece of equipment to another, or if the materials of construction are different, F_M, F_p, and F_T must take weighted average values.

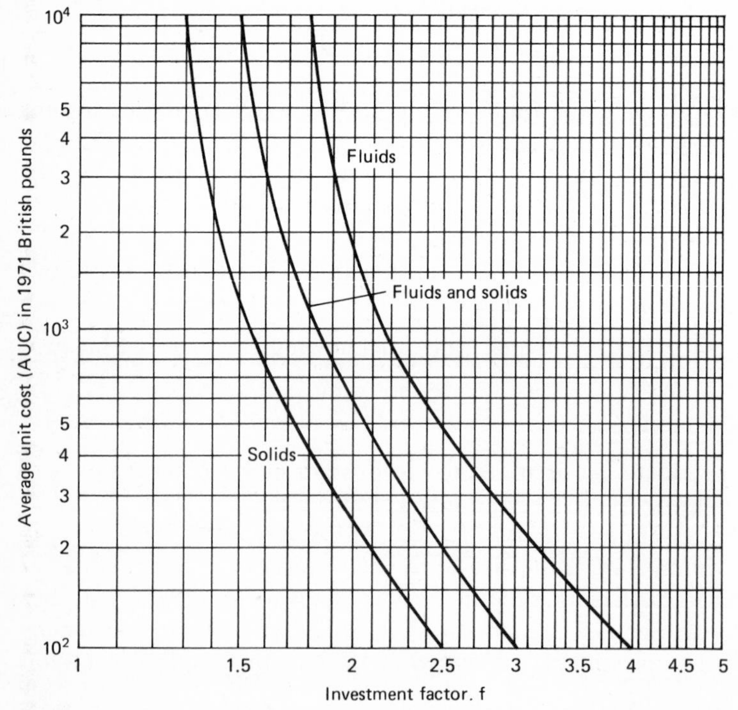

Fig. 3.4 Relation of average unit cost of primary equipment to battery-limits investment.

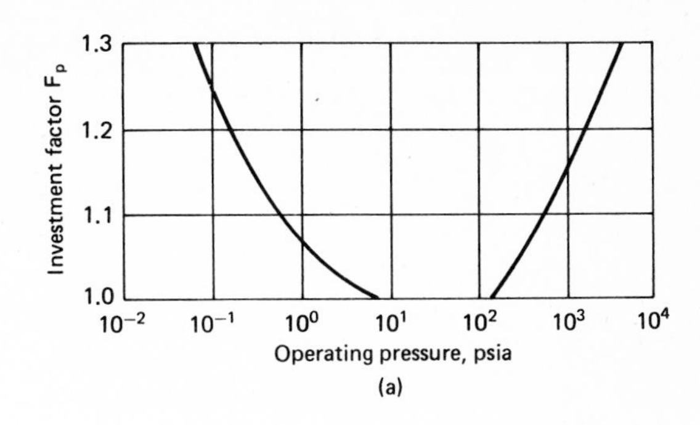

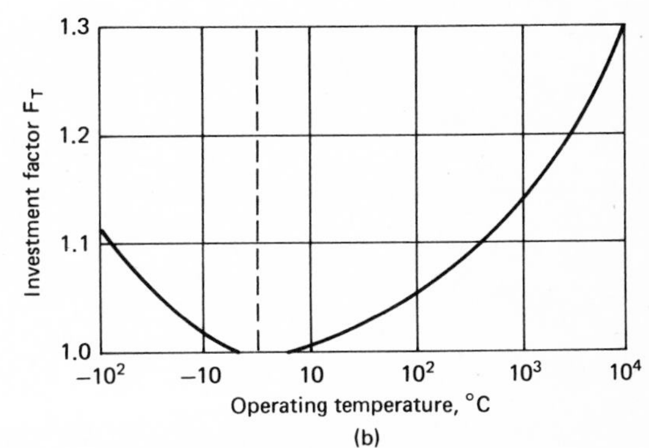

Fig. 3.5 (*a*) Relation of pressure to battery-limits investment. (*b*) Relation of temperature to battery-limits investment.

According to G. T. Wilson, the accuracy of this method is such that the investment can be calculated to $\pm 30\%$, as long as V is between 10,000 and 1,000,000 t/yr. In practice, the accuracy of the method is tied to how much can be known about the unit and its material balances—not only overall but line by line, including the general operating characteristics of the primary equipment.

3.5.1.2e The Method of D. H. Allen and R. C. Page

According to the authors, this method, which uses some of the parameters calculated by G. T. Wilson, allows for determining investments to within an accuracy of -20% and $+25\%$. It applies only to plants handling fluids.

The principle consists of determining the delivered equipment cost (DEC) as accurately as possible. The battery-limits, or even grass-roots, investment is estimated from this with statistical coefficients such as those introduced by J. E. Haselbarth and J. M. Berk (Sec. 3.5.3). The DEC varies between 15 and 30% of the grass-roots investment, depending on the situation; it is calculated as

$$\text{DEC} = N(SF)(\text{BIC})$$

where N = number of pieces of primary equipment (including pumps) as determined by a flowsheet for the process

SF = complexity factor related to the operating conditions and the materials of construction, as

$$SF = F_T \times F_p \times F_M$$

with F_T and F_p representing maximum conditions of temperature and pressure and F_M average materials requirements, and all three factors determined from Fig. 3.5*a* and *b* and Table 3.10,

BIC = the basic item cost, a function of the throughput of product (TP) and calculated as

$$\frac{\text{BIC}}{(\text{BIC})_0} = \left(\frac{\text{TP}}{(\text{TP})_0}\right)^{(\text{EXP})}$$

with BIC = 1972 dollars; TP = throughput for the unit, lb-mol/yr; $(\text{BIC})_0$ = \$7,000 in June 1972; $(\text{TP})_0$ = 2.5 million pound-moles per year; and EXP = an exponent characteristic of the type of unit. The authors recommend the Marshall and Stevens Index for updating costs.

Also, the average throughput (TP) is calculated as

$$(\text{TP}) = (\text{CAP})(\text{FF})(\text{PF})$$

where CAP is the total charge to the unit in pound-moles per year; FF is a flow factor defined as the total number of lines entering and leaving each piece of primary equipment divided by the total number of pieces of primary equipment; and PF is a phase factor defined as 0.0075 + (VI)N in which 0.0075 represents the average ratio of vapor-to-liquid densities for hydrocarbons and the vapor index (VI) is the fraction of the unit's primary equipment sized on the basis of vapors (more than 2% of the fluids vaporized in a stream).

Finally, the exponent EXP is calculated as the average weighted exponent for all the categories of equipment included in the plant, based on unit equip-

TABLE 3.10 Effect of Materials of Construction on Battery-Limits Investment

Material	Materials factor F_M
Mild steel	1.00
Bronze	1.05
High-temperature steel	1.07
Aluminum	1.08
Low-alloy steel	1.28
Austenitic stainless steel	1.41
High-alloy steel	1.50
Hastelloy C	1.54
Monel	1.65
Nickel	1.71
Titanium	2.00

ment cost and an extrapolation exponent for each category; the authors recommend data by K. M. Guthrie (Sec. 3.5.4.5) for this calculation, which data are summarized in Table 3.11.

This method offers a means for rapidly estimating investments to higher accuracy than an order of magnitude estimate. However, it suffers from the same inconveniences as the methods using constant multiplying factors (see Sec. 3.5.2).

3.5.2 METHODS USING CONSTANT MULTIPLYING FACTORS

In these methods the plant investment is obtained by means of a constant average multiplying factor for the cost of the primary equipment, which is in turn determined from the characteristics of each piece of equipment included in the plant, according to

Published data (as in Apps. 1 through 10)

Direct information from vendors

If the cost of a piece of equipment at a given capacity is available, the cost at a different capacity can be estimated with an exponential factor characteristic of the equipment (Sec. 3.5.1.1). When the characteristics of the equipment are not known, they must be established through chemical engineering calculations based on material and energy balances, and a flow scheme that includes recycles, refluxes, purges, etc. In addition to equipment sizes, such calculations must give utilities consumption, furnace and exchanger duties, pump and compressor horsepowers, etc., using theoretical values times yield factors for the equipment when necessary.

When the estimate involves research projects with only preliminary data that often are based on fragmentary results of laboratory or pilot plant experiments, the chemical engineering calculations should conform to certain precautions. This aspect of estimating is taken up in Sec. 3.5.2. The constant-factor methods bring an increased precision to profitability calculations, relative to those of Sec. 3.5.1.2, due to a better determination of the investments and utilities consumptions. Their credibility rests on the use of correct values of the multiplying factors, for which correlations have been proposed by several authors, including J. J. Lang, N. G. Bach, W. E. Hand, and C. H. Chilton.

3.5.2.1 The Lang Factor

J. J. Lang was the first to state the empirical law that the relation between the cost of a plant and its primary equipment is a constant—the result of analyzing the capital investment for the construction of a number of plants. The value

TABLE 3.11 Estimating Exponents for Various Types of Process Equipment

Equipment	Exponent	Base Cost,* $1,000
Reaction furnace	0.85	135.0
Fired heater	0.85	103.5
Steam generator		
15 psig	0.50	92.0
150 psig	0.50	101.2
300 psig	0.50	115.0
600 psig	0.50	138.0
Packaged boiler	0.70	60.0
Tubular exchanger	0.65	6.5
Reboiler	0.65	8.80
U-tube exchanger	0.65	5.5
Cooler	0.66	6.8
Cooling tower	0.60	9.9
Plate column	0.73	33.5
Packed tower	0.65	35.2
Vertical tank	0.65	7.6
Horizontal tank	0.60	5.0
Storage tank	0.30	6.0
Pressurized storage tank		
Horizontal	0.65	4.8
Spherical	0.70	8.0
Centrifugal pump with		
Motor	0.52	1.5
Turbine	0.52	3.0
Reciprocating pump		
With motor	0.70	6.0
Steam driven	0.70	1.1
Compressor		
Gases to 1,000 psig	0.82	85.0
Air to 125 psig	0.28	36.5
Crusher		
Cone	0.85	12.0
Gyratory	1.20	3.0
Jaw	1.20	4.7
Pulverizer	0.35	23.4
Grinding mill		
Ball	0.65	4.4
Bar	0.65	40.0
Hammer	0.85	8.0
Evaporator		
Forced circulation	0.70	270.0
Vertical tube	0.53	37.2
Horizontal tube	0.53	30.4
Jacketed (glass lined)	0.60	32.0
Screen		
Conical	0.68	0.1
Silo	0.90	0.4
Blower; fan	0.68	9.5

TABLE 3.11 Estimating Exponents for Various Types of Process Equipment *(Continued)*

Equipment	Exponent	Base Cost,* $1,000
Crystallizer		
Growth	0.65	385.0
Forced circulation	0.55	276.5
Batch	0.70	32.5
Filter		
Plate and frame	0.58	4.3
Pressure leaf		
wet	0.58	5.3
dry	0.53	15.1
Rotary drum	0.63	17.5
Rotary leaf	0.78	31.0
Dryer		
drum	0.45	30.0
pan	0.38	12.5
rotary vacuum	0.45	43.4

*In 1968, United States.

of this constant, or *Lang factor,* depends on the type of process, particularly the products treated or manufactured (Table 3.12).

The original intent of the author should be emphasized: The Lang factor gives the total investment, including off-sites. When the plant is limited to the treatment of fluids, Lang's values seem low, and modifications introduced by N. G. Bach lead to better results.

3.5.2.2 The Modified Lang Factor (or Bach Factor)

The factors proposed by N. G. Bach permit direct calculation of the investment for a fluids-processing plant and its associated utilities generation and storage facilities. The processing plant, the utilities production, and the storage facilities are treated as independent units (Table 3.13) for which the investment is calculated directly by multiplying a factor times the cost of the primary equipment. These factors do not include the engineering fees.

3.5.2.3 Hand's Method

Derived by W. E. Hand from Bach's method, this method consists of multiplying a specific factor times each category of primary equipment and adding the

TABLE 3.12 Lang Factors

Type of Process	Factor
Solid products	3.10
Mixed solids and fluids	3.63
Fluid products	4.74

TABLE 3.13 Modified Lang Factors

Type of Facility	Factor
Products manufacturing or treating	2.3–4.2
Utilities generation	1.7–2.6
Storage facilities	2.8–4.8

TABLE 3.14 Hand Factors

Type of Equipment	Factor
Distillation columns	4
Pressure vessels	4
Heat exchangers	3.5
Furnaces	2
Pumps	4
Compressors	2.5
Instruments	4
Miscellaneous	2.5

multiples to arrive at the total battery-limits investment. The principal factors are assembled in Table 3.14.

The author points out that the need for special steels of operating pressures above 180 psia increases the importance of the primary equipment relative to other parts of the investment and thus slightly reduces the normal multiplying factors.

Hand's method, like the others using constant multiplying factors, relies on statistical data and leads to average values based on the analysis of the investments for a great number of various kinds of plants. However, these methods are not realistic for plants in which special technology or materials cause the cost of the primary equipment to vary from the normal. The need to refine the analysis and adapt the Lang factor for each project is thus apparent.

3.5.3 METHODS USING VARIABLE MULTIPLYING FACTORS

In order to adapt an estimate to individual conditions, variable multiplying factors are required. The following elements are identified

Primary equipment delivered at the site

Erection of the primary equipment

Instrumentation

Underground piping

Aboveground piping

Structures

Buildings
Site preparation
Foundations
Electrical installations
Insulation
Painting
Access roads, fences, etc.

Each one of these elements is estimated proportionally to the primary equipment by choosing precentage factors that best correspond to the process. For example, the cost of piping can vary from 52 to 125% of the cost of the primary equipment, according to the technology.

Already in 1949, C. H. Chilton presented some cost percentages for certain items in relation to the primary equipment. Examples include valves and piping (10–40%), instruments (5–15%), buildings (0–8%), auxiliaries (0–75%), engineering (30–40%), and contingencies (10–40%). These percentages vary in relation to some simple characteristics of the complexity or level of experience for the plant. N. G. Bach, J. E. Haselbarth and J. M. Berk, H. E. Bauman, and M. S. Peters and K. D. Timmerhaus have published similar factors that can be used in widely varied cases.

The latter have established a cost distribution that takes into account the type of process in order to determine *direct costs,* which are described as the investment for the production unit and its associated facilities (Table 3.15). It is possible, then, to proceed from these direct costs to the *indirect costs* and obtain the complete structure of capital for a plant (Table 3.16). However, estimating a realistic value for each division of a particular project is often complex and calls for foresight and judgment plus experience.

TABLE 3.15 Distribution of Direct Investment Costs*

Type of Material Figuring in Direct Costs	Percentage of Primary Equipment According to Form of Material Handled		
	Solids	Mixed Solids and Fluids	Fluids
Delivered primary equipment	100	100	100
Erection of equipment	45	39	47
Instrumentation, installed	9	13	18
Piping, installed	16	31	66
Electrical system, installed	10	10	11
Buildings	25	29	18
Site preparation	13	10	10
General services and utilities	40	55	70
Land	6	6	6
TOTAL DIRECT INVESTMENT	264	293	346

*This classification does not correspond to that described in Chap. 2.

3.5.4 METHODS ADAPTED TO SPECIFIC PROJECTS

A critical analysis of the structure of investments reveals the weaknesses of both methods using either constant or variable multiplying factors: The factors must differ from one plant to another according to the following:

- The structure of the primary equipment costs, i.e., the relative importance of the various categories. When, for example, the equipment operates at high pressure, its fraction of the total investment is greater than when the process operates at low pressures.
- The capacity of the unit. When capacity increases, the investment for secondary equipment, installation, civil engineering, etc., does not increase proportionally with the cost of the primary equipment. In fact, none of the costs increase in the same way with an increase in capacity, some holding practically constant (instrumentation, buildings, etc.), others changing slightly (civil engineering, electrical installations, etc.). As a consequence, the Lang factor must decrease with the increase in capacity.
- The operating pressure. When it increases, the cost of the primary equipment increases while certain other items are not changed or are changed but little (instrumentation, electrical wiring, insulation, etc.). As a consequence a higher pressure means a lower Lang factor.

TABLE 3.16 Distribution of Fixed Capital Investment*

	Proportion, %		
	Fixed Capital		Primary Equipment
Type of Material	Range	Average	Ratio
Direct costs:			
Primary equipment	20–40	22.8	100
Erection of equipment	7.3–26.0	8.7	38
Instrumentation, installed	2.5–7.0	3.0	13
Piping, installed	3.5–15.0	6.6	29
Electricals, installed	2.5–9.0	4.1	18
Buildings	6.0–20.0	8.0	35
Site preparation	1.5–5.0	2.3	10
General services and utilities	8.1–35.0	12.7	56
Land	1.0–2.0	1.1	5
Indirect costs:			
Engineering and supervision	4.0–21.0	9.1	40
Construction	4.8–22.0	10.2	45
Contractor's fee	1.5–5.0	2.1	9
Contingincies	6.0–18.0	9.3	41
TOTAL FIXED CAPITAL		100.0	439

*This classification does not correspond to that described in Chap. 2.

• The nature of the material of construction. The presence of corrosive products requires special materials of construction, either for the equipment itself or for linings. It follows that the absolute value of the cost of the primary equipment is increased while the cost of the secondary is not altered, so that the Lang factor should normally be lower.

Among the published methods for taking all these things into account are those of J. Clerk and J. T. Gallagher (materials of construction), the New York Section of the American Association of Cost Engineers (capacity), J. H. Hirsch and E. M. Glazier (materials and capacity), C. A. Miller (complexity and capacity), E. M. Guthrie (complexity and capacity), and Stanford Research Institute (materials, complexity, and capacity).

3.5.4.1 The Influence of Materials of Construction

J. Clerk has described a method to correct the price of each type of primary equipment with a multiplying coefficient based on the relationship of the costs in special steels or materials to the cost in mild steel. Curves showing the variation in installed equipment costs with this coefficient are then used to determine a modified Lang factor (excluding engineering fees). J. T. Gallagher used this method to develop a method for measuring the influence of corrosive reactants on the economics of a system.

3.5.4.2 The Influence of Capacity

The New York Section of the American Association of Cost Engineers has developed a method, called *module estimating technique,* which assigns installation costs for associated secondary equipment (foundations, structures, piping, electrical systems, insulation and labor) to each piece of primary equipment. Curves show the cost of the associated equipment as a function of the cost of the primary equipment. Since this cost has been determined according to the size of the equipment, the curves show the effects of capacity on the associated equipment. The distribution of costs for an entire plant can be obtained by assembling the costs for each category as determined by the pieces of primary equipment; and the cost-capacity relations can be studied in this manner. However, the method is limited to:

1. The types of equipment shown by the authors (towers, tanks, exchangers and pumps). Furnaces and compressors, among other things, are not handled. (It is true that prices for the latter equipment often include associated equipment.)
2. Equipment made of common steel.

3.5.4.3 The Influence of Materials and Capacity

J. H. Hirsch and E. M. Glazier have introduced a method based on a breakdown of the Lang factor into three coefficients, excluding indirect costs such as contractors fees and contingencies. The factors are

Field factor, F_L, which covers costs of the construction site
Piping factor, F_P, for piping valves, supports, etc.
Miscellaneous factor, F_M, for foundations, structures, buildings, electrical installation, instrumentation, insulation, field supervision, etc.

The battery-limits investment I is then defined by the following:

$$I = E[A(1 + F_L + F_P + F_M) + B + C]$$

This approach differentiates between equipment for which the installation cost is and is not known, and identifies the additional costs resulting from the use of special steels for certain equipment. Thus,

A = total cost of the equipment, assumed to be entirely of mild steel and for which the installation cost is not exactly known, including both shop fabricated and field fabricated items
B = cost of equipment, such as furnaces, for which installation is included in the price
C = additional cost of special steels required for some equipment, representing the difference between the actual cost and the assumed cost in mild steel, A
E = coefficient, close to 1.4, for the indirect costs (J. P. O'Donnel has furnished a method for determining this factor more accurately)

The coefficients F_L, F_P, and F_M are obtained by means of empirical relationships between A and the cost of exchangers in common steel (e), the cost of field-assembled vessels* (f), the cost of pumps and motors (p), and the cost of tower shells (t). When the value of the equipment is expressed in thousands of dollars, these relations are

$$\log F_L = 0.127 - 0.154 \log A - 0.992\left(\frac{e}{A}\right) + 0.506\left(\frac{f}{A}\right)$$

$$\log F_P = -0.308 - 0.014 \log A - 0.156\left(\frac{e}{A}\right) + 0.556\left(\frac{p}{A}\right)$$

$$F_M = 0.443 + 0.033 \log A - 1.194\left(\frac{t}{A}\right)$$

*Except for high-pressure, multiwalled vessels, vessels larger than 13-ft diameter are assumed to be field erected.

Also, F_L, F_P, and F_M can be found directly from Fig. 3.6*a* through *c*. A practical application of this method is presented at the end of this section.

3.5.4.4 C. A. Miller's Method of Statistical Averages

This method consists of determining the estimated average unit cost of a piece of equipment taken to be representative of the primary equipment in the plant under consideration. To arrive at this, the total cost of the process equipment is divided by the number of equipment items, i.e., by the total number of towers, tanks, exchangers, pumps, compressors, furnaces, etc. Instrumentation is not included.

The average cost thus obtained, which is characteristic of the complexity and size of the plant, is then compared to a schedule based on the author's experience and made up of several (seven altogether) cost categories. Veritable oceans of percentages are attached to each category, as well as to a certain number of items other than those included among the equipment entering into his definition of a battery-limits unit. These percentages are the result of statistical calculations and thus present high and low value limits. Inside these oceans one can choose multiplying factors for calculating the cost of

Field erection of basic equipment
Foundations and structural supports
Piping
Insulation of equipment and piping
Electrical
Instrumentation
Miscellaneous
Buildings

When the average cost of the representative piece of equipment passes from the lowest to the highest level, the values of the percentages and thus of the multiplying factors decrease, so that the influence of capacity is taken into account. Since the range that can be taken for the choice of these factors is relatively wide, C. A. Miller was forced to make his method more precise by defining various possible cases for each item. This allows him to make a finer cut. Table 3.17 illustrates the method for foundations and structural supports. The method stays the same for other items, although certain nuances are sometimes introduced, particularly for buildings.

The battery-limits investment can be calculated by adding the results obtained from the different items. A procedure for estimating costs of general

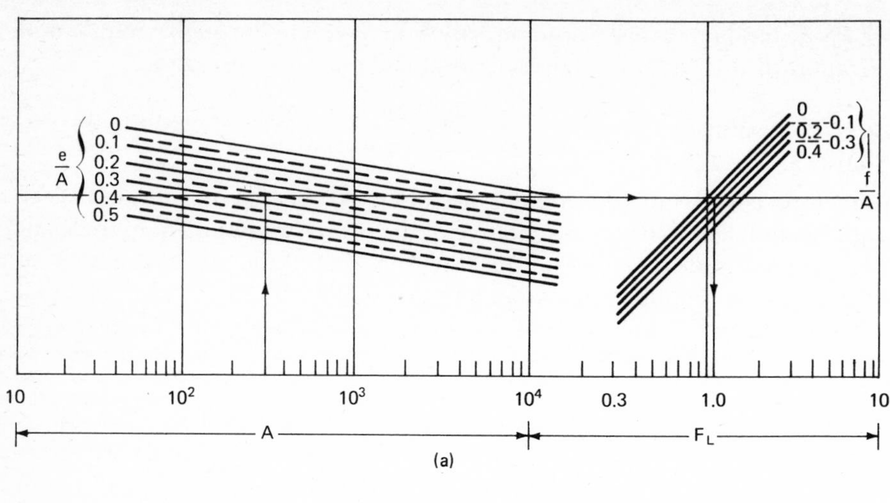

(a)

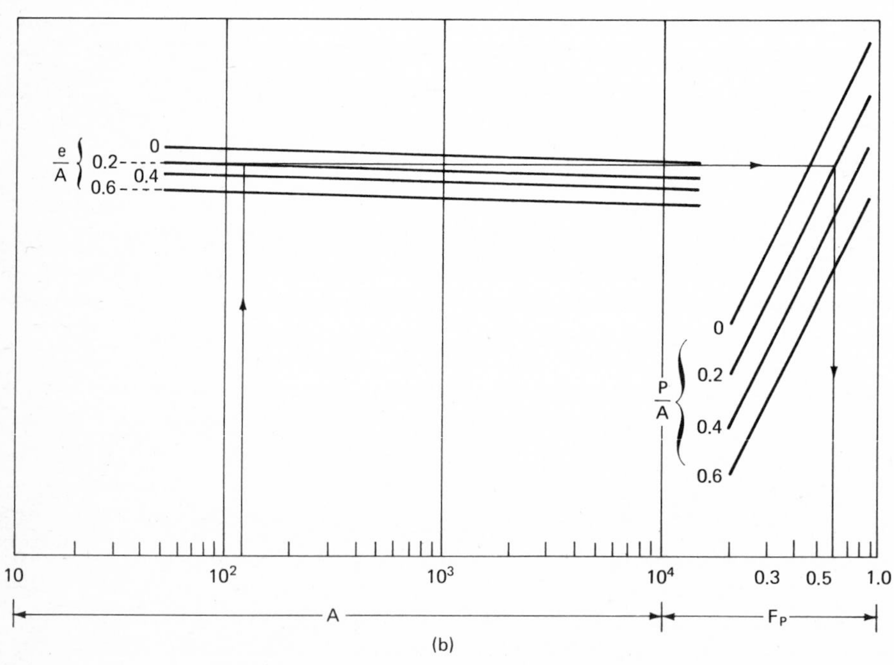

(b)

e	=	cost of exchangers in mild steel, \$
f	=	cost of field erected vessels, \$
p	=	cost of pumps and motors, \$
t	=	cost of tower shells, \$
A	=	total cost of equipment when in mild steel, \$
F_L	=	coefficient for construction costs
F_P	=	coefficient for costs of piping valves and supports
F_M	=	coefficient for costs of foundations, structures, buildings, electrical installation, instrumentation, insulation, field supervision, etc.

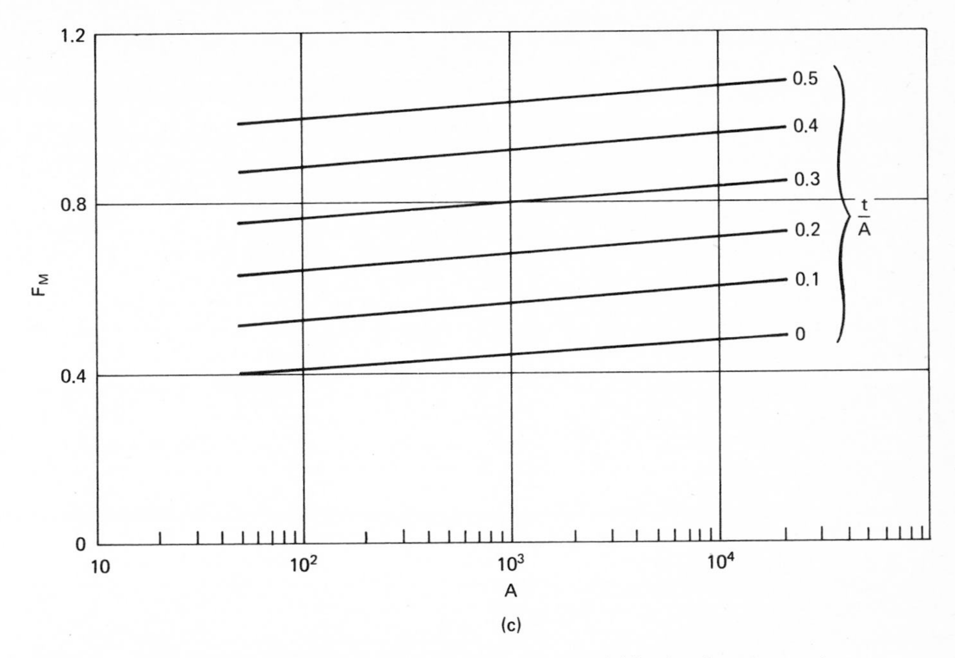

Fig. 3.6 Coefficients for (*a*) field costs, (*b*) piping costs, and (*c*) miscellaneous costs.

services, utilities, and supplies is included. The basis for this calculation is the battery-limits investment and no longer the primary equipment. A distinction should be made between plants built on developed sites and those built on new sites for which the offsites cost is higher.

It should be remembered that Miller's article gives only a suggestion of the potential percentages and really furnishes extreme values for multiplying factors; in cases where it is not possible to make a choice, he recommends adding 10% to the lowest number and taking 10% off the highest.

The Lang factor, whose most probable intermediate value is not the arithmetic average of the higher and lower limits (unless by chance) is thus found to be in a narrower range.

One of the principal advantages of this method is that it furnishes results with a known precision. But it requires broad industrial experience for those who would use it, thus limiting its applicability.

3.5.4.5 K. M. Guthrie's Method

As with the preceding, this method relies on statistics from examining investments for a number of plants. Being relatively complex, it takes into account (1) *the composition of the primary equipment list,* (2) *capacity,* (3) *operating conditions,* and (4) *the materials of construction* as described hereafter.

TABLE 3.17 Factors (%) for Relating Costs of Foundations and Structural Supports to Battery-Limits Investment

	Average Cost of One Foundation, $(1958)						
	Under 3,000	3,000 to 5,000	5,000 to 7,000	7,000 to 10,000	10,000 to 13,000	13,000 to 17,000	Over 17,000
High: Predominance of compressors and mild steel equipment requiring heavy foundations	—	—	17–12	15–10	14–9	12–8	10.5–6
Average: Typical of foundations supporting usual equipment in mild steel	—	—	12.5–7	11–6	9.5–5	8–4	7–3
Average: Typical of foundations supporting equipment fabricated of expensive alloys	7–3	8–3	8.5–3	7.5–3	6.5–2.5	5.5–2	4.5–1.5
Low: Predominance of light equipment requiring only light foundations	5–0	4–0	3–0	2.5–0	2–0	1.5–0	1–0
For pilings or rock excavation:	Increase the above values by 25–100%.						

The composition of the primary equipment list The items are divided into broad categories; and installation costs can be calculated with multiplying factors given for each of these categories (or if the installed cost is known, the factors can be used to get the item cost). The multiplying factors can be used to calculate the cost of associated equipment for each category of primary equipment, and these costs of secondary equipment can be added to arrive at a total investment cost.

Capacity Its influence is taken into account on two levels. First, the author recommends a group of curves or relationships that afford a "base cost" for each item of equipment as a function of its size. Second, analogously to C. A. Miller's method, the costs of the different equipment categories are separated into different levels; and the multiplying factors vary from one level to another.

Operating conditions A correction for the base cost of primary equipment can be obtained by using factors related to the operating conditions.

The materials of construction Here, too, correction factors can modify the base cost which is generally established by assuming the equipment is made of carbon steel.

This method is used for calculations of battery-limits investment as well as the investment for general services and supplies, or even for the various costs that go with grass-roots installations. Contrary to procedures followed in the methods studied up to this point, where the primary equipment is referred to as an associated whole, Guthrie distinguishes two broad types of equipment that he treats separately, as responsible for direct costs or indirect costs. Six modules make up the total investment, five dealing with direct costs and the sixth with indirect costs. The five dealing with direct costs are:

1. Chemical process equipment, i.e., the furnaces, exchangers, vessels, pumps and drivers, compressors, and miscellaneous primary equipment, plus secondary equipment such as piping, foundations, structures, instrumentation, electricity, insulation and painting
2. Solids-handling equipment, i.e., the mills, blenders, centrifuges, conveyors, crushers, dryers, evaporators, filters, presses, screens, hoppers, scales and other primary equipment, plus secondary equipment such as the piping, foundations, structures, instrumentation, and electricity
3. Site preparation, i.e., the purchase of land including surveyor's fees, drainage, clearing, excavation, grading, sewers, piling, parking lots, landscaping, fences, fire protection, and access ways.
4. Industrial buildings, i.e., administrative offices, laboratories, medical personnel offices, shops, warehouses, garages, cafeteria, and various steel structures

5. Off-site facilities, i.e., steam and electricity generation and distribution, cooling towers and the circulation network for this water, fuel systems, blowdown and flares, pollution control facilities, waste-water water handling, lighting, communications systems, and receiving and shipping facilities

The five modules of direct cost include not only the cost of the primary and secondary equipment but also the costs of installation. Thus they lead to installed costs.

The single module for indirect costs covers freight, taxes and insurance, construction overhead, compulsory labor benefits, field supervision, temporary foundations, construction equipment, engineering, and so forth. Directly dependent on the other modules, the total indirect cost is determined from the total installed cost. In performing the calculations, therefore, the direct costs are calculated independently, and the indirect costs are derived from their sum. The total investment is obtained by adding contingencies and contractors' fees.

It is impossible to show Guthrie's entire method here. In order to better understand his way of applying the method, however, his use of the modules can be demonstrated in a calculation for conventional chemical process equipment, i.e., shell-and-tube heat exchangers.

3.5.4.5a Calculating the Expected Cost of an Exchanger

The method relies on calculating a base cost for each item of equipment. This cost is obtained from a number of graphs of cost versus capacity for standard, mild-steel equipment. Then a series of corrections provides an "expected cost" by taking into account differences between selected equipment and that used for the cost references, operating conditions, materials of construction, and the variation in cost indexes between the current date and his reference date (mid-1968). A graph in Guthrie's article shows the following linear relation:

$$\log(\text{FOB price before taxes, \$1{,}000}) = 0.65\log(\text{exchanger surface, ft}^2) - 1.0$$

This base cost assumes a mild-steel, floating-head exchanger good for 150 psig. From the base cost, the expected cost is calculated as

$$\text{Expected cost, \$} = (\text{base cost, \$})(F_d + F_p)F_m(\text{index ratio})$$

where F_d, F_p, and F_m are correction coefficients obtained from Table 3.18.

3.5.4.5b Calculating the Module Cost for Exchangers

The equipment is next grouped into the broad categories of process furnaces, fired heaters, shell-and-tube exchangers, air coolers, vertical and horizontal pressure vessels, pumps and drivers, and compressors and drivers. For each

TABLE 3.18 Correction Factors for Heat-Exchanger Costs

Complexity Factor F_d		Pressure Factor F_p	
Type of exchanger	F_d	Design pressure, psia	F_p
Reboiler	1.35	Under 150	0.00
Floating head	1.00	150–300	0.10
U-tube	0.85	300–400	0.25
Fixed tube sheet	0.80	400–800	0.52
		800–1,000	0.55

Materials-of-Construction Factor F_m
Shell/tube materials*: F_m

Heat-exchange surface, ft^2	CS/CS	CS/brass	CS/Mo	CS/SS	CS/Monel	Monel/Monel	CS/Ti	Ti/Ti	SS/SS
Under 100	1.00	1.05	1.60	1.54	2.00	3.20	4.10	10.28	2.50
100–500	1.00	1.10	1.75	1.78	2.30	3.50	5.20	10.60	3.10
500–1,000	1.00	1.15	1.82	2.25	2.50	3.65	6.15	10.75	3.26
1,000–5,000	1.00	1.30	2.15	2.81	3.10	4.25	8.95	13.05	3.75
5,000–10,000	1.00	1.52	2.50	3.52	3.75	4.95	11.10	16.60	4.50

*CS = carbon steel; Mo = molybdenum steel; SS = stainless steel type 18-8; Monel = Monel; Ti = titanium.

of these categories a bare module cost is determined by adding the costs for each item.

The costs of secondary equipment are obtained by applying multiplying factors to the base cost for each category of equipment, even though these latter generally come from a module that has been treated separately. (Also, these coefficients vary with the size of the base cost, which is divided into levels.) In Guthrie's example the different multiplying factors are assembled in tables that accompany each of the curves for base cost. In our example of shell-and-tube exchangers, Table 3.19 is obtained.

Using such tables for all categories of standard equipment in mild steel, more precise calculations are made for (1) the direct costs of equipment and labor, including primary equipment, secondary equipment, direct field costs, and (2) the indirect costs. The sum of the direct and indirect costs constitutes the *bare module cost.* Finally, by adding contingencies and contractors' fees, the total module cost is obtained.

This bare module cost assumes the use of standard equipment. In order to take into account the complexity of the equipment, as well as the actual materials of construction and current prices, the expected cost for the module is calculated as described for the individual item.

3.5.4.5c Calculating the Indirect Costs Module

The indirect field costs are calculated from

$$\text{Indirect field costs, \$} = 0.178 F_{co} F_{mo} \text{ (index ratio)}$$

where the indirect field costs are in U.S. dollars, F_{co} is a correction factor depending on the labor-to-material ratio (Table 3.19), F_{mo} is a coefficient for the size of the direct cost; both coefficients are determined from the following tables:

Ratio of direct construction labor costs to direct material costs	0.1	0.3	0.5	0.7	1.0
F_{co}	0.3	1.0	1.3	1.6	2.0

Direct cost of material plus construction labor, \$ millions (mid-1968)	1	3	5	7	10
F_{mo}	1.07	1.02	1.0	0.97	0.95

The engineering fees are determined from

$$\text{Engineering fees, \$} = \text{(direct cost of material plus labor, mid-1968 \$)} (0.10)(F_{ce})(F_{me})(F_{pt})\text{(index ratio)}$$

TABLE 3.19 Calculations for the Bare Module Cost: Heat Exchangers

	Base Dollar Magnitude, $100,000									
	Under 2		2–4		4–6		6–8		8–10	
Equipment f.o.b. cost	100.0		100.0		100.0		100.0		100.0	
Secondary equipment cost	71.4		70.5		69.9		69.5		69.3	
Piping		45.6		45.1		44.7		44.4		44.3
Concrete		5.1		5.0		5.0		5.0		5.0
Steel		3.1		3.0		3.0		3.0		3.0
Instruments		10.2		10.1		10.0		9.9		9.8
Electrical		2.0		2.0		2.0		2.0		2.0
Insulation		4.9		4.8		4.7		4.7		4.7
Paint		0.5		0.5		0.5		0.5		0.5
Direct materials cost	171.4		170.5		169.9		169.5		169.3	
Construction labor costs	63.0		61.2		60.1		59.4		59.0	
Installation		55.4		54.7		54.2		53.9		53.8
Equipment setting		7.6		6.5		5.9		5.5		5.2
Direct costs: materials and labor	234.4		231.7		230.0		228.9		228.3	
Indirect costs	94.7		86.8		83.9		83.5		81.0	
Freight, insurance, taxes		8.0		8.0		8.0		8.0		8.0
Miscellaneous		86.7		78.8		75.9		75.5		73.0
TOTAL BARE MODULE COST	329.1		318.5		313.9		312.4		309.3	

Note: Unlike the other multiplying coefficients, which come from statistics, indirect costs are obtained through a formula developed for each module.

where F_{pt} is dependent on the type of project under consideration and is 1.4 for a chemical complex, 1.0 for a chemical processing plant, 0.8 for a solids and fluids treating plant, 0.6 for solids handling, and 0.4 for buildings only; and F_{ce} and F_{me} are determined from the labor-to-materials ratio and direct materials cost analogously to F_{co} and F_{mo}, using the following tables:

Ratio of direct construction labor costs to direct material costs	0.1	0.3	0.5	0.7	1.0
F_{ce}	1.6	1.1	0.8	0.7	0.6

Direct cost of material plus construction labor, $ millions (mid-1968)	1	3	5	7	10
F_{me}	1.15	1.05	1.0	0.95	0.90

Table 3.19 illustrates the degree of accuracy one can expect from Guthrie's method. Applied in its entirety, it is better adapted to estimating; it requires simplification for project evaluation, which is why a simplified version has been retained in the second part of this book for predicting the cost of secondary equipment, material erection and indirect field costs.

However, the accuracy of this method depends on the reliability of the curves for determining the base cost of the equipment; and it is impossible to sum up all the actual costs for one category of equipment in a single curve and a couple of correction factors. Because of this, Guthrie's method can be better applied either to the costs quoted by fabricators or to equipment costs resulting from a network of more complex curves.

3.5.4.6 The Method of Stanford Research Institute

Stanford Research Institute has developed a computer program called *PEP-COST* (Process Economic Program) for rapid calculation of plant costs. It uses a group of coefficients similar to those recommended by J. H. Hirsch and E. M. Glazier, C. A. Miller, and K. M. Guthrie, but it differs from those methods in the way it takes into account the effects of materials of construction on piping. In the other methods (J. H. Hirsch and E. M. Glazier, and K. M. Guthrie in particular), the piping is a fixed percentage of the cost of the primary equipment in mild steel and that approach overlooks the fact that the stainless piping gets longer as the stainless-steel equipment gets more numerous; it can be compensated for by either (1) increasing the increment due to materials of construction (parameter C in the method of J. H. Hirsch and E. M. Glazier) or (2) modifying the coefficient for the corresponding item. SRI has chosen the second and modifies its piping factor by the following:

$$f'_p = f_p\left(1 + \frac{X - A}{A}\right) = f_p\left(\frac{X}{A}\right)$$

where f_p = initial piping factor
f'_p = modified piping factor
A = cost of equipment in mild steel
X = actual cost of equipment

Similarly to SRI, most engineering companies have developed computer programs for estimating plant costs. Also, numerous organizations offer either the results of such programs or the programs themselves. The American Association of Cost Engineers offers such a tool. Finally, there are numerous others who use analogous methods in very precise studies for comparing processes.

Example 3

Using Cost Indexes to Calculate the Current Investment for a Cumene Plant

E.3.1 THE PROBLEM

Information is available on a French plant for obtaining cumene by alkylating benzene with propylene in the presence of a phosphoric acid catalyst. Production capacity for this plant is 30,000 tons/yr, and details of the costs of the primary equipment are known for 1968. Most of the equipment is mild steel; the reactor and some of the less important equipment (the filter among others) is of stainless steel. Thus a breakdown in French francs is obtained as shown in Table E3.1.

Now, the following information is required for a plant built in Western Europe:

- The battery-limits investment in 1968 and 1974 for a production capacity of 30,000 tons/yr, using the method recommended by J. H. Hirsch and E. M. Glazier.
- The battery-limits investment in 1974 for a production capacity of 70,000 tons/yr.
- The value of the modified Lang factor for the 30,000 tons/yr capacity.

E3.2 THE RESOLUTION

The calculations employ the method of J. H. Hirsch and E. M. Glazier described in Sec. 3.5.4.3, particularly Fig. 3.6*a* to *c*, which are established for

TABLE E3.1 Cost of Primary Equipment in 1968

Equipment category	Cost in Mild Steel, fr	Cost Supplement for Stainless Steel, fr	Total
Towers	189,000		189,000
Reactor	145,000	415,000	560,000
Tanks	52,000		52,000
Exchangers	326,000		326,000
Pumps	103,000		103,000
Motors	21,000		21,000
Miscellaneous	65,000	160,000	225,000
PRIMARY EQUIPMENT TOTAL	901,000	575,000	1,476,000

dollars in 1957. A critical part of the calculation thus becomes that of prorating costs to and from 1957 and of converting francs to and from dollars. The equipment cost data are prorated to 1957; the method of Hirsch and Glazier is used to calculate the battery-limits investment, and this is then adapted to francs in 1968 and 1974. Finally, the investment for 70,000 tons/yr and the modified Lang factor are calculated.

E3.2.1 DETERMINING THE INVESTMENT IN 1968 FOR A 30,000-TONS-PER-YEAR CUMENE PLANT

E3.2.1.1 Adapting the Data

The problem is twofold:

1. Find a rate of exchange between the franc and the dollar at an effective date.

2. Use the conversions to adapt the equipment costs in French francs in 1968 to U.S. dollars in 1957.

First, a period can be chosen when the parity between the franc and dollar was relatively constant and, if possible, close to the study values. Thus the curve in Fig. 4.6 shows that the greatest changes in parity occur in 1969; and up through 1968 a relatively constant parity of 5 francs to the dollar can be adopted as a first approximation.

Second, cost indexes of various categories of equipment, and not overall indexes, should be used to convert the data (see Sec. 3.3.2). The Nelson index is the one that best lends itself to this type of calculation. This index (Sec. 3.3.2.2c), particularly the trends of certain component indexes (Tables 3.3 and 3.4, and Figs. 4.4 and 4.7), is used to calculate equipment costs in 1957.

The method of Hirsch and Glazier can then be applied, and the battery-limits investment determined in 1957 dollars. A comprehensive index, i.e., the CE cost index (Table 3.5 or Fig. 4.5) is best suited for petrochemical plants and provides the cost of the investment in dollars, and thus in francs, for 1968. Table E3.2 is established.

E3.2.1.2 Applying the Method of Hirsch and Glazier (Sec. 3.5.4.3)

The different parameters have the following values:

$$A = \$158{,}000 \qquad B = 0 \qquad C = \$99{,}800$$

$$e = \$59{,}300 \qquad f = 0$$

$$p = 15{,}100 + 4{,}400 = \$19{,}500$$

$$t = 33{,}600 + 25{,}800 = \$59{,}400$$

Accordingly,

$$\frac{e}{A} = 0.375 \qquad \frac{p}{A} = 0.123$$

$$\frac{f}{A} = 0 \qquad \frac{t}{A} = 0.376$$

From Fig. 3.6*a* to *c*, the multiplying coefficients are determined as

$$F_L = 0.82 \qquad F_P = 0.52 \qquad F_M = 0.87$$

Similarly, the formulas give

$$F_L = 0.84 \qquad F_P = 0.515 \qquad F_M = 0.865$$

The multiplying factors can then be substituted in the formula for the battery-limits investment based on the cost of primary equipment, and

TABLE E3.2 Converting Primary Equipment Costs for a French Cumene Plant Built in 1968 to Dollars in 1957

Equipment Type	Nelson Component Indexes			Cost of the Equipment, $ (1957)		
	1968	1957	Ratio	Mild Steel	Stainless Steel	Total
Towers	199	177	0.889	33,600	—	33,600
Reactor	199	177	0.889	25,800	73,800	99,600
Tanks	199	177	0.889	9,200	—	9,200
Exchangers	223	203	0.910	59,300	—	59,300
Pumps	284	208	0.732	15,100	—	15,100
Miscellaneous	229	186	0.812	10,600	26,000	36,600
Motors	175	182	1.040	4,400	—	4,400
TOTAL				158,000	99,800	257,800

$$\frac{I}{E} = A(1 + F_L + F_P + F_M) + B + C$$

$$\frac{I}{E} = \begin{array}{l} \$606{,}000 \text{ (using coefficients from the charts)} \\ \$608{,}000 \text{ (using coefficients from the formulas)} \end{array}$$

These are the battery-limits investments in 1957.

E3.2.1.3 Calculating the Battery-Limits Investment in 1968 and 1974

Table E3.3a is established using the *Chemical Engineering* cost index (Table 3.5) and the Nelson index (Table 3.3). This comparison shows that the two different indexes give results that differ by 40.5%:

$$\frac{100(7{,}713{,}000 - 5{,}107{,}000)}{0.5(7{,}713{,}000 + 5{,}107{,}000)} = 40.5\%$$

This difference reflects the difference between the two curves for the indexes in Fig. 3.2*a*, where they are both based on 100.

In western Europe, where the problem plant is situated, it is preferable to choose a cost index more representative of that situation; and the WEBCI index (Table 3.5) is selected for results as shown in Table E3.3b.

E3.2.2 CALCULATING THE BATTERY-LIMITS INVESTMENT IN 1974, FOR A CAPACITY OF 70,000 TONS PER YEAR

The exponential extrapolation equation can be used (Sec. 3.5.1.1) as follows:

$$\frac{I_2}{I_1} = \left(\frac{C_2}{C_1}\right)^f$$

where I_1 now equals 7,435,000 fr, C_2 = 70,000 tons/yr, and C_1 = 30,000 tons/yr. The coefficient f falls between 0.5 and 0.7 for conventional equip-

TABLE E3.3a Battery-Limits Investments Prorated by U.S. Cost Indexes

Cost Indexes							
1957		1968		Ratio 1968/57		Investment in 1968	
CE	Nelson	CE	Nelson	CE	Nelson	CE	Nelson
98.5	206	114	304	1.157	1.476	3,517,000	4,487,000
Cost Indexes							
1968		1974		Ratio 1974/68		Investment in 1974	
CE	Nelson	CE	Nelson	CE	Nelson	CE	Nelson
114	304	165.5	522.5	1.452	1.719	5,107,000	7,713,000

TABLE E3.3b Battery-Limits Investments Prorated by the WEBCI Cost Index

Cost Indexes (for Indicated Years) and Ratios					Investments	
1957	1968	Ratio 1957/68	1974	Ratio 1974/68	1968	1974
67.5*	104	1.541	165†	1.587	4,685,000	7,435,000

*Calculated with the West German Index, Table 3.5.
†Estimated average.

ment; and taking into account the large proportion of distillation towers and the like in the primary equipment of this unit, a coefficient of 0.6 seems reasonable. Thus

$$I_2 = 7{,}435{,}000 \times 2.333^{0.6} = 7{,}435{,}000 \times 1.662$$

$$I_2 = 12{,}500{,}000 \text{ fr}$$

E3.2.3 CALCULATING THE MODIFIED LANG FACTOR

The modified Lang factor (Sec. 3.5.2.2) relates the total cost of primary equipment to the battery-limits investment, and usually falls between 2.3 and 4.2. We now have available the original data for the total cost of equipment in 1968, plus the calculated battery-limits investment for 1968, as 1,476,000 and 4,-685,000 fr, respectively. Thus the modified Lang factor for this cumene plant is about 3.2.

Part 2

Applications: Evaluating the Principal Types of Project

The aim of this second part is to explain and give procedures for the practical calculations of profitability, which is to say estimating in many instances. We want to emphasize that

- The information assembled here does not allow for evaluating all types of projects. It would be utopian to be able to describe methods that applied equally well to such different technologies as organic chemicals, mineral chemicals, and polymers or to plants considered singly at an already industrialized site and a grass-roots complex, or to such different sizes as pilot plants and high-capacity units. Thus, the reader should not follow this manual like a cook book but should understand and use the information intelligently.
- The methods developed here are not the only ones recommended, and the reader might be interested in looking further into other more accurate and thus more complicated models that sometimes might be adapted to his or her problem.
- The indications of price, both for equipment and for materials, are only guesses, and a fabricator whose working conditions are set will be the best source of information on this point. Here again, the reader should adapt the approach to the specific problem.

Depending on the kind, the amount, and the degree of credibility of the information, any method for measuring profitability can be made up of several variants. Those methods used will differ both in complexity and in accuracy; however, the most complex method is not necessarily the most accurate.

For our present purposes there are two categories of project: industrial projects and research projects. We will discuss industrial projects in Chap. 4 and research projects in Chap. 5.

Chapter 4

Cost Estimating for Industrial Projects

Whenever a company undertakes an expansion of its manufacturing facilities, it must resolve three essential problems:

1. Where the new manufacturing facilities should be located.
2. How the financial investment for the facilities should be handled.
3. What manufacturing technology should be employed.

Any study to resolve these problems requires information, of which a certain amount already available will fix the general objective along with the framework of the project. This information, which is conventionally gathered from services within the company, usually covers markets, raw materials, and site. Market studies fix the annual sales program and the possible price of the products. Production balance sheets show the available raw materials either directly from units already operating within the enterprise or by means of revamping existing units. The status of land for the new plant is usually set either because the land is already owned by the company or because it can be acquired in one of several already determined industrial zones.

4.1 CHARACTERISTICS OF COST ESTIMATING FOR INDUSTRIAL PROJECTS

The assembled information can indicate many different possibilities to be weighed. Generally, a computer program can be used to find the optimum among those possibilities, depending on the preliminary criteria. In many cases, however, the limitations are such that it is easy to reject certain combinations that would lead to an obviously uneconomic or impossible situation. When there are a great many viable possibilities, they can be classified according to relative advantage. Thus, a choice would lean toward a site where the supply of raw materials is easy because of available transport and large capacity; and another choice would decline, at least on first analysis, a process requiring purchase of all the feedstocks in deference to a process that used available by-products from manufacturing already taking place in the company.

After the first analysis is completed, each possibility is studied independently, and comparison of the results indicates which offers the best value within the chosen economic criteria. As shown in Chap. 2, this best value is revealed by a profitability calculation involving the fixed investment costs and the proportional operating costs. Before detailing a practical approach to this calculation, we should elaborate on the complex nature of the relationship between the profitability parameters and characteristics of the project such as location, the investment program, and the technology.

4.1.1 EFFECTS OF GEOGRAPHICAL LOCATION

The site selection exerts an influence on all aspects of the economic calculation, and most especially on the costs of the investment, of the raw materials and chemicals, of the energy, of the labor, and of the environmental protection.

4.1.1a The Costs of the Investment

The cost of plants can vary noticeably from one country to another, or from one region to another, usually because the necessary materials are or are not available nearby (Sec. 3.4). Although various authors have tried to establish indexes to account for the effects of location on the investment or its components, it is virtually impossible to form simple and logical rules for establishing and evolving these indexes over the course of time, such as is done for cost indexes. Actually, indexes for the effects of location have only a momentary value, since they depend on currency exchanges, the economic situation of one country as compared to others, and the political situation.

On the other hand, those considerations do not change among regions within the same country. Correlations can be suggested; and certain disparaties can be discerned among those correlations, especially in the United States. At this level, the most significant variables are labor and transportation.

Land is a significant part of the investment. Not only does it provide the necessary area, but also the situation for the plant in an industrial zone or out in the open country, as well as the soil with its bearing characteristics and contours. When a totally new site is considered, the required leveling and possible stabilization of the subsoil will also have to be considered, along with connections to highways, railroads, or rivers. In an industrial zone the existing installations should be taken into account.

4.1.1b Cost of Raw Materials and Chemicals

The choice of a location is influenced by the effects of transportation on the cost of raw materials and chemicals. Shipping rates can easily be obtained from the shippers and specialized companies, and entered into the economic calculations. These shipping rates depend on the type of carrier, as well as the value and weight of the products. Table 4.1 indicates the relative costs of different types of carrier in 1974. Also, shipping consumes energy. Table 4.2 gives some examples of the energy consumed by various carriers.

Handling charges must be added to the cost of shipping. They come, for example, to about 1 dollar per ton for unloading a ship onto a train.

TABLE 4.1 Costs of Different Shipping Methods

Type of Transport	Cost, ¢/(t)(km)
Pipeline	0.21
Barge carrying 4,000 t	0.42
Barge carrying 800–2,000 t	0.84
Railroad (full train consignment)	0.63–0.84
Truck	4.2–5.2
Coastal tanker	Less than 0.2
Oil tanker, ocean going	Less than 0.4

TABLE 4.2 Energy Consumption of Different Shipping Methods

Type of Transport	Consumption, therms/(t)(km)
Pipeline	0.015
Railroad	0.03
Canal	0.04
Truck	0.15–0.20

Industrial gases such as oxygen, hydrogen, or nitrogen are often manufactured for a specific consumer; and the availability of these commodities at an attractive price can be an incentive to locate a plant at one site rather than another. In the north of France and the Great Lakes area of the United States, for example, there are steel mills using oxygen shipped in by pipeline; this permits nearby chemical companies to benefit from the favorable rates, even though they would use less than the steel mills. A nearby refinery could also offer the advantage of inexpensive hydrogen.

4.1.1c The Availability and Cost of Energy and General Utilities

Plants that consume a great deal of energy (manufacture of aluminum, chlorine/caustic, ethylene, etc.) find an incontestable advantage in being located near large sources of energy, such as refineries, hydroelectric plants, coal mines, natural-gas fields, and crude oil.

Companies can also choose a site according to the dependence on water, both potable and industrial—its availability, quality (especially hardness), regularity of supply, cost, and restrictions or authorizations. Water can be taken directly from the ocean, from a river, from a water table (well), or from local installations such as municipal water.

4.1.1d Cost of Labor

Labor includes not only the operating personnel but also the construction and start-up personnel. In certain cases permanent lodging extending to entire villages must be provided for the operating personnel. The construction and start-up personnel must be displaced for more or less long periods and need convenient housing.

The availability and demographic evolution of a labor force enters into the choice of a site through the effects on delays in construction and planned future expansion programs. Other criteria should also be investigated, including the kind of personnel (male, female, foreign), their qualifications, their capacity for training, the prevalent salary level, the amount of unionization, the frequency of strikes, and the local work habits, especially concerning willingness to accept shift work.

4.1.1e Cost of Environmental Protection

Depending on the local rules, it is sometimes necessary to add equipment or even whole processing units to satisfy pollution control standards. Examples might be additional height to a smokestack, landscaping, a water-treatment plant, treatment of gaseous effluents, solids disposal, and noise prevention.

4.1.1f The Final Choice

In practice, the choice of an industrial site is often made quite empirically. Instead of a decision based on minimum cost or maximum profit, as might be supposed from the various parameters just examined, other factors can dominate. The existing industrial environment, with construction companies, services, and equipment already present might be the deciding factor. The sociocultural environment (lodgings, administrative centers, schools, universities, sports and cultural activities, etc.) might dominate, or financial incentives manifested in the prime lending rates, exemptions, or aids may affect the decision. As an example, Table 4.3 shows relative weights of location-affecting criteria of international concerns in northwestern Europe.

4.1.2 EFFECTS OF AN INVESTMENT PROGRAM

The amount of investment in manufacturing capacity should be graduated to reflect as nearly as possible an expanding market for the product, as forecast by a market study. The manufacturing capacity can be built all at once or in successive stages. In the latter case there are two possibilities, either to add production lines as they are needed or to make modifications that increase production progressively without duplicating essential equipment.

When a plant is integrated into a complex, its capacity may be planned for downstream units instead of the market. Under those conditions, it is essential that the program of investment be related to the entire complex.

Thus when a generating plant for steam or electricity runs the chance of being overloaded by the added demand of a new process unit, the necessary generating capacity must be forecast and installed. In fact, in order to calculate a new process unit the overall development for the entire complex would be taken into account and not only the immediate needs of the process unit. Such production management calls for an initial overdesign. Subsidiary investments, although not entirely accountable to the plant under consideration, are none the less related to its construction.

The investment program thus exercises its influence on

- The market-price quotations and contract-purchase variations for raw materials not available within the company
- The method of financing, either through the company's own funds or with the help of loans for which quantities and schedules must be spelled out
- The extent of depreciation of the appropriated capital and the finance charges tied to the investment
- The overall profitability of the project, and the risks incurred in completing it

4.1.3 EFFECTS OF THE TECHNOLOGY

The choice of technology has an effect on the investment, the feedstocks, energy consumption, labor, and necessary pollution control.

The size of the investment can differ markedly from one process to another due to variations in optimization, the kind and size of equipment, energy consumption, etc. Also, the same product can be manufactured from different feedstocks by competitive technology that differs by the amount of investment

TABLE 4.3 Criteria of Plant Location in Northwestern Europe

Criterion	Weight of Importance, 100 = deciding factor
Choice of country	
Taxes	80.3
Delays	69.5
Available financing	76.4
Financial assistance	72.2
Choice of a region	
Elements of cost	
Markets	68.1
Resources	
Raw materials	51.4
Energy	55.6
Water	66.7
Labor	
Availability	63.8
Quality	75.0
Schools	65.3
Wages	45.8
Economic structures	
Personnel transportation	58.3
Material transportation	79.2
Governmental liaisons	56.0
Communications	80.6
Terrain and industrial zones	
Cost	72.2
Availability	71.9
The environment	
Economic	
Existing industry	59.4
Industrial services	44.9
Universities	53.6
Social	
Strikes	50.0
Unions	51.4
Cultural	
Schools	42.4
Leisure	40.6
Reception	40.6

as well as other features. It may be necessary to buy certain feedstocks if they are not available within the company.

Depending on the reactions, the operating temperature levels and the possibilities for energy recovery, different processes will require different amounts of energy per unit of product.

The amount and qualifications of the labor force will be affected by the operations of the process, whether complex machinery is used, whether hazardous chemicals or expensive catalysts are handled, and so forth. Finally, the waste effluents from a process will affect the amount of additional investment required to satisfy pollution control standards.

Another less directly felt effect is the financing required to cover purchase of imported materials or special equipment.

4.2 CHOICE OF THE ECONOMIC CRITERIA

The economic criteria on which to base any conclusions should be established before undertaking the estimate for an industrial project. Those criteria most often used have been described in Chap. 2 and are listed below with their section references in order of increasing complexity:

Payout time (Sec. 2.6.2)
Return on investment (Sec. 2.6.4.3)
Turnover ratio (Sec. 2.6.4.1)
Cumulative net present value (Sec. 2.6.3.2.a.)
Capital growth (Sec. 2.6.4.2)
Discounted rate of return (Sec. 2.6.3.2b)
Payout time discounted (Sec. 2.6.3.2c)

Each of these criteria can be complicated or simplified, depending on the available data. It would be illusory to use elaborate profitability calculations when numerous hypotheses were still necessary to fix basic elements of the project, such as location, method of financing, procurement and sales programs, evolution of prices, and so forth. Accordingly, the criterion of profitability (or better, the form in which it is used) should be established in keeping with how well the project is defined, beginning with the simplest methods so as to eliminate the misleading answers and using more complicated and more exact methods in the final stages.

Also, since each criterion has limitations, it is important to judge the profitability on the basis of two or more complementary criteria, such as payout time with cumulative net present value. This becomes even more important as the project advances and becomes more clearly defined.

4.3 PROFITABILITY CALCULATION

As seen in Chap. 2, the economic calculation does not in itself offer any particular difficulties; it can be put into standardized procedures using calculation form sheets. When the task is to compare competitive processes with different technologies and economics or to compare variants of the same process or the influence of variations in certain economic parameters, the calculation becomes repetitive, and a computer can be used to obtain a rapid answer with the help of curves. However, such a study is possible only with a complete set of consistent data.

4.3.1 PREPARING THE DATA

The first step is to assemble a consistent set of data. This step is important and requires good judgment; its results depend on the information that is first encountered. Basically, the function of the analyst is to adapt the available information, to make it consistent and complete it if necessary through educated guesses and complementary calculations.

The information for a proposal project can be divided into three stages described as preliminary, detailed, and engineered (Fig. 4.1).

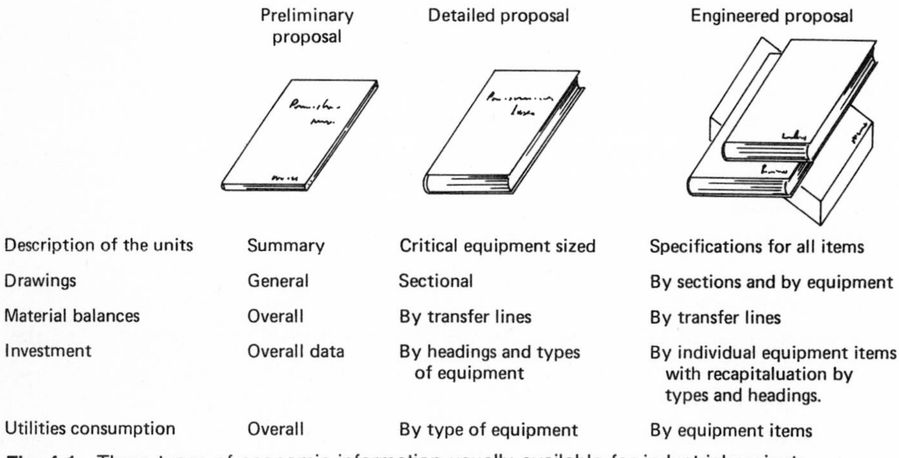

	Preliminary proposal	Detailed proposal	Engineered proposal
Description of the units	Summary	Critical equipment sized	Specifications for all items
Drawings	General	Sectional	By sections and by equipment
Material balances	Overall	By transfer lines	By transfer lines
Investment	Overall data	By headings and types of equipment	By individual equipment items with recapitaluation by types and headings.
Utilities consumption	Overall	By type of equipment	By equipment items

Fig. 4.1 Three types of economic information usually available for industrial projects.

Preliminary proposals In a well-defined situation (production capacity, year, site, etc.) the following data are available:

- An overall material balance with specifications for the feedstocks and principal products
- The investment costs, including battery-limits investment, initial charges of catalysts and solvents, royalties, process data book, and the start-up expenses

- Consumption of utilities and chemicals
- Quantity and quality of operating labor

These data, to which is added a sketch of the proposed unit and a summary of its operation, usually make a preliminary proposal. It is also the sort of information frequently found in publications or conference papers.

Detailed proposal The process flow sheet can be more detailed, showing pieces of primary equipment with operating conditions, as well as the essential transfer lines. The material balance is often more detailed, coded to the transfer lines and shown along with them on the process flow scheme. The battery-limits investment can be divided into primary equipment, secondary equipment, field installation and setting, indirect field costs, and even a division of the primary equipment into its different types such as reactors, towers, tanks, and so forth. Sometimes the same can also be done for the secondary equipment.

It is rare to find information corresponding to detailed proposals available in current publications. On the other hand, there are organizations such as Stanford Research Institute, which specialize in economic comparisons of different processes; those studies generally furnish sizes of the primary equipment and details of its cost.

Engineered study A detailed material balance, exact operating conditions, operating characteristics of the primary equipment, and the detailed utility consumption for each piece of equipment are all provided as part of the engineered study. We call *engineered proposals* those in which all equipment has been sized and where the instrumentation and piping characteristics are given.

The associated cost estimate may offer the same scope and detail as the engineering information. Generally, it is very rare to have the same quality of information in the engineering study and the cost estimate. Because of this the analyst is forced to prepare a detailed estimate in order to make profitability studies, using wherever possible the advice of the equipment fabricators.

Economic data alone, published as they are in a fragmentary way, do not permit construction of a schedule exact enough for an operating plant, nor do they permit one to fix the exact nature of the necessary investments. Consequently, it is important to imagine and to calculate the missing elements, especially in analyzing research projects. The published data serve only as an indication.

Detailed information for a project does not necessarily assure improved credibility. Thus a detailed publication, established from other published information (patents, for example) is not necessarily as reliable as a preliminary engineering proposal or a proposal from the licensor of a process, or even as reliable as a research-project estimate made on the basis of personal technical information.

4.3.2 THE ACTUAL ECONOMIC CALCULATION

As has been shown in Chap. 2, numerous methods exist for determining the profitability of a plant. There are also a large number of methods for analyzing the elements of the economic calculation. The problem becomes a matter of choosing a method and adopting a plan that meets the needs of the analysis. Thus a relatively practical and simple method is chosen because it can be modified to suit the immediate circumstances as well as to take into account nearly all of the economic guidelines.

Only three of the numerous profitability criteria have been chosen: payout time, cumulative net present value, and discounted rate of return. All of these can be easily adapted to the accuracy and credibility of the data.

This approach is particularly suited for engineers, or those in research or development, or even for management people who would like to add an economic analysis to a project they are studying, even though these methods are too rudimentary for accounting and handling concrete business problems.

In our chosen method, three stages come into the economic calculation: determining the overall investment, establishing the operating cost, performing the profitability calculations.

4.3.2.1 Determining the Amount of Investment

Section 2.1 described the principal parts of an investment that are considered when analyzing a project. Figure 4.2 relates these parts to economic analyses. When the information is detailed enough, each part can be calculated separately. However, only the battery-limits investment and sometimes the off-sites are most often available. The other components must be figured from the first of these two.

Table 4.4 assembles some numbers that permit such figuring. These numbers are only approximations and have to do with petrochemicals or bulk organics. It should be noted that royalties, payment for know-how or for an invention, are extremely variable from one product to another or from one process to another. The typical value of 0.07 times the battery-limits investment, shown in Table 4.4, is thus approximate, and it is often preferable to make the calculation without considering royalties when no better data are available. Some examples of specific royalties might be helpful. The know-how fee for an ethylene pyrolysis plant with capacity for 1 million tons per year of naphtha was about $1.05 million in mid-1975. The paid-up royalty for a catalytic reforming unit treating a million tons per year of naphtha was about $0.70 million in mid-1975.

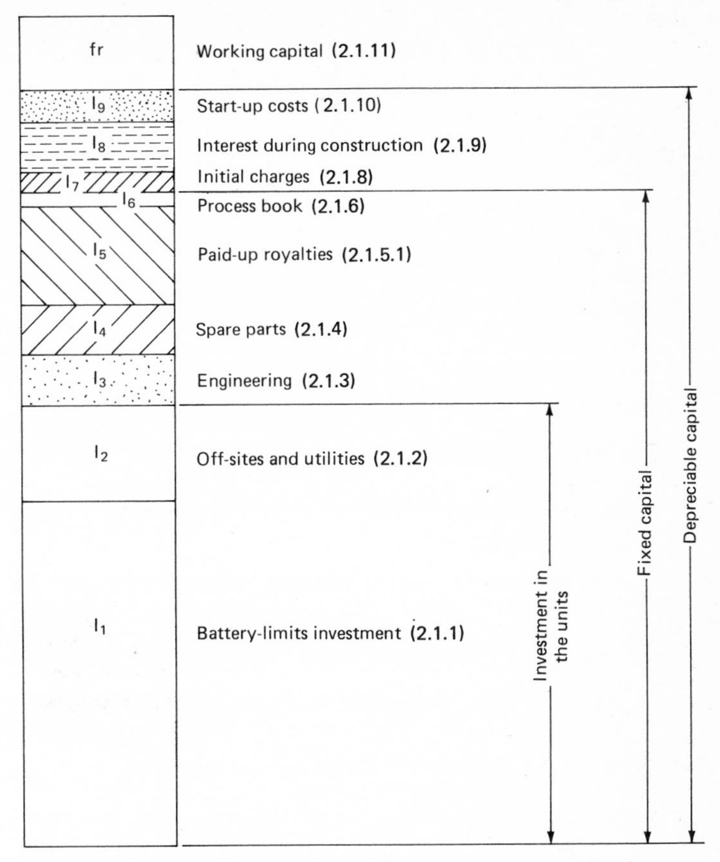

Fig. 4.2 The different segments of the investment cost (with references to sections of this book).

Also, the cost of the initial charges of catalysts and solvents depends on the required quantities of those materials, as well as their unit prices. The tonnage required for solvents is determined by the size of the absorption or extraction tower and its holdup (App. 1). Typical costs in mid-1975 for the principal industrial solvents are shown in Table 4.5. In practice, the cost of these products is highly dependent on their availability at the site of the plant.

The initial charges of catalyst involves only heterogeneous catalysts and depends on the space velocity, $ft^3/(ft^3)(h)$, and the production capacity (App. 2). The unit price of the catalyst or the overall cost of the charge is often part of the basic data, since it is difficult if not impossible to include catalyst manufacture in the analysis of a processing plant. Even when the catalyst-manufacturing technology is described in detail, the new estimate is much more delicate. In fact, it is impossible to envisage a catalyst manufacturing unit dedicated solely to the needs of a processing project. In addition to other problems there

TABLE 4.4 Formulas for Relating Segments of the Investment Cost

Investment Cost Segment	Observed Relationships
Battery-limits investment	I_1
Off-sites (production and distribution of utilities not included)	$I_2 \cong I_1$ (0.3 to 0.4)
Process unit investment	$I_1 + I_2$
Engineering	$I_3 = (I_1 + I_2)$ (0.08 to 0.20), typically $0.12\,(I_1 + I_2)$
Spare parts	$I_4 \cong$ zero in industrialized countries.
Paid-up royalties	$I_5 =$ variable (see text), with the average $\cong 0.07 I_1$
Process book	$I_6 \cong$ \$40,000 to \$160,000 in early 1975
Fixed capital	$FC = I_1 + I_2 + I_3 + I_4 + I_5 + I_6$
Initial charge of feedstocks	I_7
Interest during construction	$I_8 =$ about 0.07 FC for a 2-year long project (see text).
Start-up costs	$I_9 =$ 0.5–4 months of operating costs either with or without raw materials (see Table 4.7)
Depreciable capital	$DC = FC + I_7 + I_8 + I_9$
Working capital	$f =$ according to the needs, or generally, $2I_9$ plus the cost of any precious metals for catalysts

TABLE 4.5 Typical Prices for the Principal Industrial Solvents

Solvent	FOB Price in mid-1975, \$/t
Acetonitrile	1,000
Dimethylacetamide	1,670
Dimethylformamide	1,000
Dimethylsulfoxide	940
Furfural	830
n-Methylpyrolidone	1,670
Sulfolane	1,460

are choice of optimum capacity, markets for by-products, and small equipment of pilot-plant size that is operated batchwise and used from time to time for other manufacturing operations.

In the case of catalyst containing a precious metal, cost I_7 of Table 4.4 includes only the cost of the support, of its manufacture, and of the promoter; the cost of the precious metal, which can be recovered and resold or reused, is charged as working capital. The posted prices of these metals vary considerably, depending on the competitive market. Thus for industrial uses, there is a value that is established on the basis of the actual cost by long-term contracts and not subject to fluctuations. In general, the producer's value for such contracts is lower than the free-market value. In 1975, however, this situation was reversed; Table 4.6 shows, as examples, the prices of some principal precious metals at the end of May 1975. Gold and silver have no producer's values, only the market prices.

The interest paid during construction (Sec. 2.1.9.) affects the capital appropriations even before the plant produces and is, as a first approximation, part of the fixed capital. In actual plants, however, the spare-parts inventory is often made up only after start-up and experience with operating the equipment, while engineering fees and royalties are not wholly paid until after the acceptance of the units. Consequently, the number given for I_8 in Table 4.4 should be modified not only as a function of the duration of construction but also according to agreements with the various lendors and users, assuming of course that such information is available and that the accuracy of the calculation allows for it.

4.3.2.2 Calculating the Operating Cost

Figure 4.3 furnishes a summary of principal segments of the operating cost, which have been separated according to the realistic needs of project analysis. This breakdown avoids becoming inflexible and permits changing cost categories as well as more general calculations. As shown in Sec. 2.5, the operating cost can be expressed in various forms according to the circumstances:

- Annually (from the accounting point of view), when it is spoken of as operating cost
- Proportionally by ton of product or of feedstock, when the manufacturing or product cost is determined for comparison by plots or curves
- Currently, at the actual time of operation, when the cost is expressed from the point of view of manufacturing personnel

4.3.2.2a Choosing a Stream Factor and a Feed Rate

The values of stream factor and feed rate should be identified first, since they characterize the operation of the plant (Sec. 2.5). In actual practice, however, these two notions present ambiguities, such as the difference between a theo-

TABLE 4.6 Producer's Price and Market Price for Precious Metals
(at the end of May 1975)

Metal	Producer's Price, $/lb	Free Market Price, $/lb
Iridium	53,800	26,900–30,500
Osmium	14,150–15,900	9,910–11,680
Palladium	8,480–8,870	4,300–4,630
Platinum	10,960–11,680	10,460–10,630
Rhodium	24,780–25,500	19,830–21,970
Ruthenium	4,240–4,630	3,190–3,520

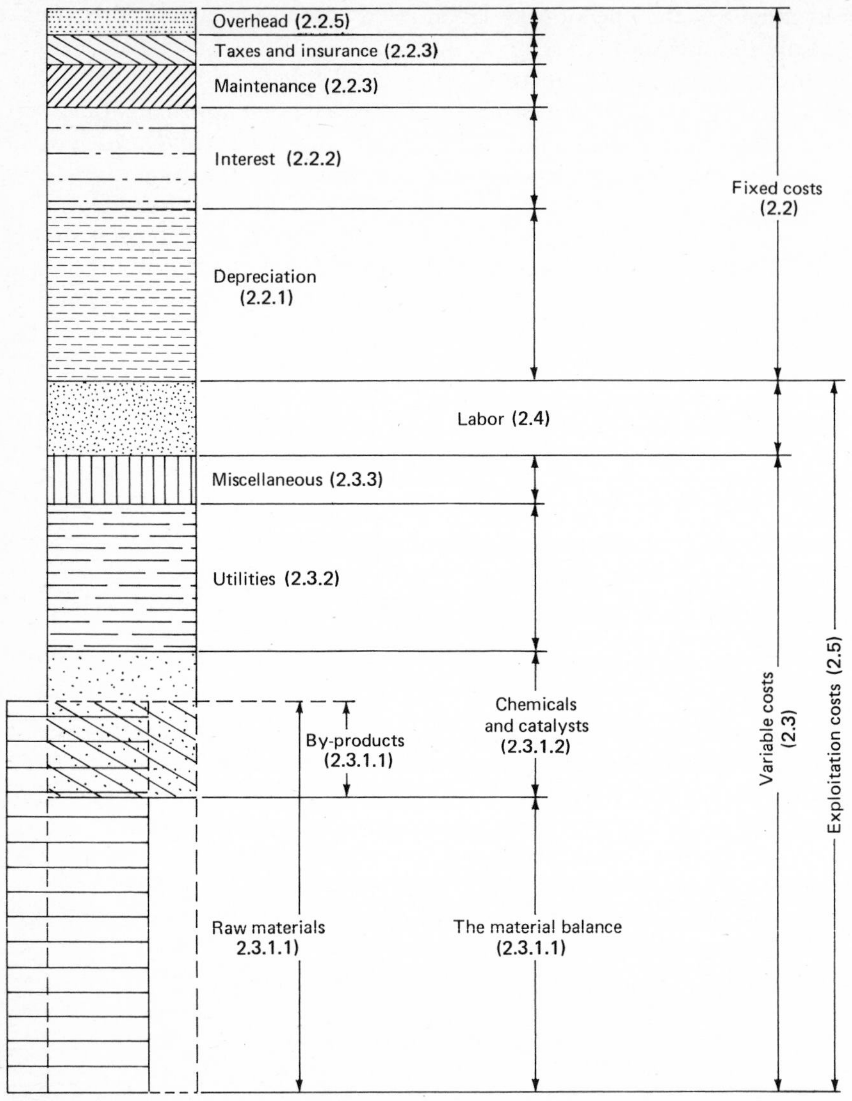

Fig. 4.3 Operating costs, their composition, and their importances (with reference to sections of this book).

retical and an actual stream factor. The theoretical stream factor corresponds to the annual operating time calculated on the basis of planned shutdowns (for example, 8,000 hours per year or 340 days per year), whereas the actual stream factor takes into account the unforeseen shutdowns and expresses the annual operating time as, for example, 6,000 hours per year.

The theoretical factor has the character of a forecast, whereas the actual factor expresses operating experience. When the two are compared, for example, $100(\frac{6,000}{8,000}) = 75\%$, the resulting use factor indicates the realistic production possibilities for the plant.

In contrast to both these factors, actual production is fixed by the market, which is often smaller than the plant's capacity; this actual production rate is an essential feature of the plant's accounting. For example, if the current operating rate is 100, 90, or 80% of the annual design capacity, the following relation is intended:

$$\text{Actual annual production} = (\text{theoretical annual production})\,(\text{rate of use}) \int_{t=0}^{t=1\text{ year}} (\text{operating rate})\, dt$$

This relation expresses the accountant's point of view, which assumes knowing the results of operating the plant for a year. In feasibility studies, where the studies are forecasts, the operating data are not available; so the estimates are based on the theoretical stream factor (8,000 hours per year or 333 days per year) and an equally theoretical feed rate, which varies with market forecasts but is generally taken as constant over the span of a year. The notion of a *rate of use* disappears and becomes mixed with the notion of the *feed rate,* thereby giving only one coefficient that directly represents the relation between the forecast production and the calculated capacity.

4.3.2.2b Method of Calculation

Our calculation plan includes the three basic headings of the classification in Chap. 2: variable costs, labor, and fixed costs. Table 4.7 and the following comments offer values and coefficients that can be used quickly to get the operating data from the investment costs. These numbers apply particularly to petrochemical and bulk organic chemicals, taking into account, where necessary, the effects of the cost of energy. The calculation can be generalized and applied to other types of production, such as pharmaceuticals, minerals, nuclear energy, and steel by modifying the various coefficients in Tables 4.4 and 4.7. However, the following comments should be noted with respect to the variable costs, feedstocks, by-products, catalysts, utilities, labor, depreciation, and coproducts stated in Table 4.7 as follows:

Variable costs These are most often expressed per unit of feedstock or product. Sometimes the only known data are the overall costs and the overall revenue (i.e., production times sales price).

Feedstocks As has been seen, it is difficult to come by an exact price for a chemical product without actually buying or selling it under contract. The best alternative is association with a person or company department that is actively trading the chemical and thus has the data. Barring that:

Prices for small quantities or for fine chemicals may be estimated from published market prices (for references and descriptions, see Sec. 2.3.1.1).

Prices for large quantities of bulk chemicals can be determined according to average transaction prices.

Average transaction prices are the results of weighing all available data and keeping those data up to date. Transaction prices fluctuate extremely in the competitive market and serve only as a preliminary approach. For this reason it is useful to maintain a consistent price list that affords comparisons on a

TABLE 4.7 Coefficients for Calculating Total Operating Costs, Including Fixed Costs, Labor, and Variable Costs

Cost Category	Approximate Values
Variable Costs	
Feedstocks	See text and Table 4.9
By-products	Generally a credit, see text
Raw materials	Generally feedstocks minus byproducts
Chemicals and solvents	See text
Catalysts	See text
Utilities	For mid-1975 costs, see text and as follows:
Steam	0.31 ¢/lb or 0.30–0.33 ¢/lb according to pressure
Electricity	2.5 ¢/kWh
Fuel	3.1 ¢/lb
Cooling water	0.40 ¢/1,000 gal
Boiler feedwater	9.5 ¢/1,000 gal
Process water	14.2 ¢/1,000 gal
Refrigeration	$25.00 per million kilocalories at −20°C.
Miscellaneous costs	Allow 3% of sales
TOTAL VARIABLE COSTS	See Fig. 4.3
Labor	$83,333/yr per station (see text)
TOTAL OPERATING COSTS	See Fig. 4.3
Fixed costs	See Table 4.4
Depreciation	Take 12.5%/yr of depreciable capital; see Table 4.4 and text
Financing costs	
On depreciable capital	7% average (mid-1975)
On working capital	9% average (mid-1975)
Maintenance	
On battery-limits units	4% of the investment
On off-sites	3–4% of the investment
Taxes and insurance	2% of battery limits plus off-sites
Overhead	1% of battery limits plus off-sites
TOTAL FIXED COSTS	See Fig. 4.3
Operating (or bare manufacturing) cost	See Fig. 4.3
Coproducts	See text
Adjusted operating or manufacturing costs	See Fig. 4.3

TABLE 4.8 Sales Prices of Some Intermediates as a Function of the Price of Crude Oil

BASIS: $P_{fuel} = P_{crude}$ $P_{naphtha} = 1.6P_{crude}$
BASIC EQUATION: $A{\cdot}P_{crude} + B =$ dollars per metric ton

Product	*A*	*B*	Example: Price, \$/t (when $P_{crude} = \$79/t$)
Benzene	2.30	83	265
Toluene*	2.00	38	196
o-Xylene	2.75	125	343
p-Xylene	2.90	181	411
Ethylbenzene	2.86	173	395
Ethylene	2.75	108	326
Propylene	1.65	108	239
Chlorine†	1.25	68	167

*The price of a mixture of xylenes can be taken as equal to the price of toluene.

†The price of caustic soda can be assumed equal to that of chlorine. The sales price of butadiene can be calculated from that of styrene on the assumption that $P_{polybutadiene} = P_{SBR}$. Accordingly,

$$P_{butadiene} = 0.79P_{styrene} - 0.70P_{crude} - 48 = \text{dollars per metric ton}$$

more realistic basis and avoids the extreme fluctuations of the market. Such a list can be easily maintained for the principal petrochemicals whose costs at any given moment are relatively well known. If the price of naphtha is known, for example, and the battery-limits investments for the various process units are known, the formulas in Tables 4.4 and 4.7 can be used to determine either (1) the effect of the cost of energy on the selling price of a naphtha derivative, since Table 4.8 gives coefficients for calculating energy costs that cause price fluctuations, or (2) the selling price required to allow for given energy cost, when it is necessary merely to apply a reasonable profit margin of, say, a return on investment of 30% in order to obtain a relatively consistent price suitable for a preliminary estimate (Table 4.9).

By-products A pricing coefficient can usually set the effects of by-products, if it accounts for the amount produced, the quality or value, and the available markets. Sometimes it is necessary to add a cost of disposal or secondary treatment. Without better information, a first guess at a by-product value should hardly place it higher than 50% of its price as a prime product.

Catalysts The amount of catalyst consumed in homogeneous reactions, like consumptions of chemicals, is generally a technical characteristic of the process. The amount of catalyst consumed in heterogeneous reactions depends on its life; the theoretical annual consumption is reported in relation to the production or consumption of product or feedstock. When the catalyst is

TABLE 4.9 Comparing Calculated Average Sales Prices* and Market Prices for Chemicals

Product	Price, dollars per metric ton in 1975		
	Calculated*	June	December
Acetaldehyde	304	290	333–354
Vinyl acetate	617	385	458–429
Acetone	352	292	333–354
Acetic acid	471	312–375	479
Acrylic acid	579	617	708
Adipic acid	696	675	729
Terephthalic acid (pure)	579	—	—
Maleic anhydride	535	583–792	792–854
Phthalic anhydride	504	481	542
Cumene	300	308	333
Cyclohexane	319	263	292
Cyclohexanol/one	483	—	583
Cyclohexanone	540	694	750
Diethylene glycol	492	500	625
Dimethyl terephthalate	660	713	771
Ethanol (95%)	276	288	323
Formaldehyde	221	167–229	250
Isopropanol	246	204	254
Methanol	146	125–167	158
Ethylene oxide	448	490	542
Propylene oxide	558	458	521
Phenol	554	521	563–625
Styrene	525	423	438

*Based on crude oil at $79/metric ton.

a precious metal, its costs cover the support, any promoters or inhibitors, forming and impregnating the support, recovery and treatment of the precious metal, and any loss of the precious metal.

Utilities Unit prices for utilities have come to fluctuate as much as the prices of products and feedstocks, particularly for steam, electricity, fuel, and refrigeration, which all depend largely on the cost of energy. Costs for electricity and steam can be examined with respect to the type of fuel and source of supply.

Electricity from large utilities should be distinguished from that by-produced by large plants like petroleum refineries. *L'Électricité de France,* for example, scheduled its price per kilowatthour in 1974 according to the fuel value expressed in the price of a therm, t, as about the following:

Producing cost for fossil-fuel–fired generating stations, ¢/kWh: $0.48 + 2.3t$

Producing cost for nuclear-fueled generating stations, ¢/kWh: $0.8 + 1$(price of nuclear fuel)

The cost of electricity from refineries, by contrast, depends essentially on the type of equipment and on whether steam is by-produced. Thus 5.5 megawatts of electricity and 100 tons/h of 150 psig steam from a 450 psig boiler and a noncondensing turbine can yield electricity priced at $(0.8 + 1.1t)$ cents per kilowatthour, where t is again the fuel value in cents per therm.

The cost of steam, which depends on the cost of fuel, makes up one part of fixed costs depending on the size of the plant, the plant's distribution network, and the steam pressure. One ton of fuel furnishes from 13 to 14 tons of steam. For a refinery, the cost of producing steam in dollars per ton is close to the following as t becomes cents per therm:

Low pressure (60 psig) steam: $1.6 + 7.3t$
Medium pressure (300 psig) steam: $1.6 + 7.5t$
High pressure (750 psig) steam: $1.8 + 7.7t$

Refrigeration The cost of cooling below ambient conditions depends essentially on the temperature level. As a first approximation, the following mid-1975 values can be used:

Temperature, °C	Cost of refrigeration, \$/10⁶ kcal
0	12
−10	18
−20	24
−30	24
−40	36
−50	44

The average mid-1975 cost of utilities can be assumed as shown in Table 4.7.

Labor In the absence of other information, an average crew of 4.5 people per day can be assumed in most cases for the continuous operation of a plant. The cost shown in Table 4.7 for one shift or for one operator per shift is equivalent to the wages for 4.5 people, benefits, and 20% additional for supervision included.

For batch operation or a plant that has diversified manufacturing, the work hours should be allocated to each type of product, with benefits and supervision included in the hourly rate. Also, transition periods for start-up and shutdown should not be forgotten in establishing the schedule.

Depreciation A period of 8–10 years, with straight-line depreciation of 10–12.5% per year, is usual for organic chemicals and continuously operating

TABLE 4.10 Procedure for Calculating Payout Time

Heading, $/yr or $/unit of product	Symbol	Remarks (see Sec. 2.6.2)
Sales volume	V	Market study
Total operating cost	C	(see Table 4.7)
Gross profit	B	$B = V - C$
Tax rate	a	taxes $a.B$; $a \simeq 0.5$
Net profit		$B(1 - a)$
Depreciation	A	(see Table 4.7)
Gross cash flow	(CF)	$(CF) = B(1-a) + A$
Depreciable capital	(DC)	(see Table 4.4)
Payout time	(POT)	$(POT) = (DC)(CF)$

plants. Smaller units with batch operations or units with diversified production usually have shorter depreciation periods on the order of 5 years, with 20% of the capital recovered annually.

Products, coproducts, and by-products Just as the distinction between raw materials and treating chemicals is sometimes awkward, so is the distinction between coproducts and by-products. On a first analysis, they can be differentiated according to the tonnages relative to the tonnage of principal product. On a second analysis, one should examine the specifications for each and the equipment needed to have those specifications meet the requirements of normal commercialization. Usually a by-product will need additional treatment in order to satisfy commercial standards. Also, organometallic compounds are almost always the primary products in organic chemicals plants.

4.3.2.3 Profitability Study

Payout time, cumulative net present value, and rate of return with discounting, as discussed in Secs. 2.6.2, 2.6.3.2a, and 2.6.3.2b, respectively, can be applied as follows:

Payout time Two questions can be posed: What is the payout time in years, given the selling price and volume of sales for the product? What is the minimum product price needed to give a predetermined payout time? Table 4.10 shows the steps of this calculation with references to the appropriate sections of discussion.

Cumulative net present value Since the source of financing is usually unknown during project analysis, we here assume the general case where the company provides its own financing. The source of capital is often unknown for feasibility studies; so the general case of self-financing is most often used. Even in this context different formulations are possible, and we have selected the two that take into account taxation, as follows:

1. A general approach in which the receipts, costs, and depreciation vary from year to year. Table 4.11a shows the steps of this calculation, which consists of creating a balance sheet and using Table 4.12a to determine $\alpha = 1/(1 + i)^n$ for various values of i and n.

2. A simplified calculation, illustrated in Table 4.11b, where receipts, costs and straight-line depreciation are assumed to be identical from year to year. Tables 4.12a and 4.12b allow for determining α and β for values of i and n.

It is important to distinguish that the operating cost, as it is presented in Sec. 4.3.2.2 and used in applying payout time, includes the finance charges, whereas in calculating cumulative net present value, the payback of capital, borrowed or not, is taken into account by the discounting rate; so the operating cost must be determined without the finance charges, and they must be left out, if given.

Rate of return with discounting A discounted rate of return can be determined for each of the two calculations for cumulative net present value by finding the rate of return, i_r, corresponding to a discounting rate, i, which will make the cumulative net present value equal to zero (Sec. 2.6.3.2b). It suffices to follow the steps summarized in Tables 4.11a and 4.11b for several values of i. In Table 4.13, the discounted rate of return is determined for the case where a simplified expression is used to obtain the cumulative net present value with taxes included and the financing handled by the company.

4.3.3 HANDLING RAW DATA: PRACTICAL METHODS FOR INVESTMENT COSTS

Once the procedure for doing the economic calculation is decided on and the choice of economic criteria has been made, the feasibility study rests solely on the data at hand, i.e., on the battery-limits investment, its associated investments, and the material balance. Since the study will involve comparisons, these basic data must be consistent and the raw reference data adapted accordingly.

The distinction made in Sec. 4.3.1 among three wide categories of plant information, particularly in the manner of presenting the investment for a plant, is reflected in the following three methods for determining the battery-limits investment: overall extrapolation (Sec. 4.3.3.1), extrapolating item by item (Sec. 4.3.3.2), and sizing the primary equipment (Sec. 4.3.3.3).

4.3.3.1 Overall Extrapolations for Capacity, Date, and Location

When overall data such as are found in preliminary proposals or the technical literature are available, including the battery-limits investment for an isolated

TABLE 4.11a Procedure for Calculating Cumulative Net Present Value:

General Case, with Financing from The Company's Treasury (see Sec. 2.6.3 for derivations of formulas and symbols)

Item	Symbols	Formulas	Calculations by year			
Year	n		0	1	2	n
Investment	I_p	Usually $I_p = 0$ after year 0	$-(I + f)$	$-I_1$	$-I_2$	$I_r + f$
Sales volume	V_p			V_1	V_2	V_n
Exploitation costs	D_p	See Sec. 2.5		D_1	D_2	D_n
Depreciation	A_p			A_1	A_2	A_n
Financing costs	F_p	Assume $F_p = 0$		0	0	0
Operating cost	C_p	$C_p = D_p + A_p + F_p = D_p + A_p$		C_1	C_2	C_n
Net profits, given a		$(V_p - C_p)(1 - a)$		$(V_1 - C_1)(1 - a)$	$(V_2 - C_2)(1 - a)$	$(V_n - C_n)(1 - a)$
Depreciation	A_p			A_1	A_2	A_n
Cash flow	$(CF)_p$	$(CF)_p = (V_p - C_p)(1 - a) + A_p - I_p$	$(CF)_0$	$(CF)_1$	$(CF)_2$	$(CF)_n$
Discounting factor, given i	α	$1/(1 + i)^p$	1	$1/(1 + i)$	$1/(1 + i)^2$	$1/(1 + i)^n$
Discounted revenue	$(CF)_p/(1 + i)^p$		$(CF)_0$	$(CF)_1/(1 + i)$	$(CF)_2/(1 + i)^2$	$(CF)_n/(1 + i)^n$
Cumulative net present value	B	$B = \sum_{p=0}^{p=n} \frac{(CF)_p}{(1 + i)^p}$				

Note: Verify that the financing costs are zero in calculating the operating cost.

TABLE 4.11b Procedure for Simplified Calculation of Cumulative Net Present Value:

General Case, with Financing from The Company's Treasury (see Sec. 2.6.3 for derivations of formulas and symbols)

Item	Symbol	Remarks	Calculation
Depreciable capital	I		$-I$
Working capital	f		$-f$
Life of project, years	n		n
Discounting factor, given rate i	α		$1/(1+i)^n$
Total investments to year p	ΣI_p	No supplementary investments	$-[I + f(1-\alpha)]$
Total of yearly sales revenues	V_p		
Total of yearly exploitation costs	D_p		
Annual depreciation*	A_p	Linear depreciation	A
Financing costs	F_p	Assumed zero	0
Annual operating cost	C_p	$C_p = D_p + A_p + F_p$	$C = D + A$
Net annual profits, given a		$(V_p - C_p)(1-a)$	$(V - C)(1-a)$
Annual depreciation	A_p		A
Revenues for year p†	$(CF)_p$	$(V_p - C_p)(1-a) + A_p$	$(V - C)(1-a) + A$
Cumulative discounting factor, assuming i	β		$\dfrac{(1+i)^n - 1}{i(1+i)^n}$
Discounted annual returns through year p			$\beta[(V - C)(1-a) + A]$
Cumulative net present value	B		$-[I + f(1-\alpha)] + \beta[(V - C)(1-a) + A]$

*Optional calculation when the operating cost is known.
†Working capital at its present worth is included in the investments.
Note: Verify that the financing costs are zero in the calculation of operating cost.

TABLE 4.12a Values of Coefficient α as a Function of $1/(1 + i)^n$

Year	Discounting rate i									Year
n	0.04	0.05	0.06	0.08	0.10	0.12	0.15	0.20	0.25	n
1	0.9615	0.9524	0.9434	0.9259	0.9091	0.8929	0.8696	0.8333	0.8000	1
2	0.9246	0.9070	0.8900	0.8573	0.8264	0.7972	0.7561	0.6944	0.6400	2
3	0.8890	0.8638	0.8396	0.7938	0.7513	0.7118	0.6575	0.5787	0.5120	3
4	0.8548	0.8227	0.7921	0.7350	0.6830	0.6355	0.5718	0.4823	0.4096	4
5	0.8219	0.7835	0.7473	0.6806	0.6209	0.5674	0.4972	0.4019	0.3277	5
6	0.7903	0.7462	0.7050	0.6302	0.5645	0.5066	0.4323	0.3349	0.2621	6
7	0.7599	0.7107	0.6651	0.5835	0.5132	0.4523	0.3759	0.2701	0.2097	7
8	0.7307	0.6768	0.6274	0.5403	0.4665	0.4039	0.3269	0.2326	0.1678	8
9	0.7026	0.6446	0.5919	0.5002	0.4241	0.3606	0.2843	0.1938	0.1342	9
10	0.6756	0.6139	0.5584	0.4632	0.3855	0.3220	0.2472	0.1615	0.1074	10
11	0.6496	0.5847	0.5268	0.4289	0.3505	0.2875	0.2149	0.1346	0.0859	11
12	0.6246	0.5568	0.4970	0.3971	0.3186	0.2567	0.1869	0.1122	0.0687	12
13	0.6006	0.5303	0.4688	0.3677	0.2897	0.2292	0.1625	0.0935	0.0550	13
14	0.5775	0.5051	0.4423	0.3405	0.2633	0.2046	0.1413	0.0779	0.0440	14
15	0.5553	0.4810	0.4173	0.3152	0.2394	0.1827	0.1229	0.0649	0.0352	15
16	0.5339	0.4581	0.3936	0.2919	0.2176	0.1631	0.1069	0.0541	0.0281	16
17	0.5134	0.4363	0.3714	0.2703	0.1978	0.1456	0.0929	0.0451	0.0225	17
18	0.4936	0.4155	0.3503	0.2502	0.1799	0.1300	0.0808	0.0376	0.0180	18
19	0.4746	0.3957	0.3305	0.2317	0.1635	0.1161	0.0703	0.0313	0.0144	19
20	0.4564	0.3769	0.3118	0.2145	0.1486	0.1037	0.0611	0.0261	0.0115	20
21	0.4388	0.3589	0.2942	0.1987	0.1351	0.0926	0.0531	0.0217	0.0092	21
22	0.4220	0.3419	0.2775	0.1839	0.1228	0.0826	0.0462	0.0181	0.0074	22
23	0.4057	0.3256	0.2618	0.1703	0.1117	0.0738	0.0402	0.0151	0.0059	23
24	0.3901	0.3101	0.2470	0.1577	0.1015	0.0658	0.0349	0.0126	0.0047	24

TABLE 4.12a Values of Coefficient α as a Function of $1/(1 + i)^n$ *(Continued)*

Year	Discounting rate i									Year
n	0.04	0.05	0.06	0.08	0.10	0.12	0.15	0.20	0.25	n
25	0.3751	0.2953	0.2330	0.1460	0.0923	0.0588	0.0304	0.0105	0.0038	25
26	0.3607	0.2812	0.2198	0.1352	0.0839	0.0525	0.0264	0.0087	0.0030	26
27	0.3468	0.2678	0.2074	0.1252	0.0763	0.0469	0.0230	0.0073	0.0024	27
28	0.3335	0.2551	0.1956	0.1159	0.0693	0.0419	0.0200	0.0061	0.0019	28
29	0.3207	0.2429	0.1846	0.1073	0.0630	0.0374	0.0174	0.0051	0.0015	29
30	0.3083	0.2314	0.1741	0.0994	0.0573	0.0334	0.0151	0.0042	0.0012	30
31	0.2965	0.2204	0.1643	0.0920	0.0521	0.0298	0.0131	0.0035	0.0010	31
32	0.2851	0.2099	0.1550	0.0852	0.0474	0.0266	0.0114	0.0029	0.0008	32
33	0.2741	0.1999	0.1462	0.0789	0.0431	0.0238	0.0099	0.0024	0.0006	33
34	0.2636	0.1904	0.1379	0.0730	0.0391	0.0212	0.0086	0.0020	0.0005	34
35	0.2534	0.1813	0.1301	0.0676	0.0356	0.0189	0.0075	0.0017	0.0004	35
36	0.2437	0.1727	0.1227	0.0626	0.0323	0.0169	0.0065	0.0014	0.0003	36
37	0.2343	0.1644	0.1158	0.0580	0.0294	0.0151	0.0057	0.0012	0.0003	37
38	0.2253	0.1566	0.1092	0.0537	0.0267	0.0135	0.0049	0.0010	0.0002	38
39	0.2166	0.1491	0.1031	0.0497	0.0243	0.0120	0.0043	0.0008	0.0002	39
40	0.2083	0.1420	0.0972	0.0460	0.0221	0.0107	0.0037	0.0007	0.0001	40
41	0.2003	0.1353	0.0917	0.0426	0.0201	0.0096	0.0032	0.0006	0.0001	41
42	0.1926	0.1288	0.0865	0.0395	0.0183	0.0086	0.0028	0.0005	0.0001	42
43	0.1852	0.1227	0.0816	0.0365	0.0166	0.0076	0.0025	0.0004	0.0001	43
44	0.1780	0.1169	0.0770	0.0338	0.0151	0.0068	0.0021	0.0003	0.0001	44
45	0.1712	0.1113	0.0727	0.0313	0.0137	0.0061	0.0019	0.0003	0.0000	45
46	0.1646	0.1060	0.0685	0.0290	0.0125	0.0054	0.0016	0.0002	0.0000	46
47	0.1583	0.1009	0.0647	0.0269	0.0113	0.0049	0.0014	0.0002	0.0000	47
48	0.1522	0.0961	0.0610	0.0249	0.0103	0.0043	0.0012	0.0002	0.0000	48

TABLE 4.12b Values of Cumulative Discounting factor β as $[(1+i)^n - 1]/i(1+i)^n$

Year n	Discounting rate i									Year n
	0.04	0.05	0.06	0.08	0.10	0.12	0.15	0.20	0.25	
1	0.962	0.952	0.943	0.926	0.909	0.893	0.870	0.833	0.800	1
2	1.886	1.859	1.833	1.783	1.736	1.690	1.626	1.528	1.440	2
3	2.775	2.723	2.673	2.577	2.487	2.402	2.283	2.106	1.952	3
4	3.630	3.546	3.465	3.312	3.170	3.037	2.855	2.589	2.362	4
5	4.452	4.329	4.212	3.993	3.791	3.605	3.352	2.991	2.689	5
6	5.242	5.076	4.917	4.623	4.355	4.111	3.784	3.326	2.951	6
7	6.002	5.786	5.582	5.206	4.868	4.564	4.160	3.605	3.161	7
8	6.733	6.463	6.210	5.747	5.335	4.968	4.487	3.837	3.329	8
9	7.435	7.108	6.802	6.247	5.759	5.328	4.772	4.031	3.463	9
10	8.111	7.722	7.360	6.710	6.145	5.650	5.019	4.192	3.571	10
11	8.760	8.306	7.887	7.139	6.495	5.938	5.234	4.327	3.656	11
12	9.385	8.863	8.384	7.536	6.814	6.194	5.421	4.439	3.725	12
13	9.986	9.394	8.853	7.904	7.103	6.424	5.583	4.533	3.780	13
14	10.563	9.899	9.295	8.244	7.367	6.628	5.724	4.611	3.824	14
15	11.118	10.380	9.712	8.559	7.606	6.811	5.847	4.675	3.859	15
16	11.652	10.838	10.106	8.851	7.824	6.974	5.954	4.730	3.887	16
17	12.166	11.274	10.477	9.122	8.022	7.120	6.047	4.775	3.910	17
18	12.659	11.690	10.828	9.372	8.201	7.250	6.128	4.812	3.928	18
19	13.134	12.085	11.158	9.604	8.365	7.366	6.198	4.843	3.942	19
20	13.590	12.462	11.470	9.818	8.514	7.469	6.259	4.870	3.954	20
21	14.029	12.821	11.764	10.017	8.649	7.562	6.312	4.891	3.963	21
22	14.451	13.163	12.042	10.201	8.772	7.645	6.359	4.909	3.970	22
23	14.857	13.489	12.303	10.371	8.883	7.718	6.399	4.925	3.976	23
24	15.247	13.799	12.550	10.529	8.985	7.784	6.434	4.937	3.981	24

TABLE 4.12b Values of Cumulative Discounting factor β as $[(1+i)^n - 1]/i(1+i)^n$ *(Continued)*

Year n	Discounting rate									Year n
	0.04	0.05	0.06	0.08	0.10	0.12	0.15	0.20	0.25	
25	15.622	14.094	12.783	10.675	9.077	7.843	6.464	4.948	3.985	25
26	15.983	14.375	13.003	10.810	9.161	7.896	6.491	4.956	3.988	26
27	16.330	14.643	13.211	10.935	9.237	7.943	6.514	4.964	3.990	27
28	16.663	14.898	13.406	11.051	9.307	7.984	6.534	4.970	3.992	28
29	16.984	15.141	13.591	11.158	9.370	8.022	6.551	4.975	3.994	29
30	17.292	15.372	13.765	11.258	9.427	8.055	6.566	4.979	3.995	30
31	17.588	15.593	13.929	11.350	9.479	8.085	6.579	4.982	3.996	31
32	17.874	15.803	14.084	11.435	9.526	8.112	6.591	4.985	3.997	32
33	18.148	16.003	14.230	11.514	9.569	8.135	6.600	4.988	3.997	33
34	18.411	16.193	14.368	11.587	9.609	8.157	6.609	4.990	3.998	34
35	18.665	16.374	14.498	11.655	9.644	8.176	6.617	4.992	3.998	35
36	18.908	16.547	14.621	11.717	9.677	8.192	6.623	4.993	3.999	36
37	19.143	16.711	14.737	11.775	9.706	8.208	6.629	4.994	3.999	37
38	19.368	16.868	14.846	11.829	9.733	8.221	6.634	4.995	3.999	38
39	19.584	17.017	14.949	11.879	9.757	8.233	6.638	4.996	3.999	39
40	19.793	17.159	15.046	11.925	9.779	8.244	6.642	4.997	3.999	40
41	19.993	17.294	15.138	11.967	9.799	8.253	6.645	4.997	4.000	41
42	20.186	17.423	15.225	12.007	9.817	8.262	6.648	4.998	4.000	42
43	20.371	17.546	15.306	12.043	9.834	8.270	6.650	4.998	4.000	43
44	20.549	17.663	15.383	12.077	9.849	8.276	6.652	4.998	4.000	44
45	20.720	17.774	15.456	12.108	9.863	8.283	6.654	4.999	4.000	45
46	20.885	17.880	15.524	12.137	9.875	8.288	6.656	4.999	4.000	46
47	21.043	17.981	15.589	12.164	9.887	8.293	6.657	4.999	4.000	47
48	21.195	18.077	15.650	12.189	9.897	8.297	6.659	4.999	4.000	48

TABLE 4.13 Procedure for Calculating Rate of Return with Discounting: Simplified Case with Financing from Within the Company*

	Formulas	Calculations for Trial Values of i		
Discounting rate	i	i_1	i_2	i_r
Discounting factor, given n	$\alpha = 1/(1+i)^n$	α_1	α_2	α_r
Cumulative discounting factor, given n	$\beta = \frac{(1+i)^n - 1}{i(1+i)^n}$	β_1	β_2	β_r
Total investment	$-[I + f(1 - \alpha]$	$-[I + f(1 - \alpha_1)]$	$-[I + f(1 - \alpha_2)]$	$-[I + f(1 - \alpha_r)]$
Cumulative present revenues	$[(V - C)(1 - a) + A]\beta$	$[(V - C)(1 - a) + A]\beta_1$	$[(V - C)(1 - a) + A]\beta_2$	$[(V - C)(1 - a) + A]\beta_r$
Cumulative net present value	$B = -[I + f(1 - \alpha)] + [(V - C)(1 - a) + A]\beta$	B_1	B_2	B_r

*The rate of return is that discounting rate at which $B_r = 0$.

plant or a complex with a given production capacity, it is possible to prorate these data to different capacities, times, and locations. Also, the costs may be converted according to the parity of currencies.

4.3.3.1a Extrapolating According to Capacity

The exponential factor described in Sec. 3.5.1.1 is the best and simplest technique for this extrapolation. Using an average value of 0.65 for the exponent f:

$$\frac{I_1}{I_2} = \left(\frac{C_1}{C_2}\right)^{0.65}$$

In actual practice, the problems confronting this formula will involve either similar units with approximately the same capacity or units where the capacities are much different. In the first case, the exponent can be taken as 0.65; when the unknown investment corresponds to a much higher capacity where production lines will be duplicated rather than enlarged, the exponent may become 1.0 for a duplicate production line or 0.9 for certain polymerization processes; and when the unknown investment corresponds to a much lower capacity, the relatively constant costs for installation and associated equipment become more important, and the exponent would reduce to 0.5–0.55.

4.3.3.1b Bringing the Investment Costs Up to Date

In Sec. 3.3 were described a number of overall cost indexes, and a choice should now be made. When a private specialized index is not available, it is necessary to use published information. Because of their regular and frequent publication, the most convenient of the published indexes are the Nelson cost index and the *Chemical Engineering* (CE) index. As has been indicated in Sec. 3.3, the Nelson index is best adapted for projects in the petroleum industry, while the CE index applies best to petrochemicals and the chemical industry in general.

The evolution of these two indexes is shown graphically in Figs. 4.4 and 4.5 for easier and more rapid use. The plot in Fig. 4.4 shows the Nelson inflation index along with its two principal components, materials and labor. More accuracy can be had from Tables 3.3 and 3.5.

These indexes relate only to North America, however, and European plants would be better represented by a European index, even though such indexes are more difficult to find. The PE cost index for the United Kingdom has to be rejected because of the recent great increase in salaries there and the depreciation of the English pound in relation to other currencies. Nor can the Dutch index be used, since it is not published regularly. The only remaining European index is that published by *Chemische Industrie,* which is repeated in Table 4.14 in comparison with the CE cost index and the Nelson inflation index.

It should be recalled that these indexes use the following relation:

$$\text{Current cost} = \text{reference cost}\ \frac{\text{index no. for the current year}}{\text{index no. for the reference year}}$$

4.3.3.1c Correcting for Differences in Location

As was seen in Sec. 3.4, there is no accurate rule for accounting for differences in location. However, the industrialized countries exhibit comparable investment costs, with few exceptions. A good approximation for the developing countries can be obtained with the coefficients proposed by Ph. Terris (see Table 3.8).

4.3.3.1d Extrapolating and Correcting Other Reference Data

Auxiliary investment costs such as initial charges of catalysts and solvents, spare parts, royalties, and the process data book can be extrapolated as follows:

- Initial charges of catalysts and solvents proportionally to capacity, when operating conditions are identical
- Spare parts according to a ratio of the battery-limits investments
- Royalties proportionally to the capacity, except for large plants where an exponential factor of 0.6–0.8 of the ratio of capacities can be used
- Process data book relatively constant

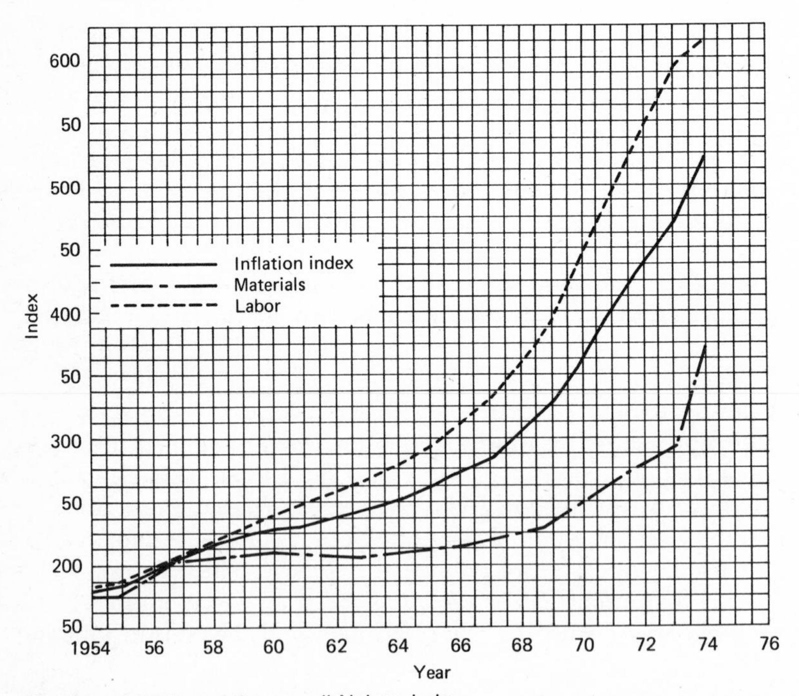

Fig. 4.4 Evolution of the overall Nelson index.

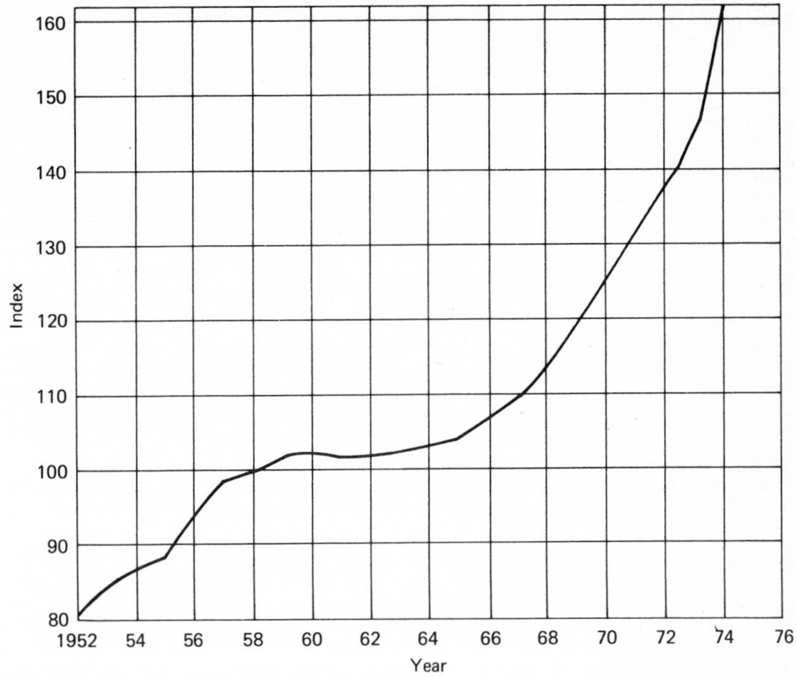

Fig. 4.5 Evolution of the CE cost index.

Miscellaneous consumptions or productions per unit of feedstock are not modified by extrapolation.

Labor costs change according to whether a change in capacity changes the number of work stations, as when there is a doubling of production lines.

4.3.3.1e Parity of Francs to Dollars

It is helpful, for making data consistent and for rapid conversions, to have available the evolution of the rate of exchange of of francs for dollars over a period of 10 years. This is the purpose of Fig. 4.6, where a certain stability can be seen up until 1969, with a parity close to 5 francs per dollar. The years 1968–1969 can thus constitute a preferred period in calculations for transposing data.

4.3.3.2 Extrapolating Item by Item

This uses information such as is furnished by detailed proposals or certain publications of restricted circulation, where an exact material balance, as well as a partial or complete breakdown of investment costs, is provided. These data are made up of one of the following:

1. The battery-limits investment and its different components (primary equipment, secondary equipment, installation, etc.) with primary equipment detailed

TABLE 4.14 Recommended Published Cost Indexes

Year	Nelson Inflation Index	*Chemical Engineering* Cost Index	*Chemische Industrie* Cost Index
1960	228.3	102.0	91.4
1961	232.7	101.5	95.5
1962	237.6	102.0	100.0
1963	243.6	102.4	101.1
1964	252.1	103.3	104.6
1965	261.4	104.2	108.7
1966	273.0	107.2	111.1
1967	286.7	109.7	107.8
1968	304.1	113.7	102.8
1969	329.0	119.0	110.1
1970	364.9	125.7	125.7
1971	406.0	132.3	135.5
1972	438.5	137.2	140.0
1973	468.0	144.1	145.5
1974	522.7	165.4	156.6
1975 (July)	578.3	182.1	167.6
1976			
1977			
1978			
1979			
1980			

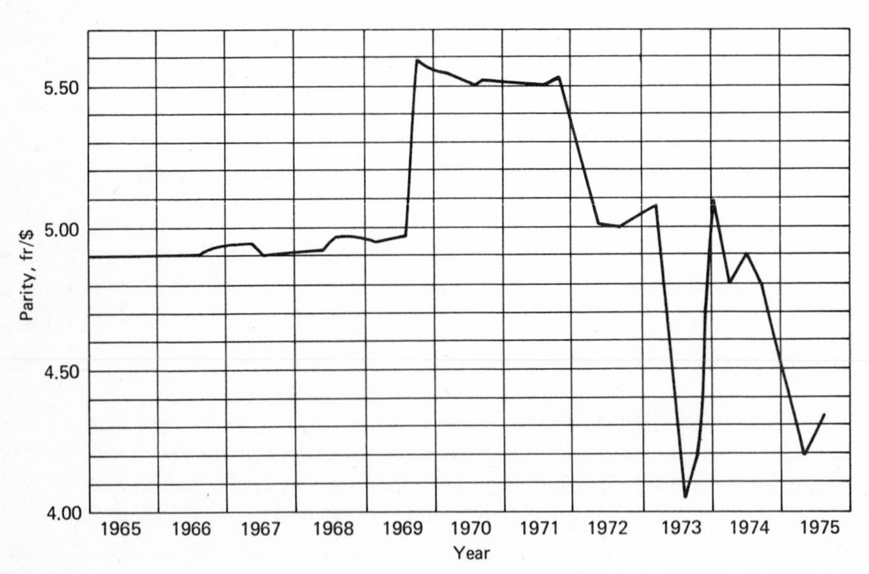

Fig. 4.6 The parity between French francs and U.S. dollars.

2. Only the primary equipment, which is defined by one of the following:
   a. Broad categories or types of equipment (towers, exchangers, compressors, etc.)
   b. Each item of equipment

This information is extrapolated for different conditions of material balance, time, and site so as to attain successively the cost of the primary equipment, the battery-limits investment, and the profitability for the new conditions of the study.

4.3.3.2a Determining the Cost of Primary Equipment

By capacity and by categories Two situations occur, depending on whether the reference data furnish the cost of the primary equipment item by item or by categories. In the first case, extrapolating is most often done in relation to the feed rate (assuming operating conditions are identical). Exceptions are distillation and extraction. In distillation, the calculation is made in terms of the overhead vapor traffic as $D(1 + R)$, with D the moles of distillate and R the reflux ratio. In extraction, the calculation is made in terms of liquid traffic as feed rate times (solvent rate plus one).

In the event information is given by categories of equipment, the extrapolation is based on the production capacity, except for special cases, and when the data are furnished for sections within a plant, partial extrapolations are done on each section according to its production capacity.

Theoretically, each piece of equipment has a unique extrapolation factor that will prorate its cost from one size to another within a given range. Most often, however, the parameter that characterizes the cost of an equipment item is not shown by the quantity of effluent leaving that piece of equipment and even less by the total production of a unit operation or a complete plant. The extrapolation is applied to the characteristic parameter (horsepower, weight, etc.) and the value of the corresponding exponent has a wide range of validity.

This is why when sizes are not available, it is difficult to use published factors that prorate equipment items costs by capacity and even more difficult to prorate broad categories of equipment. As a first approximation, however, the exponents shown in Table 4.15 can be used where only production capacities are known. These coefficients are to be used with the same restrictions as given in Sec. 4.3.3.1a for the overall exponents, particularly with respect to doubling the equipment.

By date As seen in Sec. 3.3, most of the published cost indexes only provide data for overall costs. *The Oil and Gas Journal,* which periodically furnishes the Nelson coefficients for commonly used refining and petrochemical equipment, is a rare exception; it offers indexes to update costs item by item.

TABLE 4.15 Exponential Factors for Extrapolating the Costs of Equipment Items*

Equipment Item	Factor	Equipment Item	Factor
Pressure vessels		Dryers	0.45–0.50
Columns and reactors	0.65–0.70	Filters	0.60–0.65
Tanks	0.60–0.65	Crystallizers	0.65
Exchangers		Basket-type centrifuges	1.00–1.25
Conventional	0.65	Evaporators	
Reboilers	0.50	Conventional	0.55
Air coolers	0.80	Forced circulation	0.70
Compressors	0.80–0.85	Grinders	0.65
Pumps		Storage tanks	
Centrifugal	0.40–0.50	Over 5,000 gal	0.30
Reciprocating	0.60–0.70	Under 5,000 gal	0.65
Motors	0.90–1.00	Steam generation	0.80
Furnaces	0.80–0.85	Generation of electricity	0.75
Agitated reactors	0.45–0.50	Cooling towers	0.60
Ejectors	0.45–0.55	Refrigeration	0.70
Agitators		Pressurized storage tanks	
Propeller	0.50	Horizontal	0.65
Turbine	0.30	Spherical	0.70

*For a description of applicable parameters, see text, Sec. 4.3.3.2a.

When preparing a study without specific information, the best thing to do is to use the Nelson indexes applying to pressure vessels and tanks, exchangers, pumps and compressors, motors, and furnaces. These indexes are plotted since 1954 in Fig. 4.7. Other equipment can be updated with the overall Nelson index in Table 3.3 or with the curve for miscellaneous equipment shown in Fig. 4.7. Another way around is to find the index after calculating the battery-limits investment using an overall index.

By location The influence of location on each piece of equipment cannot be accounted for without specific information. Instead, the data given on this subject in Sec. 3.4 can be used.

4.3.3.2b Determining the Battery-Limits Investment

Two kinds of raw data are generally available: either the battery-limits investment for an actual plant is known, or the costs of only primary equipment items are known.

Knowledge of the battery-limits investment assumes a more or less detailed breakdown. Costs may be differentiated for only some of the large categories such as primary equipment, secondary equipment, installation of primary equipment, installation of secondary equipment, and indirect field costs. Also, costs may sometimes be available for details of the secondary equipment, such as instrumentation and control, piping and valves, electrical, steel structures, civil engineering, insulation, painting, control room and buildings, site preparation, indirect field costs, and contingencies.

The costs of primary equipment form the basis on which to estimate other costs when extrapolating to similar process units. This is done for a new plant by means of coefficients that are applied to the primary equipment after its costs have been estimated from the reference plant. The battery-limits investment for the new plant is then obtained from the estimated components. Thus the values of the coefficients relating associated costs to primary equipment must be determined for the reference plant.

Knowledge of only the costs of primary equipment for the reference plant affords only the costs of corresponding items in the estimated plant without values for the coefficients. However, this eventuality is rather theoretical, since it is rare for a proposal or similar document to present the details of the primary equipment and not give at least the totals for secondary equipment and installation. Still, if only costs for the primary equipment is known, the problem is the same as for estimating a unit with complete sizing available or estimating a research project.

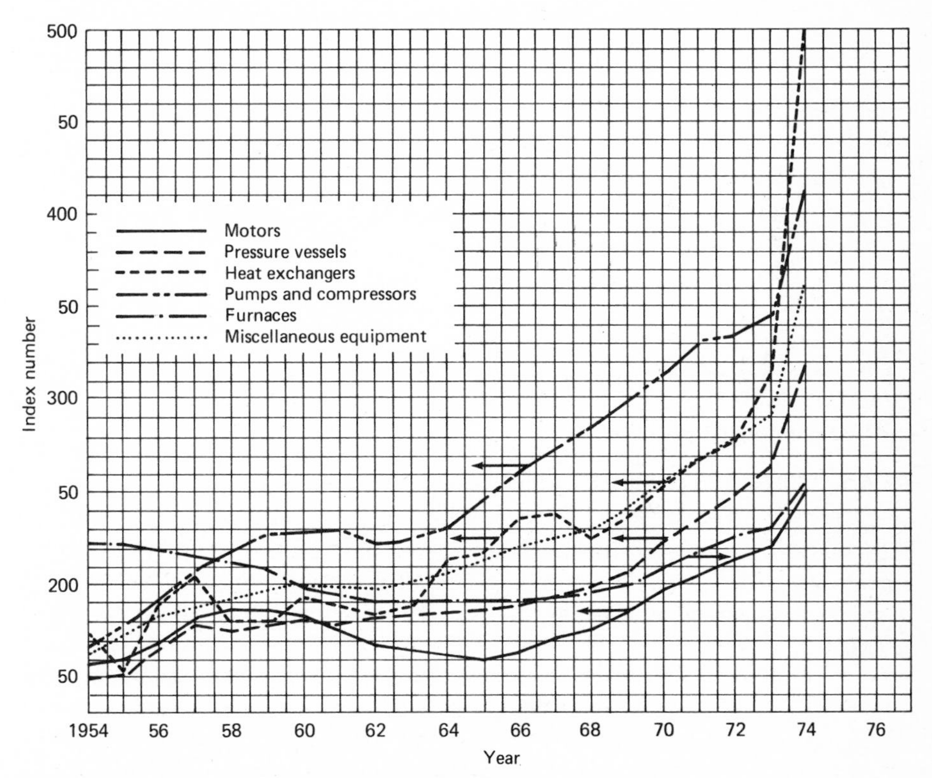

Fig. 4.7 Evolution of costs of equipment items: the Nelson indexes.

Various methods have been described in Sec. 3.5 for obtaining the battery-limits investment from knowledge of that portion corresponding to primary equipment in mild steel and under ordinary operating conditions, e.g., the methods proposed by J. H. Hirsch and E. M. Glazier, and by K. M. Guthrie. Use of these methods is assumed within the framework of calculations that involve sizing the primary equipment.

4.3.3.3 Estimates Based on Sizing the Primary Equipment

Estimators are often driven to reconciling information coming from various origins with varying accuracy and detailing. It is possible, for example, that a comparison is to be made between a process with sizes of each item of equipment available and another process with only a flow diagram and a material balance available. In the first case, the evaluation proceeds like an estimate; in the second, the primary equipment must be sized before the investment numbers are obtained. It is well understood that one does not ask an engineering service to make such sizing calculations, because of the high cost and delay in getting such an authoritative answer.

Consequently, an adequate estimating procedure should include a succinct method for rapidly calculating both the size of the primary equipment and its cost, as well as means for estimating equipment already sized.

Furthermore, it is preferable to recalculate sizes according to a consistent estimating method, even when a detailed estimate is accompanied by information on the sizing of a unit, since consistent use of a single sizing method puts everything into conditions suitable for making a comparison.

An extrapolated estimate based on a plant known in detail will differ from an estimate based on the consistent rapid sizing method, and the extrapolated estimate should be more reliable. If the difference between the two estimates is large, an error or false information is indicated. If the difference is not too great and the basic data reliable, the difference between the two estimates should be converted to coefficients that could be applied to the rapid sizing method for future estimates.

4.3.3.3a Rapid Calculations for Sizing and Pricing the Primary Equipment

This section constitutes the most extensive part of this chapter. Elements introduced here adapt themselves to the study of industrial projects as well as to research projects. Our aim is to create for each type of equipment a calculation sheet that can be completed on order or according to the available documents. In light of the role played by this information and the possibility of modifying or developing it for different applications, it has been preferable to collect it in an appendix. The methods have been limited to the equipment usually encountered in petrochemicals or in intermediate organics. Mineral chemicals, pharmaceuticals, or polymers often call for more specialized or more complicated equipment for which it is difficult to propose a simple calculation sheet. Also, one should realize that contractors define the sizing and pricing of such equipment according to their own standards.

Consequently, only a checklist is furnished for equipment of special applications. It is usually possible to indicate the characteristic size on which the extrapolation can be based; and a reference price and an extrapolation factor are then the only information needed. The difficulty rests in determining the representative size of the equipment. That problem is not taken up here.

Short-cut process design and estimating procedures are presented in the appendixes as follows.

Appendix 1: Pressure Vessels (including plate towers, packed towers, and tanks)

Appendix 2: Reactors

Appendix 3: Heat Exchangers (including shell-and-tube exchangers and coolers)

Appendix 4: Pumps and Compressors

Appendix 5: Drivers (both motors and turbines)

Appendix 6: Furnaces

Appendix 7: Ejectors

Appendix 8: Special Equipment

The data for the costs in these appendixes are generally for Western Europe, particularly France, during the first six months of 1975. This date can be made current through the indexes of Fig. 4.7 and Tables 3.3 and 3.4, combined with the most recent published versions of those indexes. Appendix 13 shows typical calculation sheets for some of the equipment. These sheets permit systematizing the methods detailed in the appendixes.

4.3.3.3b Determining the Battery-Limits Investment

Sections 3.5.2, 3.5.3, and 3.5.4 describe various authors' methods for calculating battery-limits investments from the cost of the primary equipment. It is not always necessary to resort to such procedures, since in practice either a complete sizing will be accompanied by an estimate or a detailed estimate will be available for one or more similar projects.

When a complete sizing is accompanied by an estimate, the necessary information will be contained in those documents. When a detailed estimate is available for a similar project, coefficients for its proportions can be used to determine corresponding components of the investment from the primary equipment costs for all the projects it resembles.

When the primary equipment must be sized, however, it is necessary to choose a way to find the battery-limits investment, either directly or by item. In the type of project evaluation examined here, where it is not necessary to ask for too great an accuracy, a method derived from that of K. M. Guthrie is proposed for use when the size of the primary equipment is estimated by a short-cut calculation.

The procedure consists of working with broad categories of equipment: pressure vessels, heat exchangers, pumps and drivers, compressors and drivers, furnaces, fired heaters, and miscellaneous. The calculation is performed in the following two steps:

1. Using the graphs found in the appendixes, determine
   a. The base cost P_B, which corresponds to the sum of the costs of items in a given category of equipment and excludes any adjustment for complexity or corrosion. These are the costs in mild steel, such as come from the graphs in Apps. 1 through 8. Use the general figure when choices exist.
   b. The expected cost P_R, which is the base cost P_B modified by correction factors for the complexity of the equipment and the construction materials.
2. Using Tables 4.16 and 4.17, where coefficients relating to a base cost of $P_B = 100$ are shown, calculate either one of the following.
   a. The primary equipment, the secondary equipment, and the indirect field costs for each category of equipment by applying the corresponding coefficients to the base cost P_B. If necessary, make corrections from the graph in Fig. 4.8, which gives a size-correction factor, f_c, depending on the expected cost P_R. The battery-limits investment is then the sum of the following cost components:

Cost Component	Formula or Symbol
Expected cost	P_R
Secondary equipment	$P_B(f_{ms}/100f_c)$
Installation	$P_B(f_m/100f_c)$
Indirect field costs	$P_B(f_i/100f_c)$
TOTAL	$P_R + P_B(f_{ms} + f_m + f_i)/100f_c$

   b. The battery-limits costs of the production units, directly or by type of equipment, by applying the overall correction factor f_g, adjusted or not, for the effect of size (Tables 4.15 and 4.16 and Fig. 4.8). The battery-limits cost is then $P_R - P_B + P_B f_g f_c$.

      The correction factor f_c can be used or not, depending on the accuracy of the calculation.

      It can sometimes be necessary to take into account the materials used for piping and valves. In order to calculate the overall coefficient

TABLE 4.16 Multiplying Factors, f_g, for Converting Base Costs of Primary Equipment Items to Battery-Limits Investment

	Pressure Vessels		Heat Exchangers				Furnaces	
	Towers* and Reactors	Tanks	Shell and Tube	Air Coolers	Pumps and Drivers	Centrifugal Compressors† and Drivers	Reaction	Heating
Primary equipment P_B	100	100	100	100	100	100	100	100
Secondary equipment f_{ms}	103	63	71	30	56	48	36	35
Piping and valves f_{tv}	60	40	45	14	23	17	19	17
Civil engineering	10	6	5	2	3	10	11	11
Metal structures	8	—	3	—	—	—	—	—
Instrumentation	11	6	10	4	2	7	4	5
Electrical equipment	5	5	2	9	25	11	2	2
Insulation	8	5	5	—	2	2	—	—
Painting	1	1	1	1	1	1	—	—
Erection f_m	97	58	60	29	64	54	36	35
Indirect field costs f_i	75	51	53	33	47	39	39	38
TOTAL	375	272	284	192	267	241	211	208
Multiplying coefficient f_g	3.75	2.72	2.84	1.92	2.67	2.41	2.11	2.08

*Installation of distillation trays is included for towers.

†The multiplying coefficient for reciprocating compressors is 2.95. The causes of the increase in this number, relative to that for centrifugal compressors, include piping and foundations designed to resist the vibrations.

f_g making such allowances, the factor f_{tv} of Table 4.16 should be corrected to f'_{tv} as follows:

$$f'_{tv} = f_{tv}\left(1 + \frac{P_R - P_B}{P_B}\right)$$

In certain plants such as petrochemical plants, where components like civil engineering or instrumentation are known to be more extensive, the average coefficients shown in Table 4.16 can be modified.

The battery-limits investment for the whole manufacturing complex is obtained by adding the battery-limits investments of the production units or, better, the installed costs of the corresponding categories of equipment that should have been estimated in order to apply the method.

If the costs of the primary equipment have not been brought up to date, an overall index can be applied to the erected-equipment cost by using the curves in Figs. 4.4 and 4.5 and Table 4.14.

The coefficients in Table 4.16 for converting primary equipment costs to installed costs correspond only to conditions in the petroleum and petrochemical industries, i.e., to the conditions of the cost bases indicated in the appendixes. For example, the minimum recommended thicknesses are 6 and 8 mm, respectively, for tanks and distillation towers, whereas other industries that habitually use alloy or special steels work with much thinner vessel walls and would differ markedly from the results given by the bare coefficients of Table 4.16.

4.3.3.3c Determining the Cost of Off-sites

The different situations encountered in actual practice lead either to evaluating the off-sites as a whole or as a detailed calculation. The off-sites cost can be

TABLE 4.17 Multiplying Factors f_g for Converting Costs of Miscellaneous Primary Equipment to Battery-Limits Investment

Equipment Type	Factor f_g
Mixers	1.85
Crushers and grinders	2.00
Centrifuges	1.80
Conveyors	1.90
Crystallizers	2.12
Ejectors	1.20
Evaporators	2.25
Filters	2.17
Dryers	2.05
Vibrating screens	1.50

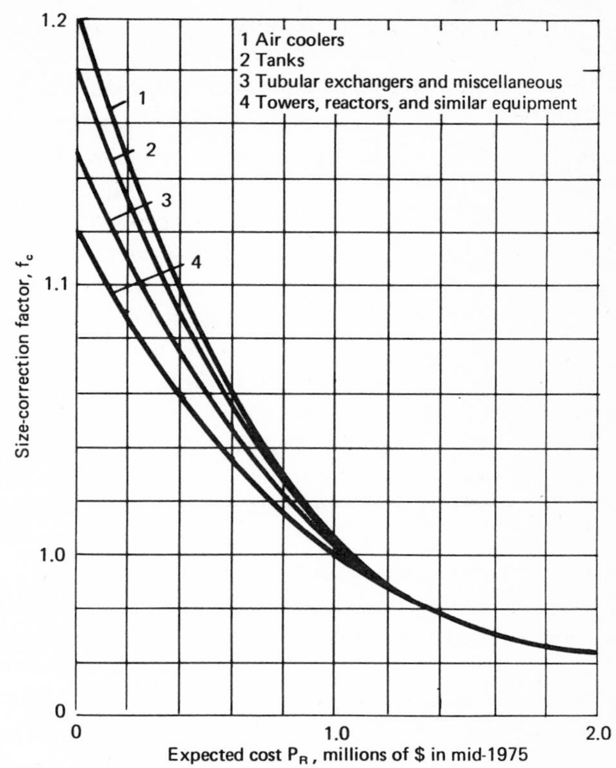

Fig. 4.8 Size-correction factors for the cost of primary equipment.

obtained directly from the battery-limits investment by multiplying by an overall coefficient of 0.3–0.4 for petrochemicals, for example, or of 3.0 for a refinery unit under the following conditions:

It is not necessary to calculate a manufacturing cost with great accuracy.

The purpose of the calculation is to compare several projects with the same or similar technologies.

The unit under study is to be installed in a complex.

However, it may be preferable to divide the off-site costs into, (1) utilities generation, (2) storage facilities, and (3) general administration services, when the study should afford greater accuracy or establish a market price for a competitive product or analyze a unit in isolation. In such situations, the utilities generation and storage facilities can be sized and estimated separately, while the administration facilities can be estimated as a fixed percentage (usually 15%) of the battery-limits investment, plus utilities generation plus storage facilities.

4.3.3.3d Determining the Investment for Utilities Generation and Distribution

The overall consumption is calculated for each utility, such as steam, electricity, cooling water, or refrigeration, that requires manufacturing or processing. Sizing is done according to the rules given in Apps. 1 through 7 and Tables 4.16 and 4.17. The cost of the complete installation of utility-generating equipment is determined according to the needs of the study through use of the graphs in App. 9. Since the utility-generating units must be linked to their corresponding distribution networks, these installation costs are converted to the equivalent of battery-limits investment by means of the multiplying factors of Table 4.18. When no distribution network is involved, the installation costs can be determined according to App. 9. These costs can then be updated by means of the CE cost index.

Other categories of utility, such as fuel, process water, air, etc., either do not involve much of an investment or are produced and transformed by primary equipment used for the basic utility or the battery-limits plant.

It is important to note that, when the utility costs amount to only 30–40% of the value of the battery-limits investment, the installations for generating such utilities as steam and electricity have not been included, and those utilities must be figured at their manufacturing cost, such as it is described in Sec. 4.3.2.2b. If the investment costs of the various units for generating utilities is determined and the investment for the necessary storage and general services is added (as for a complex), the total will be more than 30–40% of the battery-limits investment for the production unit. In such case, the cost of the generated utilities should not be included in the operating cost, except for consumptions.

TABLE 4.18 Multiplying Factor f_g for Converting Costs of Utility-Generating Equipment and Storage Equipment to Battery-Limits Investment

Equipment Type	Factor f_g
Steam generation	
Package boilers	1.50
Erected boilers	1.60
Electricity generation	1.40
Cooling towers	1.50
Refrigeration	1.40
Storage tanks	
Under 5,000 gal	1.65
Over 5,000 gal	2.05
Pressurized storage	
Horizontal	1.40
Spheres	1.85

4.3.3.3e Determining the Investment for Tank Farms

The size of a tank farm depends on how long the manufacturing units are to function without receiving raw materials or shipping products. For a refinery, the total strategic storage of finished products and crude oil, including loaded boats, is for 3 months, which puts land storage at about 2 months. For petrochemical units, an average storage of 4–5 days is allowed—8 days maximum for products and feedstocks. For other types of process, the storage time should be obtained as part of the basic data. Each manufacturing unit usually has its own storage area, which allows for differentiating the elements that make up the investment in this item.

There are three types of storage: atmospheric pressure, pressurized horizontal tanks, and pressurized spheres. The costs of the equipment are shown in the graphs in App. 10. These costs apply to the standard complete equipment item that needs only to be installed and connected. The coefficients for obtaining the battery-limits investment for tank farms from the equipment costs are assembled in Table 4.18.

Once the investment costs have been identified, the economic analysis can be carried out as described in Sec. 4.3.2.

Example 4

Calculation of the Investment Costs, Operating Costs, and Profitability for a Formaldehyde Plant

Basic data are identified, and calculations made according to the recommended methods.

E4.1 THE PROBLEM

A plant for producing formaldehyde from methanol by oxidation over a catalyst of iron oxide and molybdenum is planned for the United States. Economic data for this plant are as follows:*

Capacity: 12,500 t/yr of pure formaldehyde
Battery-limits investment: $590,000 in 1968
Initial charge of catalyst: 2.2 metric tons valued in mid-1975 at $25/kg
Life of the catalyst: 1 year
Consumptions per metric ton of pure formaldehyde:
  Methanol: 1.15 metric tons, valued at $135/ton in mid-1975
  Electricity: 256 kWh

* *Translators' note:* These data are only an example to illustrate the calculation method. In particular, they do not represent IFP/CdF-Chimie's process for production of formaldehyde.

Cooling water: 75 m^3
Boiler feedwater: 3 m^3

Steam production (at 20 bars) per ton of pure formaldehyde: 1.58 metric tons

Labor: two operators per shift

The following information is requested:

The battery-limits investment in mid-1975 for a production capacity of 25,000 metric tons per year of pure formaldehyde, assuming that doubling the reactor for this capacity will require an extrapolation factor of 0.88.

The manufacturing cost per metric ton of pure formaldehyde at both 12,500 and 25,000 metric tons per year.

The payout time when operating at 100 and 80% of design, assuming that the selling price of pure formaldehyde is $229.17/metric ton.

E4.2 THE ANSWERS

Generally, it is preferable to look into the investments first, in order to make the final comparisons easier.

E4.2.1 DETERMING THE BATTERY-LIMITS INVESTMENT IN MID-1975

The *Chemical Engineering* cost index for a unit installed at a chemical processing site in the United States is as follows:

For 1968: 113.7 or about 114
For mid-1975: about 190

$$\begin{aligned} I_{\text{mid-1975}} &= I_{1968}\left(\tfrac{190}{114}\right) = \$590{,}000(1.667) = 983{,}333 \\ &\simeq \$980{,}000 \end{aligned}$$

This mid-1975 investment for 12,500 metric tons per year capacity is then extrapolated to 25,000 tons/yr:

$$\begin{aligned} I_2 = I_1\left(\frac{C_2}{C_1}\right)^f &= \$980{,}000\left(\tfrac{25{,}000}{12{,}500}\right)^{0.88} \\ &= \$980{,}000(1.84) \\ &= \$1{,}800{,}000 \end{aligned}$$

The proportions given in Tables 4.4 and 4.7 are then used to determine the total investments, the manufacturing costs and the payout times as shown in Tables E4.1 through E4.4.

TABLE E4.1 Investment for Formaldehyde Plants of Different Capacities

	Capacity, t/yr of pure formaldehyde	
	12,500	25,000
	Investments, $ million	
Battery limits investments	0.98	1.80
General services and storage (30%)	0.29	0.54
TOTAL	1.27	2.34
Engineering (12%)	0.15	0.28
Royalties	0.10	0.20
Process data book	0.06	0.06
FIXED CAPITAL	1.58	2.88
Initial charge of catalyst	0.05	0.11
Interest on construction loan (7% of fixed capital)	0.11	0.20
Start-up costs (0.5 month of operating cost)	0.09	0.16
DEPRECIABLE CAPITAL	1.83	3.35
Working capital (1 month of operating cost)	0.17	0.31

TABLE E4.2 Manufacturing Cost for Making Formaldehyde in Plants Running at Design Capacity in mid-1975 (dollars per metric ton)

	Capacity, t/yr of pure formaldehyde		
	12,500		25,000
Variable costs			
Raw materials:			
Methanol: 1.15t at $135/t		$155.73	
Catalysts: 0.176 kg at $25/kg		4.40	
Utilities:			
Steam produced (20 bars)*: 1.58 at $6.88/t		(−)10.86	
Electricity: 256kWh at $0.025/kWh		6.40	
Cooling water: 75 m^3 at $0.0104		0.78	
Boiler feed water: 3 m^3 at $0.25/$m^3$		0.75	
TOTAL VARIABLE COSTS		157.20	
Labor: 2 operators per station at $83,333/yr per station	$13.33		6.67
Operating costs	170.53		163.87
Fixed costs			
Depreciation: 12.5% of the depreciable capital	18.30		16.75
Interest: 7% of the depreciable capital	10.24		9.38
9% of the working capital	1.27		1.12
Maintenance: 4% of battery limits investment	4.06		3.74
Taxes and insurance: 2% of battery limits investment	2.03		1.87
Overhead: 1% of battery limits investment	1.02		0.94
TOTAL FIXED COSTS	36.92		33.80
Manufacturing cost, $/t	207.45		197.67

*Although the generated steam is high pressure, it is credited as medium pressure steam because it is a by-product.

TABLE E4.3 Investment and Manufacturing Costs for Making Formaldehyde in Plants Running at 80% of Design Capacity

	Capacity, t/yr of pure formaldehyde	
	12,500	25,000
Production, metric t/yr	10,000	20,000
Variable costs, $/t produced (unchanged)	157.20	157.20
Labor, $/t produced (increased 25%)	16.67	8.33
Operating costs	173.87	165.53
Fixed costs: Investment costs for start-up, depreciable capital and working capital are increased due to the increased operating costs, so that the fixed costs are effectively increased on the order of 25%:		
Depreciation	22.88	20.94
Interest on the depreciable capital	12.80	11.73
Interest on the working capital	1.59	1.40
Maintenance, taxes and insurance, and overhead	8.89	8.19
TOTAL FIXED COSTS	46.16	42.26
Manufacturing cost, $/t produced	220.03	207.79

TABLE E4.4 Payout Time from Formaldehyde Manufacture at Two Different Capacities

	Capacity, t/yr for pure formaldehyde			
	12,500		25,000	
Feed rate, % design	100	80	100	80
Selling price of pure formaldehyde, $/t	229.17	229.17	229.17	229.17
Manufacturing cost, $/t produced	207.45	220.03	197.67	207.79
Gross income (before taxes) $/t	21.72	9.14	31.50	21.40
Net income (after taxes) $/t	10.86	4.57	15.75	10.70
Depreciation, $/t	18.30	16.75	22.88	20.94
Cash Flow, $/t	29.16	21.32	38.63	31.64
Depreciable capital, $ millions	1.83	1.83	3.35	3.35
Depreciable capital, $/t produced	146.40	183.00	134.00	167.50
Payout time, years	5.0	8.6	3.5	5.3

Example 5

Battery-Limits Investment for the Production of Cumene

This example takes up Example 3 and applies the method described in Chap. 4. In order to avoid turning back and forth, the problem along with parts of the answer is partially repeated here.

E5.1 THE PROBLEM

A unit for making cumene by alkylating benzene with propylene in the presence of a phosphoric acid catalyst is considered. Production capacity is 30,000 tons/yr and the details of the cost of the primary equipment is known for the year 1968. The major part of the equipment is of mild steel; the reactor and certain other pieces of equipment (the filter among others) are of stainless steel. Table E5.1 is obtained.

The following study is requested:

> Calculate the battery-limits investment for this unit for 1974, using an index by components. In order to do this, the data are updated by type of equipment, using the Nelson indexes; and the battery-limits investment is then calculated from the updated equipment costs with individual coefficients for each type of equipment, plus correction factors for materials of construction and size. Finally, the modified Lang factor is calculated for the unit.

TABLE E5.1 Costs of Primary Equipment for a 30,000-ton/yr Cumene Plant (Dollars, 1968)

Equipment	In Mild Steel	Supplement for Stainless Steel	Total
Towers	37,800		37,800
Reactor	29,000	83,000	112,000
Tanks	10,400		10,400
Exchangers	65,200		65,200
Pumps	20,600		20,600
Motors	4,200		4,200
Miscellaneous	13,000	32,000	45,000
Primary Equipment	180,200	115,000	295,200

Determine the 1974 battery-limits investment using an overall index. To do this, the individual coefficients are used to determine the battery-limits investment in 1968; then this battery-limits investment is updated with an overall index, and the results are compared with the previous results.

Calculate the battery-limits investment in 1974 for a production capacity of 70,000 tons/yr by

1. Extrapolating by type of equipment, using average exponents for each type, for the year 1974.

2. Determining the battery-limits investment, as well as the modified Lang factor.

3. Calculating the overall extrapolation exponent relating 30,000 tons/yr capacity to 70,000 tons/yr.

Compare results with those obtained by applying the method of J. H. Hirsch and E. M. Glazier in Example 3.

E5.2 THE ANSWER

E5.2.1 DETERMINING THE 1974 INVESTMENT FOR 30,000-TON/YR CAPACITY

E5.2.1.1 Updating from 1968

E5.2.1.1a By Type of Equipment

Table E5.2 is established using Fig. 4.7 and Tables 3.3 and 3.4.

E5.2.1.1b Calculating the Battery-Limits Investment for 1974

Table E5.3 is established using the factors from Tables 4.16 and 4.17 for the equipment in mild steel, plus the curves from Fig. 4.8 for correcting according to size.

E5.2.1.1c Calculating the Modified Lang Factor

The ratio $\frac{1,259.6}{510.6}$ is about 2.5.

E5.2.1.2 Determining the 1974 Investment with an Overall Cost Index

E5.2.1.2a Calculating the Battery-Limits Investment for 1968

Table E5.4 is calculated analogously to Table E5.3.

E5.2.1.2b Calculating the Modified Lang Factor

The ratio $\frac{740.2}{295.2}$ is about 2.5.

E5.2.1.2c Calculating the 1974 Investment Using an Overall Index

Table E5.5 is established according to the indexes used. The Nelson index seems to give a result that is closest to that calculated in Table E5.3. However, it should be noted that the cost index by categories of equipment also uses values given by Nelson. Besides, the correction factors for size, f_c, in Table E5.4 are applied to the costs at the beginning of 1975 and not to the 1968 costs, and a correction should be made from this point of view. This comparison, using the Dutch WEBCI index, shows that the accuracy of the results is at best ± 10%.

E5.2.2 INVESTMENTS IN 1974 FOR A CAPACITY OF 70,000 TONS PER YEAR

E5.2.2.1. Extrapolating by Types of Equipment

The exponents in Table E5.6 are obtained from Table 4.15. The base costs should be those for a 30,000 tons/yr-unit updated to 1974. Thus Table E5.7 is obtained.

TABLE E5.2 Prorated 1974 Costs of Primary Equipment for a 30,000-ton/yr Cumene Plant

	Nelson Equipment Indexes			*Updated Costs, $1,000*		
Equipment	1968	1974	Ratio	Mild Steel	Stainless Steel	Total
Towers	198.5	318.3	1.604	60.6		60.6
Reactor	198.5	318.3	1.604	46.6	133.2	179.8
Tanks	198.5	318.3	1.604	16.6		16.6
Exchangers	223.4	501.0	2.243	146.2		146.2
Pumps	284.5	416.3	1.463	30.2		30.2
Motors	175.3	250.4	1.428	6.0		6.0
Miscellaneous	228.8	361.8	1.581	20.6	50.6	71.2
TOTAL				326.8	183.8	510.6

TABLE E5.3 1974 Battery-Limits Investment for a 30,000-ton/yr Cumene Plant

Equipment	Cost in Mild Steel, $1,000	Multiplying Coefficient f_g	Size-correction Factor f_c	Installed Cost, $1,000
Towers	60.6	3.75	1.080	245.4
Reactor	46.6	3.75	1.080	188.8
Tanks	16.6	2.72	1.170	52.8
Exchangers	146.2	2.84	1.120	465.0
Pumps	30.2	2.67	1.135	91.6
Motors	6.0	2.67	1.135	18.2
Miscellaneous*	20.6	2.15	1.130	50.0
TOTAL	326.8			1,111.8
Addition due to the use of stainless steel, $1,000				183.8
Battery-limits investment, $1,000				1,295.6

*Such as filters.

TABLE E5.4 1968 Battery-Limits Investment for a 30,000-ton/yr Cumene Plant

Equipment	Cost in Mild Steel, $1,000	Multiplying Coefficient f_g	Size-correction Factor f_c	Installed Cost, $1,000
Towers	37.8	3.75	1.095	155.2
Reactor	29.0	3.75	1.095	119.0
Tanks	10.4	2.72	1.180	33.4
Exchangers	65.2	2.84	1.135	210.2
Pumps	20.6	2.67	1.140	62.8
Motors	4.2	2.67	1.140	12.8
Miscellaneous	13.0	2.15	1.135	31.8
TOTAL	180.2			625.2
Addition due to the use of stainless steel, $1,000				115.0
Battery-limits investment, $1,000				740.2

TABLE E5.5 1974 Battery-Limits Investment for a 30,000-ton/yr Cumene Plant

1968 Investment, $1,000	Indexes	1968	1974	Ratio	1974 Investment, $1,000	Difference from the Investment in Table E5.3, %
740.2	Overall Nelson	304	522.5	1.719	1,272.4	≅2
	Chemical Engineering	114	165.5	1.452	1,074.8	≅19
	WEBCI	104	165	1.587	1,174.7	≅10

E5.2.2.2 Calculating the Battery-Limits Investment

The battery-limits investment in 1974 for a capacity of 70,000 tons/yr is determined as in Table E5.8.

The calculation for the modified Lang factor gives

$$\frac{2,175.0}{873.8} \simeq 2.5$$

E5.2.2.3 Calculating the Overall Extrapolation Exponent between 30,000 and 70,000 Tons per Year

$$\frac{2,175.0}{1,295.6} = \left(\frac{70,000}{30,000}\right)^x$$

$$x = \frac{\ln 1.678}{\ln 2.333} = \frac{0.518}{0.847} = 0.61$$

TABLE E5.6 Capacity-Extrapolation Exponents for Types of Equipment

Equipment	Range	Selected Average
Towers	0.65–0.70	0.67
Reactor	0.65–0.70	0.67
Tanks	0.60–0.65	0.63
Exchangers*		
Conventional	0.65	0.60
Reboiler	0.50	0.60
Pumps	0.40–0.50	0.45
Motors	0.90–1.00	0.90
Miscellaneous (filters)	0.60–0.65	0.63

*The flow sheet shows that about a third of the exchangers are reboilers.

TABLE E5.7 Primary Equipment Costs in 1974 for a Cumene Plant

	Costs for 30,000-ton/yr Capacity, $1,000			Costs for 70,000-ton/yr Capacity, $1,000		
Equipment	Mild Steel	Stainless Steel	Total	Mild Steel	Stainless Steel	Total
Towers	60.6		60.6	106.8		106.8
Reactor	46.6	133.2	179.8	82.2	235.0	317.2
Tanks	16.6		16.6	28.4		
Exchangers	146.2		146.2	243.0		
Pumps	30.2		30.2	44.2		
Motors	6.0		6.0	12.8		
Miscellaneous	20.6		71.2	35.2	86.2	121.4
TOTAL	326.8	183.8	510.6	552.6	321.2	873.8

TABLE E5.8 1974 Battery-Limits Investment for a 70,000-ton/yr Cumene Plant

Equipment	Cost in Mild Steel, $1,000	Multiplying Coefficient f_g	Size-correction Factor f_c	Installed Cost, $1,000
Towers	106.8	3.75	1.055	422.6
Reactor	82.2	3.75	1.055	325.2
Tanks	28.4	2.72	1.165	90.0
Exchangers	243.0	2.84	1.100	759.2
Pumps	44.2	2.67	1.130	133.4
Motors	12.8	2.67	1.130	38.6
Miscellaneous	35.2	2.15	1.120	84.8
TOTAL	552.6			1,853.8
Addition due to the use of stainless steel, $1,000				321.2
Battery-limits investment, $1,000				2,175.0

E5.2.3 COMPARING RESULTS OF THE DIFFERENT METHODS

For a capacity of 30,000 tons/yr of cumene, the battery-limits investment in 1974 are

1. According to the method of Hirsch and Glazier, $1,487,000 (See Example 3.)

2. According to the recommended method, $1,295.6

The difference is around 15%. The number afforded by the method of Hirsch and Glazier, which has a modified Lang factor of 3.2 in 1968 and 2.9 in 1974, is considered high. A plant to produce cumene is not very complex; it has neither high erection costs nor special arrangements for the piping, civil engineering, or instrumentation.

Chapter 5

Evaluating Research Projects

Three different kinds of research results call for evaluation: inventive ideas requiring research; laboratory information that confirms, modifies, or enlarges the inventive idea; and pilot-plant information that confirms or improves on the inventions that have been tested in the laboratory. The evaluation that comes into play at these three stages should be regarded as a tool not only for rejecting or accepting, or suspending or following, the research but also for recognizing major imperfections in the technology.

5.1 OBJECTIVES AND BASIC DATA

Not only is the final result of an evaluation important, but also the economic importance of various aspects of the technology should be revealed as research opportunities (reactor design, catalyst consumption, purities, secondary products, influence of recycling, importance of pressure and compression, availability of raw materials, etc.). It is thus possible to apply research efforts to important, rather than to unimportant, problems.

To fulfill these functions, the evaluation should examine the following five subjects:

1. The manufactured products
2. The basic materials required—raw materials, reactants, solvents, catalysts, and so forth.

3. The operating material balances of the reaction section and the purification section
4. The flow diagram and the technology
5. Competitive processes

5.1.1 THE MANUFACTURED PRODUCTS

A distinction should be made between an established product from a new process and a new product.

5.1.1.1 Known Product from a New Process

Innovations of this type are characteristic of heavy industry with widely diffused products, such as bulk intermediate chemicals, chemical fertilizers, bulk inorganic chemicals, and (in a less general way) polymers. The evaluation is most often made for a well-defined market that has a predictable expansion. Since the technical characteristics of other manufacturing methods are relatively accessible, comparison studies are relatively easy to do.

5.1.1.2 New Product from New Technology

Since the product is new, it is not possible to make a direct comparison with an existing situation. Instead, evaluation consists of establishing a probable price for a product whose qualities have to be in demand during the course of a market study. The sales volume generally varies inversely to the selling price that can be assigned to such a product; and evaluation should lead to an analogous relation between the selling price and plant capacity. According to the respective positions of the two curves, an argument can be made for one of the four following cases:

1. The manufactured price is always higher than the price defined by the market study (Fig. 5.1*a*); so the technology is incompatible with introducing the product on the market.
2. The manufactured price is always lower than the market price (Fig. 5.1*b*); so the project merits development.
3. The two curves cross each other so that the manufactured price becomes lower than the market price only at a certain capacity, so that commercialization requires a plant with a high capacity from the start.
4. The two curves cross each other such that the manufactured price is lower than the market price only at low capacity (Fig. 5.1*d*), so that product development is limited to applications of low tonnage.

These various situations are encountered in industries, such as drugs, perfumes, cosmetics, pesticides, photosensitive emulsions, adhesives and glues,

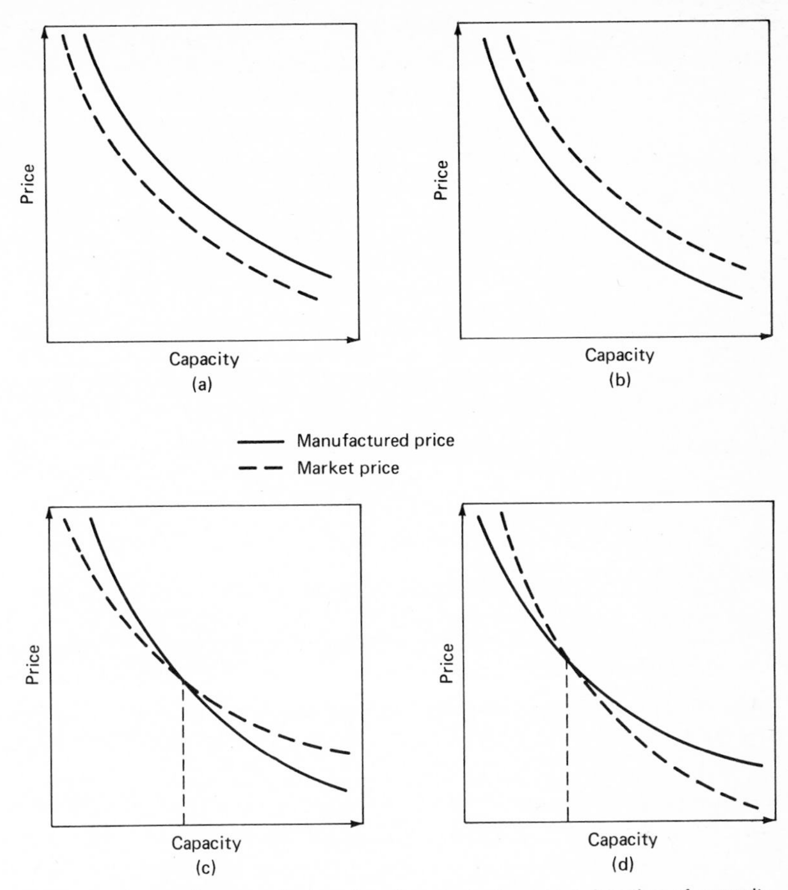

Fig. 5.1 Comparing manufactured and market prices as a function of capacity.

paints and varnishes, detergent compounds, and sometimes polymers, which put specific products on the market. In such cases, the evaluation often shows that the principal costs are for raw materials, packaging, and commercialization. Investment, energy, and personnel represent only a small part of the final cost.

5.1.2 RAW MATERIALS

No matter whether a new product or a new process is being considered, the availability, quality, and price of the basic materials (feedstocks, reactants, etc.) constitute a primary consideration.

5.1.2.1 Availbility of the Basic Materials

There are two general cases in which restrictions faced by a raw material have pronounced effects on the planning for a new plant: either the material is produced in reduced amounts and is thus not generally available, or the material is by-produced in average to low quantities with another product.

In the first case, it may be necessary to include a manufacturing unit for the unavailable raw material along with the plant for the desired product. Examples of such combinations include oxygen units for the manufacture of ethylene oxide, catalyst units for many processes, and purification units for recycling acetic acid in the manufacture of vinyl acetate.

In the second case, it may be necessary to locate the proposed manufacturing plant at a central location with respect to several sources of the desired product. Examples might be a butadiene extraction plant located centrally from several ethylene plants that by-produced the desired butadiene-rich C_4 stream, or an isoprene extraction plant similarly located to receive the C_5 stream by-produced in ethylene pyrolysis.

5.1.2.2 Quality of Raw Materials

There is usually a considerable cost advantage in using the commercial quality of raw materials, along with the implicit trace impurities or even several percentage points of equal boilers that accrue from conventional manufacturing or refining technology. Consequently, any new process should be reviewed for the effects of such impurities; laboratory studies should be carried out first with the pure products, and then with synthetic mixtures. Some examples:

- The C_3 fraction coproduced with ethylene in steam pyrolysis generally contains 93% and 98% propylene. However, the presence of propane in certain reactions, particularly oxidation, can make the purification duty heavier and adversely affect the economics.
- The C_4 fraction from steam pyrolysis of naphtha may or may not have butadiene, as well as isobutene. Some of the normal constituents of this fraction can behave as diluents or as reactants. The consequence for butadiene or isobutene utilization can be beneficial, if the reaction leads to a worthwhile product, or disastrous, if the effect of diluting makes the size of the butadiene or isobutene plant grow out of all proportion.
- The presence of sulfur in many generator gases affects the life and activity of synthesis catalysts.

Depending on the importance of the effect of the impurities, it may be necessary to add pretreatment to the process; and such treatment must be accounted for under the economics of the process being evaluated, particularly in comparisons with technology that may or may not require the same kind of pretreatment.

5.1.2.3 Prices for Raw Materials

The effect of raw material costs on research projects is close to that on industrial projects; however, potential price changes need to be taken into account, particularly if commercialization is to take place at some far future date. Such forecasting is often difficult, for it must consider both the growth in productivity of existing manufacturing plants, as well as the increase in investment and operating costs as they affect prices.

A particularly difficult problem presents itself when the new process uses products that have no current use and therefore do not have a well-established price. Examples include petroleum fractions by-produced with gasoline, isobutyraldehyde produced from oxo synthesis. and the C_4 and C_5 fractions from steam pyrolysis.

In theory, such a product has a base price depending on how its producer currently gets rid of it, e.g., fuel value for hydrocarbons and combustible products, or a price less than zero for products costing money for disposal. From the moment a demand makes itself known, however, the seller tries to charge a part of its manufacturing overhead against any by-product, and even to price that by-product so that it can yield a profit while also allowing the buyer to realize a profit.

5.1.3 MATERIAL BALANCES

Research projects, whether for a process or a product, are expressed through the design of one or more manufacturing stages, each with its own material balance, and each in turn often made up of a reaction section and a purification section. The purification section will usually consist of physical separation processes that can be studied in isolation.

5.1.3.1 Material Balances around the Reaction or Recovery Section

Generally, the material balances around a reactor or unit process give the overall yield of the operation by means of data directly accessible from the literature or laboratory results.

Most often such laboratory information comes from a batch operation, where the problems of recycling unreacted feedstocks or by-products do not enter in. Nevertheless, data from batch operations can be used to calculate conversion, which is the moles changed per mole of feed; selectivity, which is moles of product per mole of feed transformed; and yield, which is conversion times selectivity. When several reactants are involved, the conversion and selectivity can be defined in terms of each reactant. The notion of extent toward equilibrium conversion and selectivity is sometimes encountered, although it has little or no significance on the industrial level. Also, experimental results are sometimes expressed in weight percent or mole percent of feed, instead of compositions before and after reactions.

When transposing the basic information to industrial scale, the selectivity remains the same as for the batch operation, although the industrial conversion must be specified. Unreacted feedstock is recycled from either the reaction section or a subsequent purification section, so that the batch conversion becomes a conversion per pass, and the selectivity becomes equivalent to the yield around the reactor, as determined from the recovery and yields of the recycled components. The yield obtained in the batch study becomes a yield per pass.

5.1.3.2 Material Balance around a Purification Section

In organic chemistry, purification based on distillation or crystallization achieves high recovery rates. Losses are due only to design allowances (number of plates, number of stages, reflux ratio) and are most often on the order of 1%, with up to 5% allowed in some difficult cases. However the efficiency of an operation can be radically affected by phenomena such as the formation of eutectics or azeotropes due to impurities introduced with the feedstock or formed during the course of the reaction. Examples are xylenes; water and sodium chloride; propylene oxide and methyl formate; cyclohexanol, cyclohexanone, and phenol; ethanol and water; hydrochloric acid and water; etc.

Since the purification section should represent the most expedient method of achieving the required specifications for the final product, impurities should be reviewed for possibilities of removal directly in the final stage of reaction, by preliminary treatment, or yet by modifying the reaction conditions. The oxidation of *p*-xylene to terephthalic acid offers an example of the latter; operation at a higher temperature avoids formation of carboxybenzaldehyde, which is difficult to eliminate during purification.

5.1.3.3 The Use of Purges on Recycling Reactants and Solvents

Recycling either unconverted reactants, intermediate products or solvents can introduce components that are inert or deleterious to the reaction; and when the separation process for removing such components is too expensive, they may be removed in a purge. Calculating the purge should take into account the following:

1. The optimum recovery of the recycled product in terms of obtaining an overall yield or avoiding loss.

2. The permissible concentration of the impurity at the entrance to the reaction section.

These two aspects of the calculation are translated into economic data to which the cost of recycling must be added.

5.1.4 ESTABLISHING THE FLOW DIAGRAM

Determining the investment for a unit requires at least an elementary flow diagram showing the approximate primary equipment, as well as an idea of the materials of construction.

5.1.4.1 The Composition of a Flow Diagram

The diagram should describe the steps of pretreating, manufacturing, and separating the feedstock and products. If pretreatment or other feed preparations, as well as purifications, are extensive enough to be considered as self-contained process units, the flow diagram is divided into sections, for example, a reaction section and a purification section.

The reaction section would include equipment such as the reactor; the exchangers, furnaces, etc., necessary to the reaction heat balance; compressors pumps, etc., needed to maintain operating pressures; quench vessels and flash drums needed to receive and treat or separate the effluent; and sometimes special equipment such as filters, dryers, etc., needed to handle the process materials. The purification section would include all that equipment needed for separating, purifying, and transferring reaction products, coproducts, by-products, solvents, etc.

5.1.4.2 Reviewing the Technology

Technical considerations, even though general, can have considerable influence on a process and its economics, and it is good to have them in mind while developing the basic flow diagram, particularly the temperatures and heat balance around the reactor, temperatures in the purification section, pressures in the purification section requiring compression, and corrosion.

5.1.4.2a Heat Problems about the Reactor

Three fundamental characteristics of a reaction are its heat of reaction, its temperature level, and its mode of supplying and removing heat. These three characteristics control the reactor design, whether isothermal or adiabatic, with superimposed catalytic beds or multitube beds, and so forth.

Heat can be supplied either by direct contact (in a furnace, on a preheated refractory mass, or by introduction of a heated diluent) or by indirect contact (through feed-effluent heat exchange or exchange with a heat-transfer fluid).

Heat can be removed either by quenching in situ (using an excess of one reactant in a recycle stream, or a suitable component from a gaseous or liquid source) or by heat exchange (using a heat-transfer fluid or an integrated exchanger). Appendix 2 can be used to envisage and determine more precisely the type of reactor.

Calculation of the heat of reaction can be made from the values of the energy of chemical bonds, but it is preferable to treat those results with reserve, particularly when the heat balance is critical. In such case, it is preferable to use the heats of formation, such as can be calculated by the method of K. K. Verma and L. K. Doraiswamy, shown in App. 11. Tables giving the direct experimental values of heats of reaction can also be consulted.

5.1.4.2b Temperature Problems in the Purification Section

Temperature is critical to the stability of many chemical compounds, as well as to the most-used method of separating such compounds, distillation, where it is critical in both reboilers and overhead condensers.

Generally, organic chemicals should not be taken to higher than 250°C if random decomposition is to be avoided. Beyond this general rule, each of the principal compounds in a project under evaluation should have the temperature limits of its stability verified to set criteria for the separation equipment, such as has been done for the oxides of ethylene and propylene and for styrene. Those criteria may lead to distillation at reduced pressure, to steam stripping, to doubling of distillation columns, or to using stabilizing agents recommended by the laboratory.

In distillation, the highest temperature is normally encountered at the reboiler, where heat is supplied by steam, hot water, a heat-transfer fluid, or by pumping the reboiling mixture through the tubes of a fired heater, the choice among these alternatives depending on the temperature, the heat duty, and the stability of the reboiled products.

Steam is used up to 250°C, as a first approximation, and furnaces are used above that, provided the tube-wall skin temperatures are compatible with the stability of the products. If not, heat-transfer fluids are used. Sometimes it is preferable to replace the use of steam by that of a furnace at reboiling temperatures below 250°C. The steam cost for reboilers (including the fixed costs due to the investment for the reboiler) becomes higher than the cost of fuel (including the fixed costs for the investment for the furnace) when the duty gets larger than about 6–7 million kcal/h.

Hot water is often used for fragile products that cannot be heated to too high a temperature. Also, use of a heat-transfer fluid often leads to extending this method of heating to a whole purification train, since a single furnace for reheating the fluid can be the most economical system. The inconvenience of such hot-oil systems is their single-temperature level.

Steam-temperature levels are usually set by the steam available at the site. In the absence of such data, steam pressures may be assumed as low pressure (3–7 bars), medium pressure (10–15 bars), or high pressure (20–30 bars); and the minimum temperature difference across a reboiler may be assumed as 20–25°C.

The coolants most used for overhead condensers are water, air, and brine or refrigerants from a refrigeration system. Water, the least expensive coolant, is available at ambient temperature, and depending on its quality can be heated to about 50°C without forming excessive scale. Allowing for a maximum ambient temperature of 30°C, depending on location, the water may experience a temperature rise of 20°C; and depending on the condensing temperature-profile (which is a function both of the overhead composition and the tower pressure), the cooling water may be used for a minimum condensing temperature of about 40°C. Below this, cooling brine or less economic refrigerants will have to be used.

An analogous situation may also be encountered in a reaction section. Sometimes, for example, a researcher recommends a reaction temperature of 20–30°C based on laboratory results. However, this point of view can be questioned to find out if the process would not be feasible at a slightly higher temperature so as to reduce heat-exchange problems.

The relatively recent widespread use of air coolers is due to a number of technicoeconomic advantages that accrue to them, as follows:

The off-site investment for water treatment in water-short countries often makes air coolers more economical.

A large temperature difference (20–30°C) between the ambient air and the fluid to be cooled favors the use of air coolers, as does its corollary, a maximum ambient temperature less than 25–27°C.

The allowable temperature rise for cooling water is less than 6–8°C, thus increasing cooling water consumption.

The fouling factor for local water is higher than 0.0004 $(m^2)(h)(°C)/kcal$, thus increasing the design requirements for exchanger surface.

The process-side heat-transfer coefficient is less than 1,000 $kcal/(m^2)(h)(°C)$, thus reducing the effect of the poor air-side transfer coefficient.

The study pressure and the temperature are more than 100 bars and 150°C, thus increasing the cost of the shells for water coolers.

The process fluid requires special materials.

One of these conditions is usually not enough to clearly indicate the use of air cooling rather than water cooling, nor is it necessary to have all of them. An economic calculation should indicate the final choice.

5.1.4.2c Problems of Compression

Pressure is used in various ways in both the reaction section and in the purification section. Most often the problems are specific, but the following general principles may find application.

Just as it is often inconvenient to operate at a temperature level of 10–30°C, so is it often difficult to operate at a pressure between 1 and 3 bars, because of problems of control, of large gas volumes, and of the type of compression required. When the laboratory recommends near-ambient pressures, the recommendation should be questioned to see if it is not possible to work at a higher pressure without disturbing the performance recorded during research and thus avoid the need to compress the effluents.

The effect of substituting impure commercial products for laboratory reactants should be considered for its effect on pressure. If the partial pressure of the reactant is the determining parameter for the reaction, it is important to study the economic effect of an overall increase in pressure on the reactor, as well as on the reaction section and its equipment, particularly with respect to introducing compressors and pumps.

Although operating under pressure allows for separating unreacted feedstock by simple flash, the pressure drop employed for the flash should not be so much as to require a large recycle compressor. This is why it is sometimes useful to flash in several stages or to heat the effluent.

5.1.4.2d Corrosion Problems

During research is a fruitful period for directing attention to corrosion problems inherent in one or more of the constituents of a reaction. At this stage it is easier to reduce such constituents to noncorrosive concentrations, or barring that, to have them exposed to a minimum of equipment. Sometimes the corrosives can be eliminated by a complementary reaction or the addition of an inhibitor; other times they can be reduced to harmless concentrations by slightly modifying the general performance of the system or its operating conditions.

Recommended materials of construction for different compounds can be found in tables or lists in numerous publications, such as App. 12.

5.1.4.2e Calculating the Investment

When the flow diagram and the sizes of the primary equipment have been established, the cost of the equipment can be obtained through the methods and graphs for industrial projects, which were presented in Sec. 4.3.3.3 and in the appendixes. It should be kept in mind that the calculation is not exact, since it somes in large part from laboratory data that may be incomplete and may thereby overlook certain necessary operations. Consequently, a provision for contingencies is left up to the evaluator to add to the calculated costs. As a first approximation, such contingencies might come to 30% of the estimated investment.

5.1.5 COMPETING TECHNOLOGY

The feasibility study for a research project does not in itself have any meaning unless it is compared to studies of the competition, whether that competition is a product or a process; and it is important that the comparison studies be made on the same basis and with the same methods.

5.1.5a Comparing Competitive Processes

A study of competitive technology may require modifying the research project's flow scheme with added operating steps such as those imposed by the use of a particular kind of feedstock or the need to make a by-product saleable. On the other hand, it may be necessary to add elements to the competitive technology for analogous reasons. Finally, it may be necessary to optimize the competitive process with respect to a particular set of conditions that put it on the same basis as the research project.

5.1.5b Comparing Competitive Products

The comparison can be carried out on the manufacturing system, when that is available, or on the cost of the product, when that is the only element of comparison. When comparing product costs, the following points should be verified:

Plant capacity should correspond to manufactured price.

Conditions of delivery should be the same.

The current uses should be the same.

The specifications should be the same.

Implications of the prices of coproducts or by-products should be allowed for.

5.2 THE SEQUENTIAL STAGES OF A RESEARCH STUDY

From the practical point of view, the evaluation of a research project logically proceeds through the successive stages of defining the market, the technology, and the economics for the project in question, and then of comparing it with the competition.

5.2.1 THE MARKET STUDIES

The purpose of a market study is to determine six aspects of the research project:

1. The capacity for an eventual manufacturing plant
2. The location of the manufacturing site
3. Any necessary integration with existing industry
4. The impact of the competition
5. Any necessary provisions for assuring a supply of feedstock
6. The expected price trends for raw materials, products, and by-products

This study generally includes three different parts according to their emphasis on production volume, uses for the product, and prices. The production volume concerns not only the product but also the raw material and the principal countries or sites of production. This part is often based on the literature; it considers prior developments, production trends, and if possible the actual current production and consumption, including foreign trade.

The part emphasizing uses concerns the principal markets for the manufactured product, the apparent trends for the usual applications, and possibilities for new applications. The average annual growth rate is often the only data immediately accessible, but this appears to be sufficient for directing choices in the context of evaluating a research project.

The price study concerns price variations and tendencies. It takes on more importance when information about competitive production techniques is lacking or if the project is for manufacturing a new product.

5.2.2 DEFINING THE TECHNOLOGY

This stage employs the principles used by engineering services for defining the elements of a preliminary proposal; its purpose is threefold: establish a manufacturing process, determine the material balance, and select the operating conditions—all on the advice and careful consultation with both research and specialized services. Research provides the basic data on operating conditions, reactants, and performances, while the specialists provide information about peculairities of the reactions and purifications as applied to design.

The manufacturing process covers only the primary equipment and those transfer lines essential for defining the operation of the unit. The material balance usually evolves through three types, one based on laboratory data and reported per mole of feedstock, a second translated from the first and recorded as mass balance based on feedstock, and a third based on the second and reported in terms of throughput. The operating conditions generally consist of temperature and pressure levels for the equipment shown in the process flow diagram, plus approximate type selections and estimated sizes for the equipment and the energy balances necessary to set utilities.

The selection of operating conditions should be accompanied by approximate optimization of the process scheme with respect to energy consumption, the best possible sources of energy, and the equipment. For example, a feed-effluent exchanger for a reactor should allow a margin of safety for the uncertainties of the unit's operation, while recovering about 75% of the available heat. By-produced steam can serve for reboiling distillation towers or driving turbines.

A choice of the heating and cooling modes for the primary equipment evolves from such studies. Heating by furnaces or by exchangers, or by steam, or by a heat-transfer fluid, as well as cooling with water or a cooling fluid, are all adopted according to the requirements.

Various methods allow for determining the characteristics of primary equipment without detailed engineering or long calculations. Some such methods are given in the appendixes for pressure vessels, heat-exchange equipment, pumps and compressors with drivers, etc., and have already been shown for evaluating industrial projects in Chap. 4. This sizing and selecting is accompanied by calculation of the utilities consumption or production, first by type of equipment, then by category of utility.

5.2.3 DEFINING THE ECONOMICS

These calculations follow the identical steps given for industrial projects in Sec. 4.3.

Sometimes difficulties in rationalizing certain costs arise. Thus a refiner and a petrochemist will have a tendency to give two different values to the same product according to the uses they plan for it. On the other hand, each producer that has a product in surplus will use it while assigning it the lowest price, and that low price will be maintained as long as such use continues; but if for one reason or another the product comes into demand, it will be assigned the highest price compatible with the demand situation.

When such difficulties are envisaged, parameters can be set to articulate the areas of comparison for the techniques being studied.

5.3 ANALYZING THE RESULTS OF A STUDY

Economic evaluation of research projects is not performed by a simple comparison of the economics of competitive methods, as is usually the case for industrial projects. Instead, as pointed out in Sec. 5.1, the economic evaluation of research projects is a tool for the researcher to use in finding opportunities for perfecting the discovery. In order for the evaluation to become such a tool, its results must be analyzed, first, to determine the more important elements of cost, and then to determine the research opportunities.

5.3.1 FINDING THE DOMINANT ELEMENTS OF COST

Economic conclusions from industrial projects reflect the situation formed by the basic assumptions of the study. In the framework of a research project, however, the basis for the economic studies is subject to constant revision as the research progresses; and economic evaluations contribute to this evolution by indicating the most rewarding directions. Segments of the operating cost are compared with each other by analyzing their respective variations as functions of the basic data.

An analysis of operating costs for projects in general, whether industrial or research, can be approached in three ways, depending on whether the variable cost is most important, or the fixed cost is most important, or whether the variable and fixed costs are similar.

In about half the processes involving chemical reactions, the most important cost item is the variable cost, often because the needed raw material is in limited production; and in a parallel manner, the credit for coproducts and by-products often plays a determining role. The investment (and thus the fixed costs) is often most important in refining processes, which handle large volumes of inexpensive materials. Finally, investment and raw material costs often have equal importance in processes for separating a manufactured product, e.g., crystallization or adsorption of *p*-xylene, distillation of ethylbenzene, and extraction of butadiene and isoprene.

When the variable costs dominate, a closer analysis will differentiate raw materials, chemical products, catalyst, and utilities. It is a rare project where the raw materials are not the determining element, since they enter both on the level of unit price and by the amount of their consumption, i.e., by the selectivity of the process. However, the consumption of chemical products or of catalysts or even of the utilities can be of the same order of magnitude as the fixed costs.

When the investment dominates, the contributions of the various sections of the flow scheme are determined. The most important of these is often the reaction section. In such a case, the variations of reaction temperatures and pressure, of conversion per pass and its corollary, the recycling rate, and of the space velocity and its corollary, the volume of the reactors, are the factors to take into consideration. The kind and quantity of certain by-products can also have an effect on the materials of construction and thereby on the investment.

Sometimes the separation, recovery, and pretreating sections play an important role, but their importance can bring into question the conception of the unit's process scheme.

Labor has not been discussed. Its contribution to costs for the continuous production of bulk chemicals remains limited, but becomes perceptible as the production capacity is lowered. However, labor becomes important for batch operations and those needing special handling. Also, packaging and commercialization, for example, require more work and lead to relatively high labor costs.

In sum, analysis of the operating costs leads to weighing the influence of the different items, showing the technical parameters that govern the predominant items, and plotting graphs to show the effect of the variations on the most important parameters. Thus a sort of classification of the basic data is made and the areas of profitability of the research project are defined. The influence of the cost of the raw material and the pricing of the by-products and co-products along with the capacity are generally the parameters to be examined first.

5.3.2 DETERMINING THE POSSIBILITIES FOR IMPROVEMENT

Research results can be rejected solely on examination of the eventual markets and possible production volume of a product, or even on the availability of raw materials, without profitability calculations. When an economic study is undertaken, a simple statistical analysis can suffice to show that a research project has a relatively weak possibility of succeeding. It is not surprising if a comparison with the competition should often be unfavorable for a project.

However, to reduce the economic evaluation to the level of verification gives it a very limited value. Instead, it may serve as an industrial perspective and carry with it possibilities to explore. By weighing the various parameters that affect the investment and the operating cost, and by comparing these values to those obtained for the competition, possible improvements are revealed. Moreover, a preferential order can be established for studying the items for improvement, which amounts to specifying certain areas for research while recommending that others be let go. This leads to organizing research so that the means are concentrated on particular points.

For example, it can happen that either the reactor or the recycling compressor appear as the most expensive components of the investment. In such cases it may be possible to modify the reactor by varying the conversion per pass, or the operating pressure might be changed so that the rate of compression is reduced. It is also possible that the calculated cost for catalyst is excessive, so that it becomes desirable to reduce the performance a little, or to increase the temperature, or to improve recovery of the used catalyst.

Example 6

Profitability Calculation Applied to a Research Project for the Production of Heptenes

Laboratory research on organometallic catalysts has shown that the codimerization of propylene, *n*-butene-1, and *n*-butene-2, under proper conditions, can preferentially form heptenes along with some hexenes and octenes. Unlike phosphoric catalysts, the research catalysts produced dimer fractions all of which were capable of the oxo reaction.

E6.1 THE PROBLEM

First of all, it is necessary to know if such a synthesis can be of interest from the point of view of both potential market and price. Thus the search has been to determine the eventual production capacity of a European industrial market. Besides, an effort is made a priori to define the most propitious conditions with respect to raw material and operating costs.

E6.1.1 PRELIMINARY STUDY: THE HEPTENES MARKET

The conclusions of the preliminary market study having been taken into account, attention turns to optimization of heptenes production.

E6.1.2 ECONOMIC EVALUATION OF THE INITIAL EXPERIMENTAL RESULTS

The following principal reactions were considered:

$$C_3^= + C_3^= \rightarrow C_6^= \quad \text{(E6.1)}$$

$$C_3^= + C_4^= \rightarrow C_7^= \quad \text{(E6.2)}$$

$$C_4^= + C_4^= \rightarrow C_8^= \quad \text{(E6.3)}$$

Maximum heptene selectivity has been obtained for a given composition of the feed (propylene and *n*-butenes) that should be kept constant during the reaction.

E6.1.2.1 Using the Experimental Material Balances

E6.1.2.1a The Feedstock

A first series of experiments carried out on commercial propylene and *n*-butene feedstocks identified the typical fractions and the most important requirements (Table E6.1) as follows:

The concentrations of isobutane and *n*-butane should be as low as possible, since these act as diluents.

The proportion of the different *n*-butene isomers is of no matter.

Pentenes have no effect on the activity of the catalyst but can lead to undesired products.

Although the concentration of isobutene should be limited to avoid parasitic polymerization, the 5% present in commercial *n*-butenes is comfortably acceptable (Table E6.1).

These findings lead to the raw material specifications in Table E6.2.

E6.1.2.1b The Operating Conditions and Average Results

Material balances made on several runs under conditions of 40°C and 50% conversion of *n*-butene afforded the following weighted average composition for the polymerized product:

Component	Weight %
$C_6^=$	32
$C_7^=$	47
$C_8^=$	15
C_{8+}	6 ($C_8^=$s coming from 70% of the *n*-butenes)
TOTAL	100

The catalyst is introduced in solution; and hexenes amounting to about 0.25% by weight of the *n*-butenes is used as solvent.

The effluent can be separated into the different products by simple distillation, with recoveries as those shown in Table E6.3.

E6.1.2.1c Conclusions

After a preliminary study, a production capacity of 20,000 tons/yr and a stream factor of 8,000 hours per year were selected; and analyses were made to

TABLE E6.1 Composition of Dimerization Feedstocks

Feedstock	Compound	Weight %
C_3 fraction	Propane	5
	Propylene	95
C_4 fraction	*n*-Butane and isobutane	40
	Isobutene	5
	n-Butenes	55
	Pentenes	0

TABLE E6.2 Specification for Feedstock to Dimerization Process

	Weight %	
	C_3 cut*	C_4 cut†
Total acetylenes‡	$\leq$0.001	$\leq$0.001
Propadiene	$\leq$0.003	$\leq$0.003
Butadiene	$\leq$0.008	$\leq$0.003
Water	Saturated	Saturated
Total sulfur	$\leq$0.0005	$\leq$0.0005
Acetonitrile	Minimum	Minimum
Tertiary butanol	Minimum	Minimum

*These specifications correspond to commercial hydrogenated propylene.

†Tests by selective hydrogenation show that butadiene and acetylene specifications are met.

‡Including methylacetylene.

TABLE E6.3 Recoveries of Dimerization Products by Distillation

Distillation	Compound	Recovery, %
Debutaniser	Butenes	98.5
	Hexenes	99
Separation of the hexenes	Hexenes	98
	Heptenes	97
Separation of heptenes	Heptenes	99
	Octenes	99

establish a general flow scheme neglecting recycle of unconverted *n*-butenes, plus a general material balance assuming a polymer production of 100.

E6.1.2.2 Using the Kinetic Data

The purpose of these data are to select a reactor type and size it.

E6.1.2.2a The Experimental Results

The ratio of heptenes to total dimerized product is shown as a function of the propylene–to–*n*-butenes ratio in Fig. E6.1. This plot shows that maximum selectivity for heptenes occurs at a propylene–to–*n*-butenes ratio of about 0.025. This result is independent of the conversion as long as the secondary transformation reactions of $C_6^=$, $C_7^=$, and $C_8^=$ remain limited; and this is the case up to a conversion of about 50% of the *n*-butenes.

The shape of the curve suggests that a theoretical reactor should be either a perfectly agitated vessel with constant concentration and overflow, or a long small-bore tube perfectly agitated along its axis. With the vessel, *n*-butenes and propylene would be continuously introduced so as to maintain the proper proportions. With the tube *n*-butene would be introduced at one end, with the propylene injected at an infinite number of points along the axis.

If the volume of these reactors is expressed as a function of the *n*-butene conversion and the rate of reaction, which is assumed constant at a propylene–to–*n*-butenes ratio of 0.025, the variation of volume with conversion can be

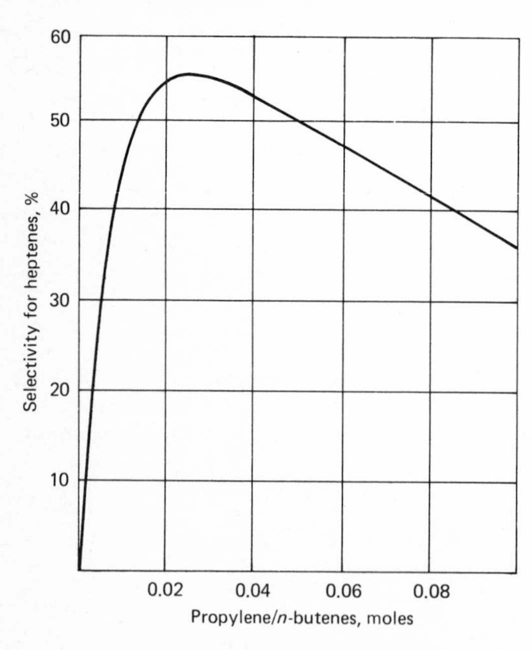

Fig. E6.1 Effect of propylene–to–*n*-butenes ratio on selectivity for heptenes.

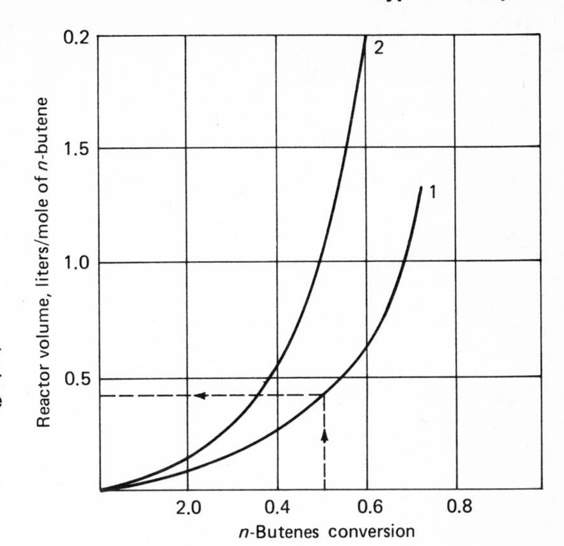

Fig. E6.2 Volume versus conversion for reactors with (1) graduated concentration and (2) single concentration.

plotted as shown in the two curves of Fig. E6.2. These curves show that the reactor with infinite propylene injections and constantly changing concentration is preferable, particularly at high conversions.

E6.1.2.2b Selecting the Reactor Design

Practically, it is not possible to have a reactor with infinite propylene injection points for a perfectly graduated concentration. However, the shape of the curve in Fig. E6.1 shows that, if the propylene-to-butenes ratio is allowed to range from a high of 0.06 to a low of 0.01, it is possible to obtain an average selectivity for $C_7^=$ above about 47%. Thus the propylene can be injected in stages. Calculating the number, position, and flow of propylene injections can be done on an analog computer. It was found, for example, that 16 injections were necessary for a conversion of 50% of the *n*-butenes.

E6.1.2.2c Significance of the Preliminary Information

The problems of defining the reactor came first, during the evaluation and sizing of the primary equipment. Then came the different separations. Thus the immediate information needs concerned the size of the reactor and its cost; the sizes and costs of the heptene separating units, including tower, condenser, reboiler, overhead accumulator, and reflux drum; and the consumption of the corresponding utilities.

From this were developed the manufacturing cost of the heptenes, with the hexenes and octenes being priced as coproducts at $83/ton. The cost of the primary equipment was estimated, and from that the battery-limits investment, plus the miscellaneous costs. To avoid all of these calculations (for which the methods have been described) data are given in Tables E6.4 and E6.5.

E6.1.3 PERFECTING THE INITIAL TECHNOLOGY

The research efforts were concerned with catalyst consumption and reactor design. Attempts to reduce catalyst consumption were made through increasing the activity. Costs of the catalysts were thus brought to 0.6 kg/h for catalyst A and 8.5 kg/h for catalyst B, which amounts to $9.50/ton of pure heptene isomers produced.

The reactor design was changed from tubular to a system of consecutive chambers with a fixed concentration in each chamber. This type of design had been rejected because of the high number of stages determined through theo-

TABLE E6.4 Mid-1975 Costs* for a Dimerization Plant to Produce 20,000 t/yr of Heptenes, $1,000

Equipment	Cost Mild Steel	Addition for Stainless Steel	Total
Reaction section			
Reactors	212.5	181.3	393.8
Tanks	5.2	8.1	13.3
Mixers	20.0	16.9	36.9
Pumps and motors	37.9		37.9
TOTAL	275.6	206.3	481.9
Fractionation section			
Towers	65.6	58.1	123.7
Tanks	12.2	13.5	25.7
Exchangers	96.7	56.3	153.0
Pumps and motors	18.5		18.5
TOTAL	193.0	127.9	320.9

*Add royalties at $10 per ton; process data book at $60,000.

TABLE E6.5 Consumption of Chemicals, Catalysts, and Utilities

Catalysts:	
Catalyst A	1.85 kg/h at $4.17/kg
Catalyst B	25.5 kg/h at $2.50/kg
Utilities	
Reaction Section	
Cooling water	110 m^3/h
Process water	25 liters/h
Electricity	28.2 kW
Fractionation section	
Steam: (12 bars)	2.8 t/h
Steam: (25 bars)	1.6 t/h
Electricity	10.2 kW
Cooling water	147 m^3/h
Labor	Two operators per station

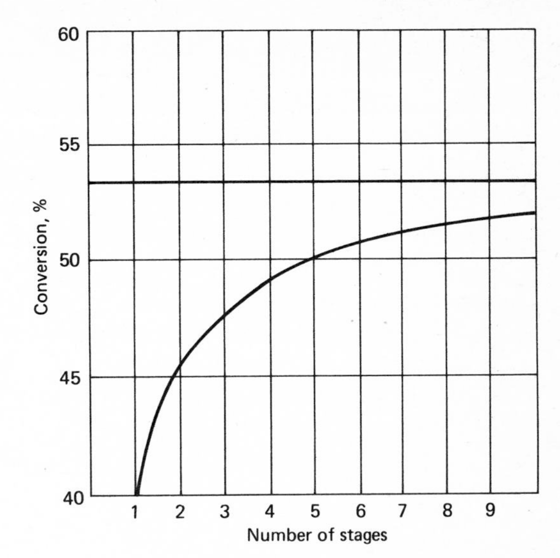

Fig. E6.3 Effect of number of equal-volume stages on conversion, where propylene-to-*n*-butenes ratio is 0.025 and the total volume is 0.5 liters.

retical calculations. However, a balance can be calculated stage by stage on the basis of an expression analogous to the one used for a reactor with stationary concentration. Two methods are possible, one involving identical volumes, the other a constant conversion for each stage. Figure E6.3 shows the influence of the number of these stages on the conversion for the first type of calculation, and thus for the total volume of the reactor. It is seen that the conversion increases rapidly with the number of stages, and if it is limited to 50%, five stages will suffice. With all calculations done, it is shown that the overall volume for the reactor is about 90 m^3. The effect of these improvements on the operating cost enter into the evaluation of the process.*

E6.2 DETAILS OF THE EVALUATION

This section and its divisions show some of the details of the market study and calculations that led to the analysis of the preceding section.

E6.2.1 THE HEPTENE MARKET

A preliminary study indicated that this project might involve two types of market, polymer gasoline and olefins for oxo synthesis. A competitive technique for making high octane polymer gasoline would be catalytic polymerization using a phosphoric catalyst.

*The information given here is furnished only as an example to illustrate the proposed calculation method. It has only a distant relationship to the actual economic and technical data of the "Dimersol" process of l'Institut Français du Pétrole (IFP).

The gasoline market is accessible only if the propylene and butene feedstocks are cheap. It is generally allowed that the butenes from steam pyrolysis are in great surplus and available at low price, but the surplus of propylene is more uncertain. In 1972–1975, certain European areas (Basse Seine, Holland) did have surpluses of propylene, because of (1) disappearance of propylene tetramers for detergents that currently find only a weak market in the manufacture of heat-transfer fluid, (2) slower growth than had been anticipated for acrylonitrile and phenol, (3) a poorly developed market for polypropylene, and (4) no economical method for shipping propylene long distances. However, this propylene surplus should not last long because of the growth of polypropylene.

Under these conditions, the various uses for propylene outside the chemical industry are classified in order of increasing importance as:

Use	Price, \$/t	Investment, \$/t
Fuel	73	—
Recycling in steam pyrolysis (priced as naphtha)	106	—
Polymer gasoline	83	4.2
LPG	83	—

The examination of the marketing situation in Western Europe indicated that the market for a new process could be

	Before 1977	1977–1982
Tons/yr	200,000	0

The market for olefins for oxo synthesis is limited by the overcapacity existing in Western Europe and Japan in 1975–1976 and the orientation of the market toward ethyl-2-hexanol. The sales objective for this version of the process could be

	Before 1977	1977–82
tons/yr.	20,000–40,000	20,000–40,000

Consequently a capacity of 20,000 tons/yr appears to be an optimum for the plasticizers market.

E6.2.2 USING THE EXPERIMENTAL RESULTS

The conclusions summarized in Sec. E6.1.2 were developed through calculations as follows:

E6.2.2.1 Using the Experimental Balances

The target capacity of 20,000 tons/yr could now be used with the experimental data on material balances to determine throughput. The reaction effluent has been determined as

Component	Weight %
$C_6^=$	32
$C_7^=$	47
$C_8^=$	15
C_{8+}	6
TOTAL	100

This, plus experimental data that the C_{8+}s come from 70% of the *n*-butenes and that the conversion of the *n*-butenes is 50%, permits calculation of the *n*-butenes and propylene consumed. Thus, for 100 units of feed,

The propylene is $32 + 47(42/98) + 0.3(6) = 53.94$.
The *n*-butenes is $15 + 47(56/98) + 0.7(6) = 46.06$.

Taking into account the initial composition of the propylene and *n*-butenes feedstocks and the rates of recovery, and after having composed a general flow diagram of the unit (Fig. E6.4), weighted material balances can be calculated as shown in Tables E6.6 and E6.7. Table E6.6 corresponds to 100 weight units of polymers, with hexenes coming from a recycling and serving as the preliminary diluent for the catalyst (0.25 weight % of *n*-butenes) added to the $C_4^=$ feedstock. Table E6.7 corresponds to a production of 20,000 tons/yr of pure heptenes, equal to 2,500 kg/h with a stream factor of 8,000 hours per year. This table is derived from Table E6.6 by multiplying through with the factor

$$\frac{2{,}500}{45.13} = 55.3955$$

E6.2.2.2 Using the Kinetic Data

E6.2.2.2a Sizing the Reaction Section

The reactor can be envisaged as a vertical tank divided inside by horizontal partitions into five chambers, the top four of which overflow into the succeeding one below through downcomers, so that the liquid entering each chamber is directed to the bottom. Both agitation and heat removal are provided by continuously withdrawing liquid from each chamber and pumping it back through an exchanger.

The characteristics of this reactor are listed on page 218.

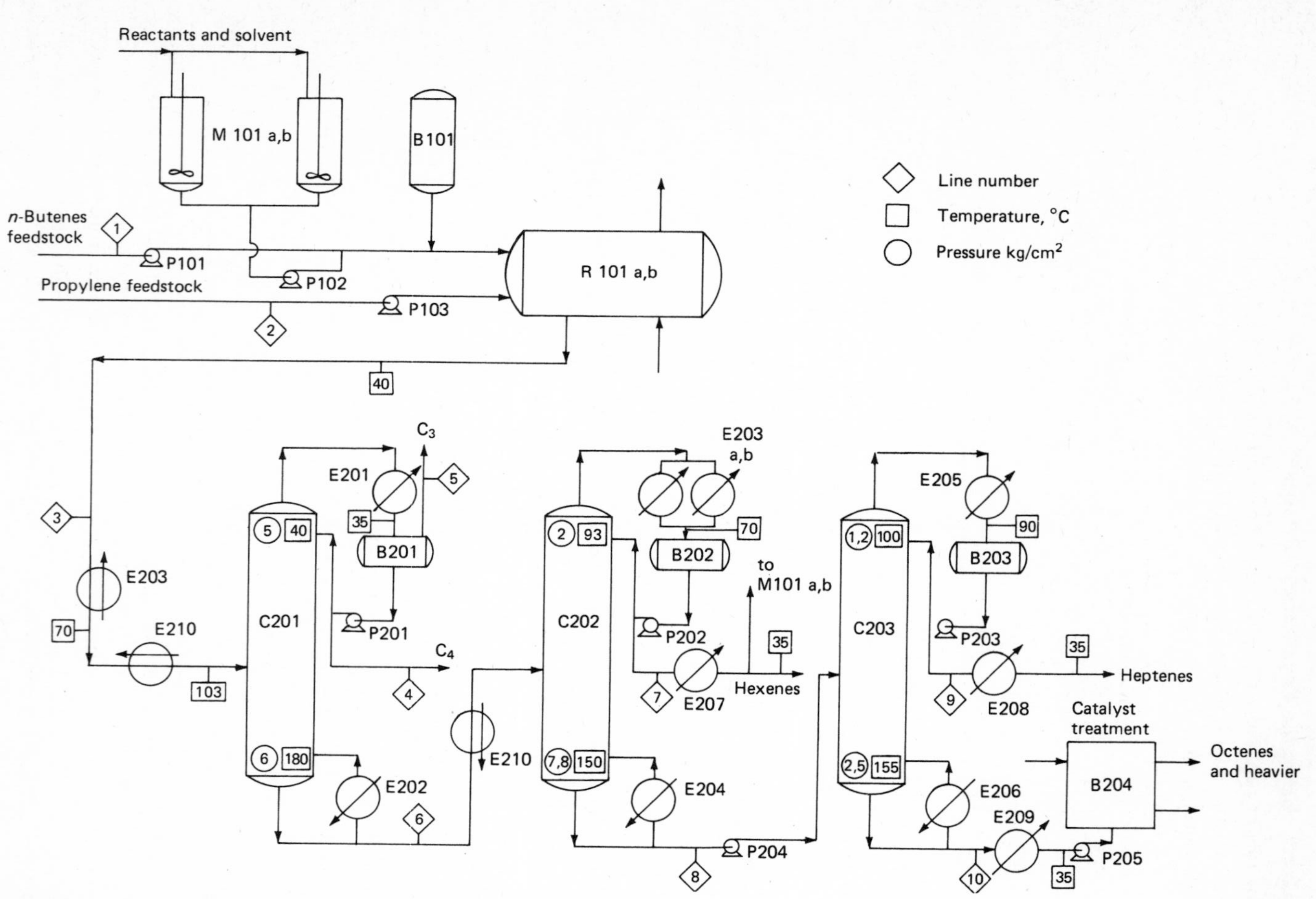

Fig. E6.4 Process flow schemes for the synthesis of heptenes.

TABLE E6.6 Material Balance for 100 Weight Units of Reaction and Product

Compound	Line Number on Flow Sheet (Fig. E6.4)									
	1	2	3	4	5	6	7	8	9	10
C_3	—	2.84	2.84	—	2.84	—	—	—	—	—
$C_3^=$	—	53.94	—	—	—	—	—	—	—	—
n-C_4 and i-C_4	67.00	—	67.00	65.99	—	1.01	1.01	—	—	—
i-$C_4^=$	8.37	—	8.37	8.24	—	0.13	0.13	—	—	—
n-$C_4^=$	92.12	—	46.06	45.37	—	0.69	0.69	—	—	—
$C_6^=$	0.22	—	32.22	0.32	—	31.90	31.26	0.64	0.64	—
$C_7^=$	—	—	47.00	—	—	47.00	1.41	45.59	45.13	0.46
$C_8^=$	—	—	15.00	—	—	15.00	—	15.00	0.15	14.85
C_{8+}	—	—	6.00	—	—	6.00	—	6.00	—	6.00
TOTAL	167.71	56.78	224.49	119.92	2.84	101.73	34.50	67.23	45.92	21.31

TABLE E6.7 Material Balance in kg/hr for a Plant to Produce 20,000 t/yr of Pure Heptenes

Compound	Line Number on Flow Sheet (Fig. E6.4)									
	1	2	3	4	5	6	7	8	9	10
C_3	—	157	157	—	157	—	—	—	—	—
$C_3^=$	—	2,988	—	—	—	—	—	—	—	—
n-C_4 and i-C_4	3,711	—	3,711	3,655	—	56	56	—	—	—
i-$C_4^=$	464	—	464	457	—	7	7	—	—	—
n-$C_4^=$	5,103	—	2,552	2,514	—	38	38	—	—	—
$C_6^=$	12	—	1,785	18	—	1,767	1,732	35	35	—
$C_7^=$	—	—	2,604	—	—	2,604	78	2,526	2,500	26
$C_8^=$	—	—	831	—	—	831	—	831	8	823
C_{8+}	—	—	832	—	—	332	—	332	—	332
TOTAL	9,290	3,145	12,436	6,644	157	5,635	1,911	3,724	2,543	1,181

Temperature: 40°C

Pressure: 8 bars (for the *n*-butenes)
10 bars (for the calculation)

Overall volume: 90 m^3

Assumed diameter: 3 m

Height: 13 m

Wall thickness:

$$e = \frac{PR}{at - 0.6P} = \frac{10(1.500)}{1.0(1{,}055) - 0.6(10)}$$

$$e = 14.3 \text{ mm} + 4 \text{ mm corrosion allowance}$$

$$e = 18 \text{ mm}$$

The weight of this reactor is determined from App. 1 as

Shell:	$P_1 =$	(24.65(3)(13)(18)	= 17,304
Heads:	$P_2 =$	160(18) (from Fig. A1.8)	= 2,880
Skirt:	$P_3 =$	24.65(3)(5)(10)	= 3,698
			23,882

The corresponding price is determined as

In low-alloy steel:

Shell and heads: 20,184(1.54)(0.85)(2.0)	≅ $52,800
Skirt: 3,698(1.54)(0.95)	≅ 5,410
Accessories: 21,880(1.8) (from Figs. A1.9, A1.10, and A1.12 and Tables A.I.9, A.I.12)	≅ 39,400
Internals, assuming each as three trays (Fig. A1.13)	≅ 12,500
Total	110,110
Add miscellaneous (15%)	127,000

In stainless steel:

$$[20{,}184(1.54)(0.85)(1) + 21{,}880(0.8) + 12(1040 - 688)]\ (1.15) = \$55{,}400$$

The exchangers are evaluated, sized, and priced according to App. 3 as follows: Since the reactor design should incorporate maximum flexibility, the exchangers should be interchangeable; and the overall duty of 1.1 million kcal/h is divided equally among the five exchangers for a duty of 220,000 kcal/h each.

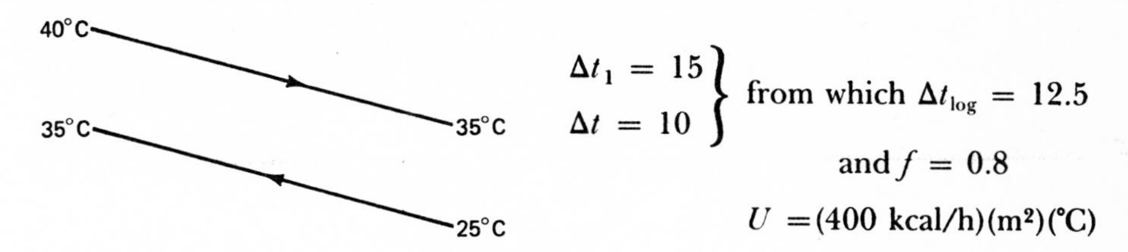

$$\left.\begin{array}{l}\Delta t_1 = 15 \\ \Delta t = 10\end{array}\right\} \text{ from which } \Delta t_{\log} = 12.5$$

$$\text{and } f = 0.8$$

$$U = (400 \text{ kcal/h})(\text{m}^2)(°\text{C})$$

From which the surface S is calculated as

$$S = \frac{220{,}000}{400(12.5)(0.8)} = 55 \text{ m}^2$$

And the consumption of cooling water is taken as 22 m³/h per exchanger.

Exchanger price

Type BES is selected from Fig. A3.4 because of the catalyst.
Tube-side pressure (reaction side): 10 bars
Base price (Fig. A3.5): $9,800
Correction factors from Table A3.6:

$$f_d = 0.92 \qquad f_p = 1.03 \qquad f_m = 1.75$$

$$f = 1.00 \qquad f_l = 1.00 \qquad f_{np} = 1.00 \qquad f_t = 1.00$$

Price per exchanger: 9,800(0.92)(1.75)(1.03) = $16,257
Total price: 16,257(5) = $81,300
Supplemental cost for materials: 9,800(0.75)(5) = $36,800

The circulating pumps are analyzed, sized, and priced according to App. 4 as follows:

Sizing The cooling duty, Q kilocalories per hour, is equal to the flow of reaction mixture, M kilograms per hour, times the specific heat of the mixture, C_p kilocalories per degree Celsius-kilogram, times the temperature drop, Δt degrees Celsius If $Q = 220{,}000$; $C_p = 0.55$ and $\Delta t = 5$, then $M = 80{,}000$ kg/h. If the density of the mixture is 0.6 g/cm³, the flow for each compartment is 133 m³/h.

Assume that pump design flow is 133(1.25) ≅ 170 m³/h, and that the pressure drop is 1.5 + 1.0 ≅ 2.5 bars. Accordingly, the pump horsepower is 0.03704(170)(2.5) = 15.7 cv. Assume a centrifugal pump with an efficiency of 76%; then the shaft horsepower = 15.7/0.76 ≅ 21 cv.

***Price** (from Fig. A4.5)* Assume a pump rotating at 1750 rpm, for a base price of $4375. Then the correction factors from Table A4.2 are $f_d = 1.00$, $f_m = 1.00$, $f_t = 1.00$, and $f_p = 0.70$, so that the price is $4,375(0.70) = $3,063.

The motors are selected and priced according to App. 5, Table A5.1, as follows:

Horsepower: 30 cv
Percent demand: 70%
Efficiency: 86%
Electrical consumption: $\simeq$ 18 kWh/h
Price (from Fig. A5.1): $1,875
Price of pump and motor: $4,938
Price for seven pumps (allowing two spares): $34,566

E6.2.2.2b Sizing the Heptene Recovery Section

Although the calculation forms in App. 13 would speed this calculation while avoiding oversights, this example will be worked out without the forms for purposes of illustration. The heptenes separation tower, no. C 203 in Fig. E6.4, is selected for this illustration. Costs for both the butenes fractionator and the hexenes fractionator, C 201 and C 202, respectively, are added to this for the investments for the fractionation section shown in Table E6.11.

Sizing the heptenes fractionator The mass balance for lines no. 8, 9, and 10 shown in Table E6.7 and Fig. E6.4 can be used to construct the material balance shown in Table E6.8. From this, the tower can be calculated according to App. 1. The key components are taken as $C_7^=$ and $C_8^=$. If the pressure at the top of the tower is taken as 1.2 bars and that at the bottom as 2.5 bars, the corresponding top and bottom temperatures will be about 100 and 155 °C, or the boiling points of heptene and octene, respectively.

These temperatures permit calculation of the relative volatilities of the keys at the top (α_1) and the bottom (α_2), as

$$\alpha_1 = \frac{1.2}{0.55} = 2.18 \qquad \alpha_2 = \frac{4.9}{2.5} = 1.96$$

So that the average relative volatility α_m is

$$\alpha_m = \sqrt{(2.18)(1.96)} \approx 2.07$$

From this average relative volatility and the compositions, the minimum theoretical plates is obtained from nomograph Fig. A1.1 as 12.5. Allowing one theoretical plate for the reboiler changes this to 11.5; and multiplying by 2 for the operating plates at operating reflux gives 23 operating plates. Similarly, the minimum reflux is obtained from nomograph Fig. A1.3, and this minimum reflux is multiplied by 1.3 to obtain an operating reflux of 1.3(1.3) = 1.70.

Next, the theoretical operating plates should be corrected for plate efficiency. Assuming 140°C as the mean column temperature, and a mean viscosity for the tray liquids of 0.15 cP, nomograph Fig. A1.2 is used to obtain a tray efficiency of about 80%; and 23/0.80 = 30 actual trays.

The diameter of the tower is calculated from its vapor traffic, the ratio of liquid density to vapor density and the assumed tray spacing by means of nomograph Fig. A1.4. Assuming constant molar travel from top to bottom of the column, the load at both top and bottom can be calculated as

$$D(1 + R) = 25.999(1 + 1.70) = 70.197 \text{ kmol/h}$$

Given this flow rate, densities for vapor and liquid can be determined top and bottom, and the tower diameter for a tray spacing of 45 cm, as shown in Table E6.9. An average diameter of 1.0 m is assumed.

Other characteristics of the tower are assumed as follows:

Distance from bottom head to bottom tray	3 m
Distance from top head to top tray	1 m
Manholes: allow two extra tray spacings	0.9 m
Tray spacing: 0.45(30) =	13.5 m
Total tower height: allow	18.5 m

Shell thickness:

$$e = \frac{PR}{at - 0.6P} = \frac{2.5(500)}{1.00(1{,}055) - 0.6(2.5)} = 1.2 \text{ mm}$$

TABLE E6.8 Material Balance around the Heptenes Fractionator

	Feed			Overhead Distillate			Bottoms Product		
Component	kg/h	kmole/h	mole %	kg/h	kmole/h	mole %	kg/h	kmole/h	mole %
$C_6^=$	35	0.417	1.17	35	0.417	1.60	—	—	—
$C_7^=$	2,526	25.776	72.06	2,500	25.510	98.12	26	0.266	2.72
$C_8^=$	831	7.420	20.74	8	0.072	0.28	823	7.348	75.21
C_{8+}	332	2.156	6.03	—	—	—	332	2.156	22.07
TOTAL	3,724	35.769	100.00	2,543	25.999	100.00	1,181	9.770	100.00

TABLE E6.9 Calculating the Diameter of the Tower

	Temperature, °C	Pressure, bar	d_v, kg/m³	d_l, kg/m³	$d_l/d_v - 1$	Vapor Flow, m³/s	Diameter, m
Top	100	1.2	3.8	625	163	0.497	0.9
Bottom	155	2.5	7.8	595	75	0.273	1.05

Assume 8-mm thickness, including corrosion allowance.

This thickness plus the dimensions permit calculating a weight, from which the tower can be priced.

The weight is calculated with Eq. (A1.5) as

Shell:	P_1 kg = 24.65 (diameter)(height)(thickness)	=	3,648 kg
Skirt:	P_2, kg = 24.65(1)(5)(8)	=	986 kg
Heads:	P_3, kg (from Fig. A1.8)	=	144 kg
Total:			4,778 kg

The corresponding price is calculated as follows: Assume the shell and heads are SA 357 and the internals of 410 stainless.

Shell and heads (Fig. A1.9 and Table A1.9): 3,792($2.02)(2.0)	≅	$15,300
Skirt (Fig. A1.9): (986)($2.02)	≅	2,000
Accessories (Fig. A1.12 and Table A1.12): $9,600(1.8)	≅	17,300
Trays (Fig. A1.13 and Table A1.13) 30($208)		6,250
Total:		$40,850
Allow 15% for contingencies:		$47,000
Cost of stainless:		
30(208 − 125)(1.15)	=	$2,900
3,792(2.02)(1) + 9,600(0.8)(1.15)	=	$17,600
Total stainless addition:	=	$20,500

Next, the overhead condenser (E.203 in Fig. E6.4) can be selected, sized, and priced as described in App. 3.

The selected construction is single-pass shell with integral channel end and a floating head with a flexible seal ring to prevent catalyst in the process fluid from leaking out (type BES, Fig. A3.4). The required surface of this exchanger depends on the variables for duty, transfer coefficient, and temperature difference, as per:

$$S = \frac{Q}{U\,\Delta t}$$

The corrected mean temperature difference is determined with the aid of Fig. A3.1 and A3.2*a:*

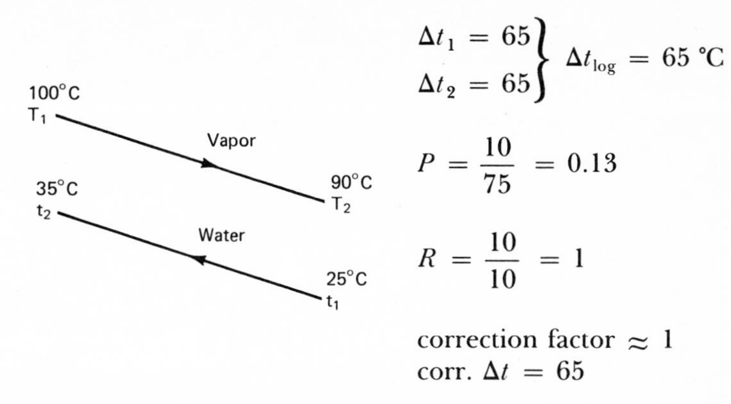

$$\left.\begin{array}{l}\Delta t_1 = 65\\ \Delta t_2 = 65\end{array}\right\} \Delta t_{\log} = 65\ ^\circ\mathrm{C}$$

$$P = \frac{10}{75} = 0.13$$

$$R = \frac{10}{10} = 1$$

correction factor $\approx$ 1
corr. $\Delta t = 65$

The overall transfer coefficient, U, is determined from Table A3.1 as being similar to propylene in the shell and water in the tubes, for about 300 kcal/h/(m²)(°C).

The duty is calculated from flow times change in heat content as

$$Q = 2{,}543(2.70)(89) = 0.611 \times 10^6\ \mathrm{kcal/h}$$

From these, then, the surface is

$$S = \frac{0.611 \times 10^6}{300(65)} = 31.3\ \mathrm{m}^2 \simeq 32\ \mathrm{m}^2$$

The cooling water consumption is calculated from the duty divided by the water-temperature rise, as

$$m = \frac{0.611 \times 10^6}{10} = 61{,}100\ \mathrm{kg/h} = 61\ \mathrm{m}^3/\mathrm{h}$$

The cost of the condenser, type BES, includes a base cost (Fig. A3.5) of \$7,290, modified by factors from Table A3.6 as follows: $f_d = 1.00$, $f_\phi = 1.00$, $f_l = 1.00$, $f_{np} = 1.00$, $f_p = 1.00$, $f_t = 1.00$, and $f_m = 1.55$ for low-alloy tubes. Thus the cost of the condenser is (\$7.290)(1.55)(1.05) $\cong$ \$11,900.

Also as determined from App. 3, the characteristics of the reboiler (E.202 in Fig. E6.4) are

Type: TS
Size:

155°C
190°C (steam 12 bars)

$\Delta t = 35\ ^\circ\mathrm{C}$
$U = 400\ \mathrm{kcal/(h)(m^2)(^\circ C)}$ (Table A3.1)

$$Q' = 1.1Q = 1.1\ (0.611 \times 10^6) = 0.672 \times 10^6\ \text{kcal/h}$$

$$S = \frac{0.672 \times 10^6}{400(35)} = 48\ \text{m}^2$$

Steam consumption: The heat of condensation of steam at 190°C is 475 kcal/kg, so that the steam consumption is

$$m = \frac{0.672 \times 10^6}{475} = 1{,}415\ \text{kg/h} \simeq 1.4\ \text{t/h}$$

Price: The base price, from Fig. A3.5, is $8,960; and the correction factors, from Table A3.6 are $f_d = 1.35, f_\phi = 1.00, f_l = 1.00, f_{np} = 1.00, f_p = 1.03,$ $f_t = 1.00$, and $f_m = 1.75$; so the corrected price of the reboiler is $8,960(1.35)(1.03)(1.75)(1.05) ≅ $22,890, and the cost of complexity is $8,960[1.35 (1.03)(1.75) − 1](1.05) ≅ $13,490.

The overhead accumulator (B203 in Fig. E6.4) can be sized and priced according to App. 1 as follows:

Size Assume 15 minutes holdup; then the capacity is 2,543/4 = 636 kg, or since the density of heptenes at 90°C is 0.63, 636/0.63 = about 1 m^3. From Table A1.7, the chosen dimensions for this tank would be 1-m diameter and 2 m long, for a total volume of 1.57 m^3. Allowing for a pressure of 1.2 bars, the wall thickness would be 6 mm, including corrosion allowance.

Price

Shell, P_1: [from Eq. (A1.5)], 24.65(1)(2)(6)	=	296 kg
Heads, P_2: (from Fig. A1.8), 17(6)	=	102 kg
Total		398 kg

Assuming low-alloy steel SA 357, the base price is obtained from Fig. A1.9 and Table A1.9, and a thickness-correction factor from Fig. A1.10, for 398 ($2.02/kg)(1.06)(2.0) ≅ $1,700. The accessories for this tank are assumed as the minimum from Fig. A1.12, modified for SA 357 per Table A1.12, or $3,440(1.8) = $6,200. This plus the cost of the tank is $6,200 + 1,700 = $7,900; and allowing 10% for contingencies brings the total cost to $8,690. Of this cost that part due to alloy steel is [398 ($2.02)(1.06)(1) + $3,440(0.8)]1.10 = ($853 + 2,752)(1.10) ≅ $4,080.

The reflux pump is sized and selected and priced according to App. 4.

Size Hydraulic horsepower, cv, is

$$PH = 0.03704\ Q_c \Delta P$$

where PH = hydraulic horsepower, cv
Q_c = flow, m³/hr
ΔP = differential pressure, bar

$$Q_c = QK = \frac{2.543(2.70)}{0.63}(1.20) = 13.078 \text{ m}^3\text{/h}$$

$$\Delta P = P_2 - P_1 + \Delta H + \delta P + (\delta\pi)$$

where $P_1 \cong P_2$
ΔH = height of the tower, or 18 m, equivalent to 1.8 bar
δP = pressure drop assumed for one valve, equal to 1 bar
$\delta\pi$ = negligible difference in pressure between suction and discharge vessels

Thus ΔP is equal to 2.8 bars.

$$PH = 0.03704(13.078)(2.8) \cong 1.36 \text{ cv}$$

Assume a centrifugal pump (Fig. A4.1), with an efficiency of 46% (Fig. A4.2), so that the actual brake horsepower is

$$P_{cv} = \frac{PH}{E} = \frac{1.36}{0.46} = 3 \text{ cv}$$

Assume, from Fig. A4.5*a,* a speed of 3,500 rpm and a base price of \$1,880, with correction factors (from Table A4.2) of $f_d = 1.00$, $f_m = 1.00$, $f_t = 1.00$, and $f_p = 0.7$, so that the corrected price is \$1,320.

The driver for this pump is assumed, from App. 5, to be an electric motor with the following characteristics (from Table A5.1):

Horsepower: 5 cv

Percentage demand: 60%

Efficiency: 81.5%

Electrical consumption: 0.735(3/0.815) = 2.7 kWh/h

The price and correction, from Fig. A5.1 and Table A5.2, respectively, are \$460(0.9) = \$410. Thus the total cost of pump and motor is \$1,730.

E6.2.3. CALCULATING THE MANUFACTURING COST

Equipment costs for the reactor and its coolers, which were calculated in Sec. E6.2.2.2*a,* are combined with other equipment for the reaction section and summarized by category in Table E6.10a. These costs for primary equipment categories are then converted to battery-limits investments in Table E6.10b, according to the method described in Sec. 4.3.3.3b.

TABLE E6.10 Investments Costs for a Propylene-Butenes Dimerizing Reaction Section

a. Primary Equipment Costs, $

Equipment	Cost in Mild Steel	Addition for Stainless Steel	Total
Reactor	71,600	55,400	127,000
Tanks	5,200	8,100	13,300
Exchangers	44,500	36,800	81,300
Mixers	20,000	19,900	39,900
Pumps and motors	74,400		74,400
TOTAL	215,700	120,200	335,900

b. Battery-Limits Investment, $

Equipment	Cost in Mild Steel	Multiplying Factor f_g (Table 4.16)	Size Factor f_d (Fig. 4.8)	Installed Costs
Reactor	71,600	3.75	1.1	292,700
Tanks	5,200	2.72	1.2	16,500
Exchangers	44,500	2.84	1.1	142,200
Mixers	20,000	3.75	1.1	81,800
Pumps	74,400	2.67	1.1	224,500
TOTAL	215,700			757,700
Addition for stainless steel				120,200
Battery-limits investment				877,900
Round figure				878,000

Similarly, the costs calculated for the heptenes fractionator in Sec. E6.2.2.2b are incorporated in battery-limits costs by category for the purification section in Tables E6.11a and E6.11b.

These investment costs, plus costs of catalyst, utilities consumption, labor, etc., are converted to operating costs by the methods described in Sec. 4.3.2.2. The results follow.

E6.2.3a Calculation of Operating Costs

Investment, dollars in mid-1975:

Battery-limits investment:	
Reaction section	878,000
Fractionation section	801,000
Total	1,679,000
General services and storage (30%)	504,000
Total	2,183,000
Engineering (12%)	262,000
Royalties	208,000

Process book	63,000
Total fixed capital	2,716,000
Construction loans (7%)	190,000
Start-up costs ($\frac{1}{2}$ month of operating costs)	308,000
Total depreciable capital	3,214,000
Working capital (1 month of operating costs)	316,000

Operating cost, dollars per metric ton of pure heptenes:

Variable costs:	
Raw materials	337.4
Catalyst	9.5
Utilities:	
Steam: 12 bars 7.7	
25 bars 4.7	
Electricity 1.3	14.8
Cooling water 1.1	
Total variable costs	361.7
Labor	8.3
Total	370.0

TABLE E6.11 Investment Costs for a Heptenes Purification Section

a. Costs of Primary Equipment, $

Equipment	Cost in mild steel	Addition for stainless steel	Total
Towers	65,600	58,100	123,700
Tanks	12,200	13,500	25,700
Exchangers	96,700	56,300	153,000
Pumps and motors	18,500		18,500
TOTAL	193,000	127,900	320,900

b. Battery-Limits Investment, $

Equipment	Cost in Mild Steel	Multiplying Factor f_g (Table 4.16)	Size Factor f_c (Fig. 4.8)	Installed Costs
Towers	65,600	3.75	1.1	270,600
Tanks	12,200	2.72	1.2	38,700
Exchangers	96,700	2.84	1.1	307,600
Pumps and motors	18,500	2.67	1.1	56,600
TOTAL	193,000			673,500
Addition for stainless steel				127,900
Battery-limits investment				801,400
Round figure				801,000

Fixed costs:	
Depreciation	20.0
Interest on depreciable capital	11.3
Interest on working capital	2.8
Maintenance, taxes and insurance, general	7.7
Gross manufacturing cost	412.0
Credit for coproducts	−103.0
Net manufacturing cost	309.0

E6.2.3b CONCLUSIONS

The attractiveness of this method for manufacturing heptenes is closely tied to the value of the coproducts.

A basic improvement would result from a higher yield of heptenes. Unfortunately, the kinetic studies have shown that a selectivity of about 50% can not be exceeded with the present catalyst system. It is consequently impossible to forsee a reduction in the cost of raw materials without a complete revision of the catalyst system.

The consumption of catalyst is, on the other hand, a basic element of the operating cost.

Appendixes

Two types of information are assembled in these appendixes: short-cut methods for the process design of equipment, and methods for estimating equipment costs. The purpose is to provide a convenient substitute for a slower and more costly engineering study. The short-cut methods relate to each other by necessity, but they can be modified or extended according to the user's experience with them. The user may note that the relative effort required by—and thus the relative accuracy of—these methods tend to reflect their relative importance to the cost of a process plant. Because of this, the methods vary in interest and complexity; and the user should know how to adapt them to the basic data available for a specific study. The LMTD corrections (Fig. A3.2), for example, have little significance when the temperatures or duty of the exchanger are uncertain.

The emphasis of these appendixes is more on pricing equipment than on sizing it, under the assumption that the pricing data and methods will be used most often. However, certain of the short-cut design methods (distillation towers, air coolers, etc.) can be used to find an order of magnitude for the first try in iterative or optimizing design calculations. Such simplifications cannot be envisaged for other types of equipment (compressors, pumps, etc.); and sometimes it is even necessary to get the advice of fabricators (extractors, filters, centrifuges, etc.). In any event the process design information gathered here is not original, but only an adaptation of data that seemed to best answer the immediate needs of evaluation. Readers are referred to Sec. 6 of the Bibliography for the references.

The cost data are more original. Indeed those curves, which are based on statistics, have been verified with references that have a high degree of credibility. The data are presented in a way that makes it easy to keep them up to date: Base prices, which are sensitive to market variations, are presented in simple curves; whereas design features, whose relative costs do not tend to vary, are accommodated through series of correction factors. Certain of the graphs do resemble data that have been published, and in such cases an appropriate reference has been included in the bibliography.

Appendix 1

Process Design Estimation: Pressure Vessels

Pressure vessels here means towers and tanks. Reactors, which may also be pressure vessels, are given separate treatment in App. 2.

Towers are used for distillation, absorption, stripping, and extraction, and their process design estimation must enter into those unit operations. Here, they are arranged in order of increasing complexity. Indeed distillation, the first, has permitted analysis through simplified nomographs, at least in the case of trays. Absorption stripping, which is more complex, cannot be analyzed by nomographs, although one or more simplified methods can be worked out. The use of packing instead of trays complicates separation problems and increases the number of calculations so that simple and direct graphic interpretation cannot be supplied. Finally, solvent extraction, which involves not only a choice between trays or packing but also a change in temperature, requires both a complicated study and (usually) consultation with the equipment fabricators in order to determine, for example, the type of mixing or shape of a particular tower. It thus appears impossible to give any short-cut methods for sizing extractors. Given the size and the power of the agitator, on the other hand, it will always be possible to arrive at an approximate cost by referring to the data for towers and reactors.

A1.1 SIZING TOWERS WITH TRAYS

A1.1.1 DISTILLATION

J. L. Gallagher's method for sizing distillation towers is recommended for immediate use when exact information is not available. It should be remembered, however, that accuracy is sacrificed for speed and that when delays are permissible, particularly in the case of complex mixtures, it is always preferable to make a more detailed calculation.

A1.1.1.1 Validity of the Recommended Method

Gallagher's method rests on the following five simplifying assumptions:

1. The feed is at its boiling point.
2. Molar flow of liquid and vapor is constant all along the tower.
3. The separation is applied to two key components in a system of several components.
4. The calculation of minimum reflux rate is rigorous only for binary mixtures giving pure distillate and bottoms.
5. The diameter is constant all along the tower.

A1.1.1.2 Determining the Theoretical Trays

The required number of theoretical trays is obtained through Fenske's equation in nomograph Fig. A1.1. The procedure consists of five steps:

1. Choose the key components of the separation as the heaviest component appearing in the overhead and the lightest component appearing in the bottoms, and set the purities desired in the distillate and bottoms.
2. Determine the average relative volatility:

$$\alpha = (\alpha_D \cdot \alpha_R)^{1/2}$$

where α_D = relative volatility at the top
α_R = relative volatility at the bottom

The relative volatility is equal to the ratio of the equilibrium constants of the two keys or, if absolutely necessary, the ratio of their vapor pressures. Thus it is necessary to first choose the temperatures and pressures at the top and bottom in conjunction with the overhead condensing temperature (light key), the bottoms boiling temperature (heavy key), along with the available pressure (tower), cooling water temperature (condenser) and steam temperature (reboiler).

A. K. Badhwar has proposed a graphic method for determining the relative volatilities based on Trouton's rule rearranged to the effect that the absolute atmospheric boiling points of normal liquids are proportional to

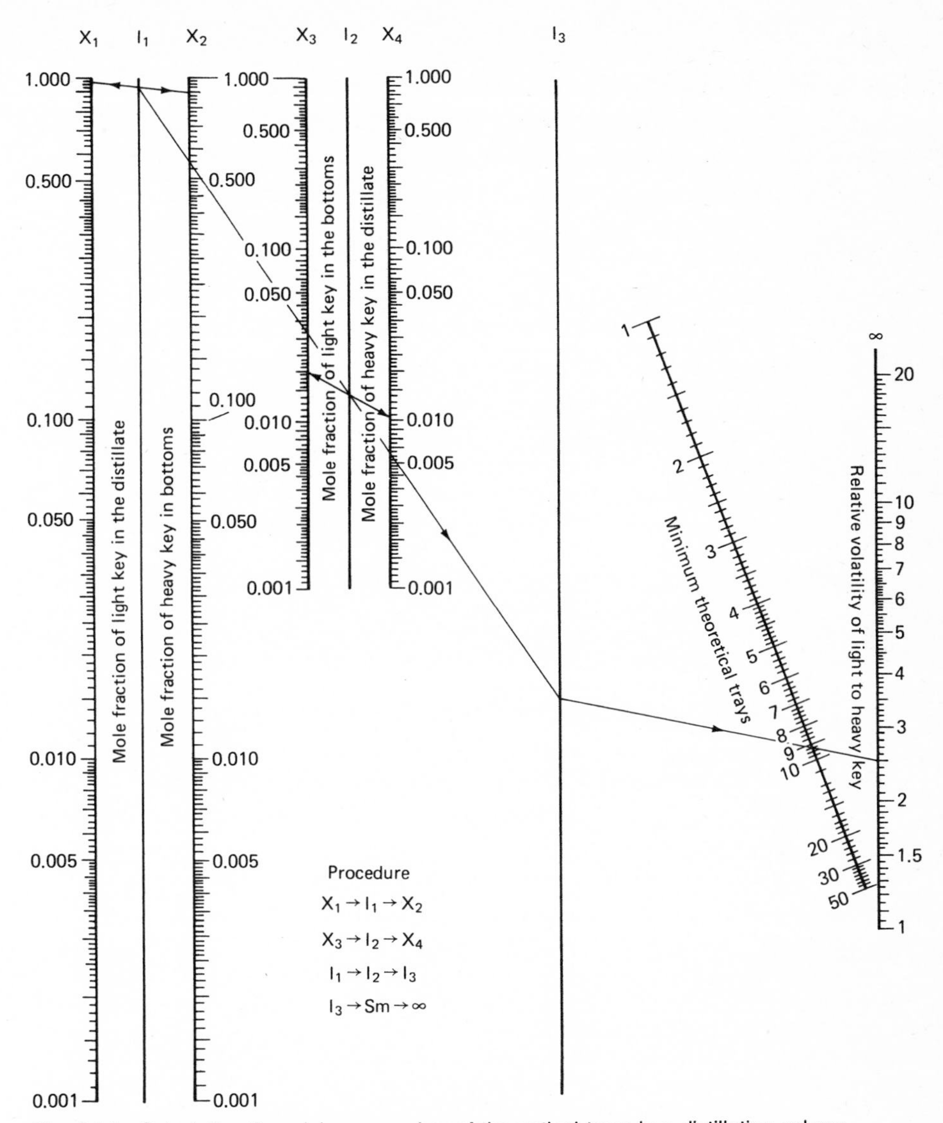

Fig. A1.1 Calculating the minimum number of theoretical trays in a distillation column.

the product of their molecular weights and heats of vaporization. It's use is of interest when vapor-pressure data are not available; but caution is needed in areas of low relative volatility. Otherwise, it is a useful complement to Gallagher's method.

3. Calculate the minimum number of trays at total reflux.

$$S_m = \frac{\log (X_1X_2/X_3X_4)}{\log \alpha}$$

where X_1 = mole fraction of the light key in the distillate
X_2 = mole fraction of heavy key in the bottoms
X_3 = mole fraction of light key in the bottoms
X_4 = mole fraction of heavy key in the distillate
S_m = the minimum number of theoretical trays

4. Calculate the minimum number of trays in the tower, N_m:

$$N_m = S_m - 1$$

This expression corresponds to the usual case, where the reboiler acts as one theoretical tray. When a partial condenser on the overhead also acts as one theoretical tray, the expression is

$$N_m = S_m - 2$$

5. To determine the theoretical operating trays, N_t, at operating reflux, assume

$$N_t = 2N_m$$

A1.1.1.3 Determining the Actual Trays

The actual number of trays depends on the overall tray efficiency, which is obtained from nomograph Fig. A1.2, based on the following equation:

$$E = 63(\alpha\mu)^{-0.212}$$

where E = overall tray efficiency
μ = viscosity of the feed liquid at the average temperature of the tower, cP

The actual number of trays, N_R, is then given by

$$N_R = \frac{N_t}{E}$$

A1.1.1.4 Calculating the Reflux Ratio

This calculation is carried out in the following two steps:

1. Obtain the minimum reflux, R_m, at infinite trays with nomograph Fig. A1.3, which solves the equation

$$R_m = \frac{1}{(\alpha - 1)X_f}$$

where X_f = mole fraction of the light key in the feed.

2. Determine the actual reflux by assuming as a first approximation:

$$R = (1.20 - 1.50)R_m$$

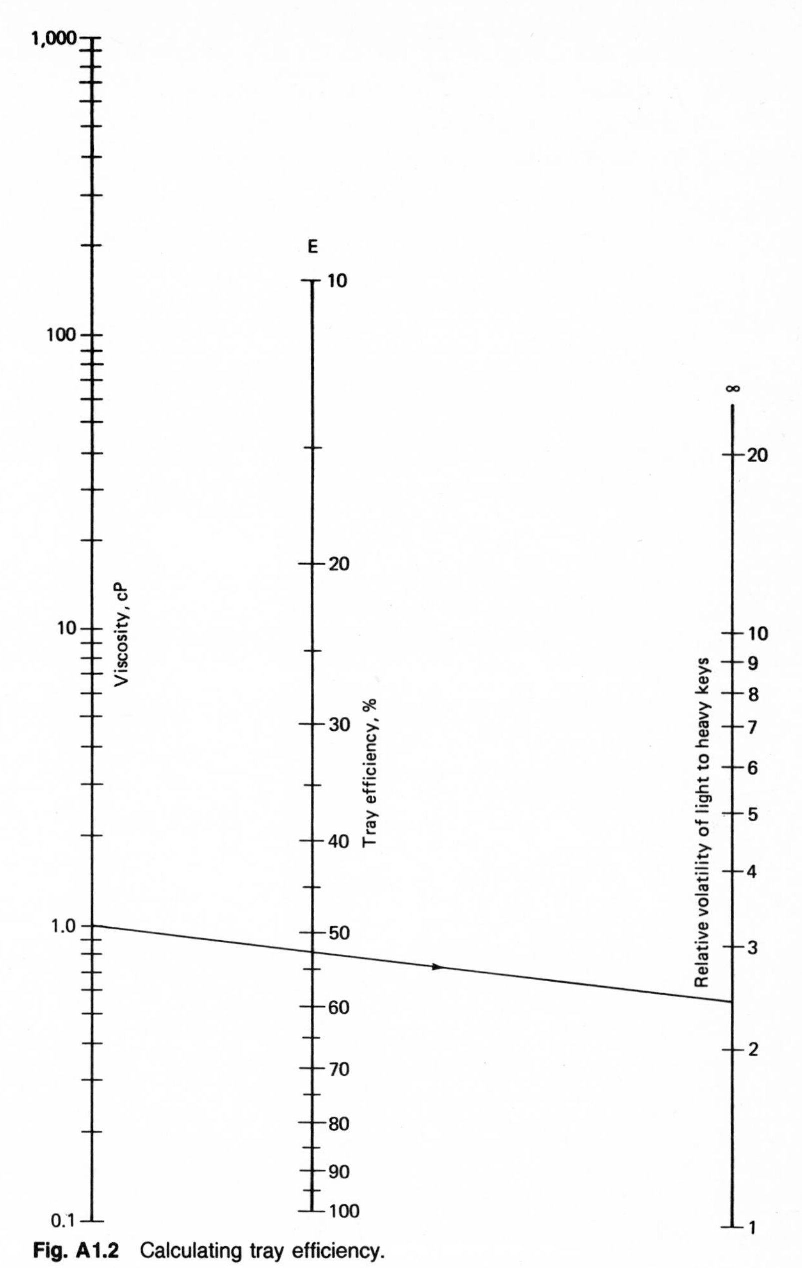

Fig. A1.2 Calculating tray efficiency.

with the average being

$$R = 1.35R_m$$

A1.1.1.5 Calculating the Vapor Traffic

The vapor load in the tower is obtained with the following relation:

$$V = D(1 + R)\ \frac{22.4}{3{,}600}\ \frac{T}{273}\ \frac{1}{P}$$

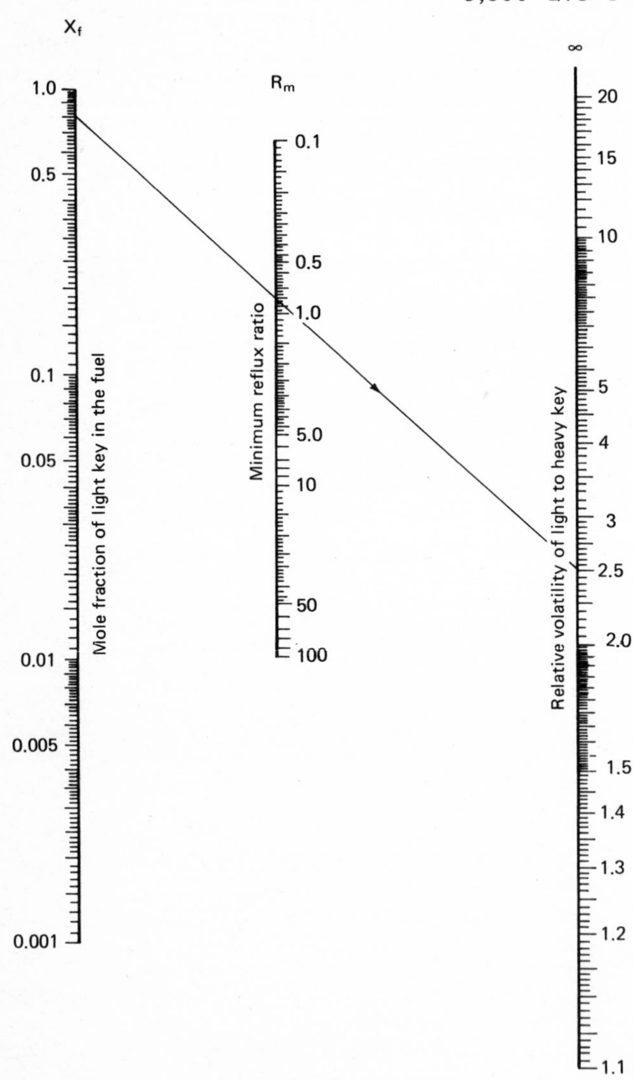

Fig. A1.3 Calculating the minimum reflux ratio.

or from $$V = 2.279 \times 10^{-5} D(1 + R) \frac{T}{P}$$

where V = vapor loading, m^3/s
D = flow of distillate, kmol/h
T = temperature of the vapor, K
P = pressure of the vapor, bar

This calculation can be done at both the overhead and bottom conditions; the molar flowrate is considered the same. Since the diameter of the tower is assumed constant, the overhead conditions, where the lowest molecular weight occurs, will usually carry the highest vapor volume and thus control the diameter of the tower. It can be different, however, when there are sidestream drawoffs.

A1.1.1.6 Determining the Tower Diameter

This calculation is performed in the following steps:

1. Determine the vapor density top and bottom.

$$d_v = \frac{M}{22.4(T/273)(1/P)}$$

or $$d_v = \frac{12.19MP}{T}$$

where M = average molecular weight of the vapor
d_v = vapor density, kg/m^3
T = vapor temperature, K
P = vapor pressure, bar

2. Determine the liquid density top and bottom, by taking the average density of the liquid at overhead and bottoms conditions from appropriate tables or graphs.

3. Choose the tray spacing. Trays are assumed to be on three spacings, each of which has its constant for the sizing calculation, as follows:

Tray Spacing	*Constant C′*
12 in (≃30 cm)	0.0229
18 in (≃45 cm)	0.0427
24 in (≃60 cm)	0.0537

In practice, an 18-in. spacing is usually adopted for a diameter of 1.5 m, and a spacing of 24 in. is adopted for diameters of 1.5 to 6 m.

4. Calculate the tower diameter. This calculation is done with the following equation:

$$D'_m = \left(\frac{4V}{\pi C'\sqrt{d_l/d_v - 1}}\right)^{1/2}$$

where D'_m = tower diameter, m
V = maximum vapor load, m^3/s
d_l = effective liquid density, kg/m^3
d_v = effective vapor density, kg/m^3
C' = tray-spacing constant

This equation can be solved by means of the nomograph Fig. A1.4.

A1.1.1.7 Determining the Height of the Tower

The following dimensions and features can be assumed for a first approximation:

The distance between the top head and top tray is 1 m.

The distance between the bottom head and the bottom tray is 2–3 m.

The vertical space occupied by the trays is the tray spacing times the number of trays.

A manhole requiring the space equivalent of one tray is located at every tenth tray.

A1.1.2 ABSORPTION AND STRIPPING

Absorption is generally a separation of the heaviest components from a gas by means of countercurrent contact with a liquid, whereas stripping is the separation of the lightest components from a mixture of liquids by means of countercurrent contact with a gas.

A1.1.2.1 The Principle

The recommended method is based on the simplified Kremser-Brown equation. The absorption of component i from a gas containing several other compounds is given by the relation

$$\left[\frac{Y^{n+1} - Y^1}{Y^{n+1}}\right]_i = \left[\frac{A^{n+1} - A}{A^{n+1} - 1}\right]_i$$

where A_i = absorption factor of component i, or L/GK_i
G = moles of gas entering the column
K_i = equilibrium constant of component i at the average temperature and pressure of the absorber
L = moles of liquid entering the column
Y_i^{n+1} = mole fraction of component i in the gas entering the absorber
Y_i^1 = mole fraction of component i in the gas leaving the absorber
n = number of theoretical absorption trays

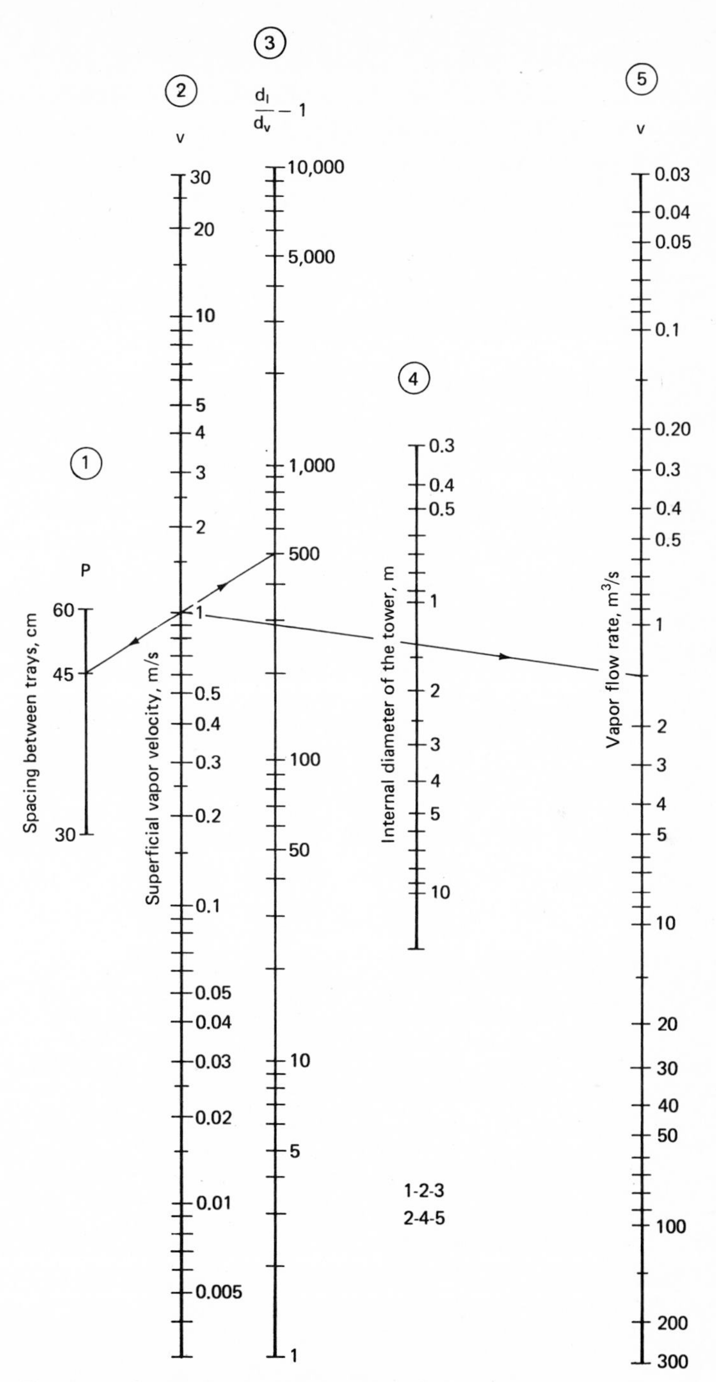

Fig. A1.4 Calculating the diameter of a distillation tower.

The stripping of component i from a liquid containing several other compounds is given by the relation:

$$\left[\frac{X^0 - X^m}{X^0}\right]_i = \left[\frac{S^{m+1} - S}{S^{m+1} - 1}\right]_i$$

where S_i = stripping factor of compound i, or GK_i/L
X_i^0 = mole fraction of compound i in the inlet at the top of the stripper
X_i^m = mole fraction of component i in the outlet at the bottom
m = number of theoretical stripping trays

A1.1.2.2 The Application to Absorption

A1.1.2.2a Calculating Liquid and Vapor Loads

Usually the gas flow, G, and the desired composition of gases entering and leaving the absorber have been set by a material balance, and calculations analogous to distillation are made prior to sizing a tower for purposes of pricing. These calculations can be made in five steps:

1. Choose a key component, i, which can be a product or the heaviest component in the gas leaving the absorber, and for this component calculate the fractional recovery, R_i, as

$$R_i = \left[\frac{Y^{n+1} - Y^1}{Y^{n+1}}\right]_i$$

2. Calculate the minimum molar ratio of absorbent to gas, $(L/G)_m$, as

$$\left(\frac{L}{G}\right)_m = R_i K_i$$

3. Choose an operating absorbent-to-gas ratio, $(L/G)_0$, as

$$\left(\frac{L}{G}\right)_0 = 1.2\text{–}1.3\left(\frac{L}{G}\right)_m$$

and calculate the operating absorbent rate, L_0, for example

$$L_0 = 1.25G\left(\frac{L}{G}\right)_m$$

4. From charts or tables of equilibrium constants, plus the average temperature and pressure in the absorber, obtain the equilibrium constant, K_i, for component i and calculate the absorption factor, A_i, as

$$A_i = \frac{L_0}{GK_i}$$

5. Determine the number of theoretical trays from Fig. A1.5.

6. With L_0, G, and n now set by the key component, it is possible to calculate A, R, and Y^1 for every other component, j, as shown in Table A1.1. This calculation verifies the material balance. If the descrepancies between the given and calculated material balance are excessive, the primary variables should be modified and the calculation repeated.

In addition to the calculated gases in the absorber overhead, a certain amount of absorbent oil will be carried over. These losses can be approximated as a function of the absorbent oil vapor pressure at the tower's top conditions by

$$M_a = G_1 \frac{P_a}{\pi}$$

where M_a = moles of absorbent lost
G_1 = moles of dry gas leaving the top of the absorber
P_a = vapor pressure of the absorbent at the absorber top temperature
π = total vapor pressure at the top

A1.1.2.2b Sizing the Absorber

The tray efficiency is obtained from Fig. A1.6 and used to convert the theoretical trays, n, to actual trays. The diameter and height of the absorber are then calculated similarly to those of a distillation tower, as described in Secs. A1.1.1.6 and A1.1.1.7.

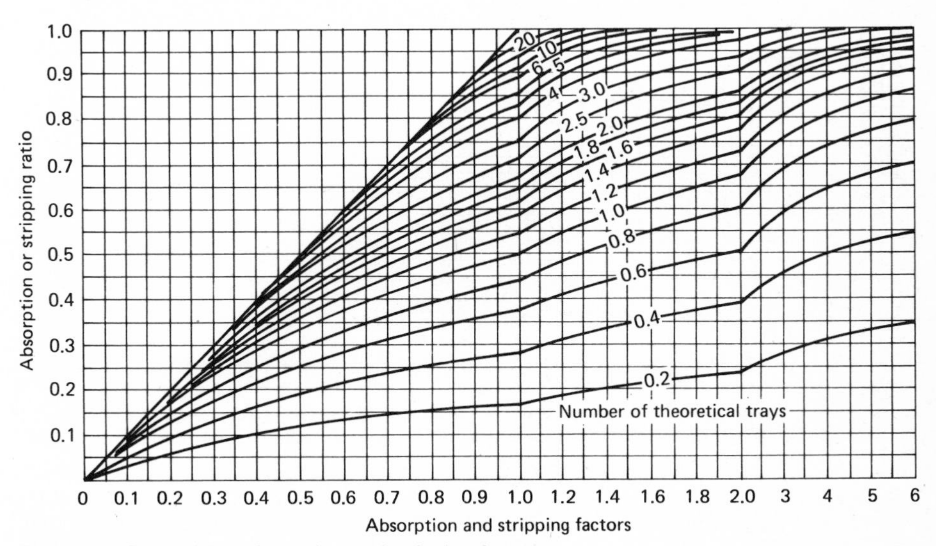

Fig. A1.5 Determining absorption and stripping factors.

A1.1.2.3 The Application to Stripping

Calculate the tower loadings in a five-step procedure analogous to that used for absorption. Figure A1.5 is used to obtain the number of stripping trays, m, and the composition of the stripped liquid is calculated according to Table A1.2, where the following symbols are found:

E: analogous to R, is the fractional removal of components from the liquid.

$(G/L)_m$: analogous to $(L/G)_m$, is the minimum ratio of gas to liquid and is equal to (E_i/K_i).

With the loading determined, the tower is sized like absorbers and distillation towers.

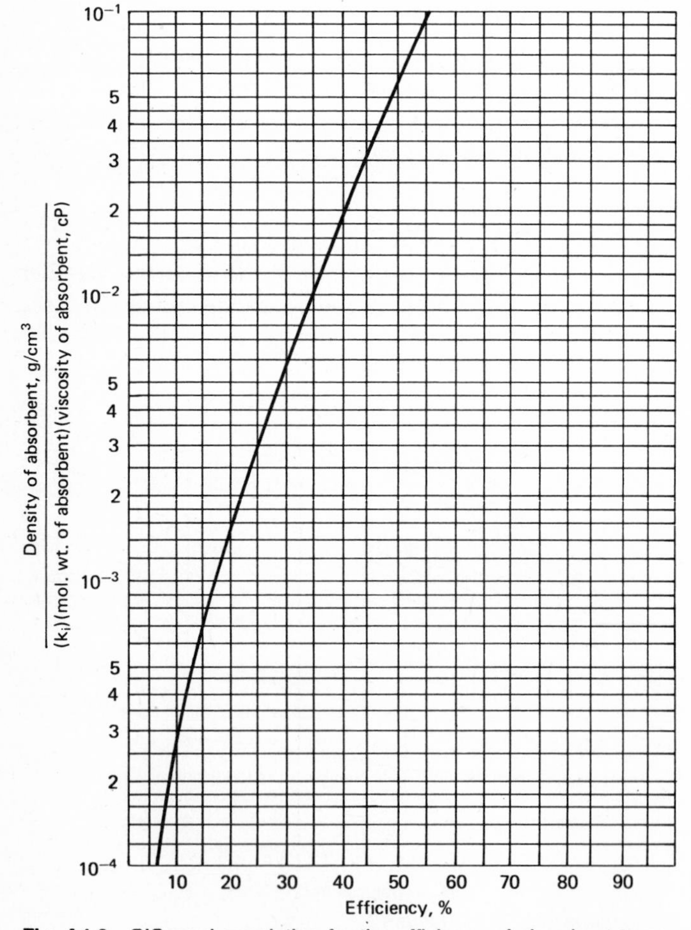

Fig. A1.6 O'Connel correlation for the efficiency of absorbent trays.

TABLE A1.1 Absorption Calculation

Component	Gas Composition $(Y^{n+1})_j$ moles	K_j	$A_j = \frac{L}{K_j G} = A_i \frac{K_i}{K_j}$	$R_j = \left(\frac{Y^{n+1} - Y^1}{Y^{n+1}} \right)$ (Fig. A1.5)	$(Y^{n+1} - Y^1)_j$ moles absorbed	$(Y^1)_j$ moles not absorbed
1	—	—	—	—	—	—
2	—	—	—	—	—	—
⋮	—	—	—	—	—	—
i	—	K_i	A_i	R_i	—	—
⋮	—	—	—	—	—	—
j	—	K_j	A_j	R_j	—	—
⋮	—	—	—	—	—	—
TOTAL	G				—	—

TABLE A1.2 Stripping Calculation

Component	Liquid Composition $(X^0)_j$ moles	K_j	$S_j = \frac{GK_j}{L} = S_i \frac{K_j}{K_i}$	$E_j = \left(\frac{X^0 - X^m}{X^0} \right)_j$ (Fig. A1.5)	$(X^0 - X^m)_j$ moles Stripped	X^m moles Not Stripped
1	—	—	—	—	—	—
2	—	—	—	—	—	—
⋮	—	—	—	—	—	—
i	—	K_i	S_i	E_i	—	—
⋮	—	—	—	—	—	—
j	—	K_j	S_j	E_j	—	—
⋮	—	—	—	—	—	—
TOTAL	L	—	—	—	—	—

A1.2 SIZING PACKED TOWERS

Packings replace trays in fractionating towers usually when the number of separation stages are few, when the capacity is small, and particularly when the circulation of liquid is low compared to the vapor. They are used especially in vacuum distillation, because of their low pressure drop, or with corrosive fluids, because common packings of stoneware are corrosion resistant.

A1.2.1 TYPES OF PACKING

Tables A1.3 through A1.6 give the characteristics of various types of packing. A choice of one or another of these types generally aims at a compromise between efficiency and cost. Raschig rings (steel or ceramic) are the most often used, often in the 1-in. size. The diameter of the tower should be at least 8 times that of the packing.

- Intalox saddles give up to 20–25% more flow.
- Pall rings (steel, ceramic and plastic) have a higher efficiency and lower pressure drop, but cost more.

A1.2.2 PACKED FRACTIONATING TOWERS

As a general rule, packing should not be used in towers more than 1 meter in diameter, without consulting a fabricator. Packing is arranged in beds, with the height of each bed equal to 3 times the tower diameter for Raschig rings and 5–10 times the diameter for Pall rings. Calculations for sizing packed towers

TABLE A1.3 Characteristics of Raschig Rings

Material	Nominal Size, in	Nominal Size, mm	Wall Thickness, mm	Number of Rings per m³	Apparent Specific Weight, kg/m³	Specific Surface, m²/m³	Void Fraction, %
Ceramic or stoneware	4	102	11	800	600	50	75
	3	76	9.5	1,900	650	70	75
	2	51	6.5	6,000	650	95	75
	1½	38	6.5	15,000	700	130	74
	1	25	3	48,000	700	200	73
	¾	19	2.5	100,000	700	240	72
	½	13	2.5	370,000	800	370	64
Steel	3	76	1.6	1,900	450	70	94
	2	51	1.2	6,000	460	100	94
	1½	38	0.9	15,000	480	140	94
	1	25	0.7	48,000	560	210	93
	¾	19	0.6	100,000	580	270	93
	½	13	0.5	370,000	700	400	91

TABLE A1.4 Characteristics of Lessing and Pall rings

Material and Type	Nominal Size, in	Nominal Size, mm	Wall Thickness, mm	Number of Rings per m³	Apparent Specific Weight, kg/m³	Specific Surface, m²/m³	Void Fraction, %
Ceramic							
Lessing rings	2	51	9.5	5,500	800	110	68
	1½	38	6.5	14,000	900	150	60
	1	25	3	46,000	800	220	66
Steel							
Lessing rings	2	51	1.2	5,500	580	120	93
	1½	38	0.9	14,000	610	170	92
	1	25	0.7	46,000	690	250	91
	¾	19	0.6	100,000	760	310	90
	½	13	0.5	370,000	880	500	89
Ceramic							
Pall rings	4	102	9.5	800	420	56	82
	2	51	5	6,000	550	125	78
	1	25	3	50,000	640	220	73
Steel							
Pall rings	2	51	1	6,000	400	105	95
	1⅜	35	0.8	19,000	430	145	95
	1	25	0.6	50,000	500	240	94
	⅝	16	0.4	200,000	550	370	93

TABLE A1.5 Characteristics of Berl Saddles and Intalox Saddles

	Size, in	Size, mm	Number per m³	Apparent Density, kg/m³	Specific Surface, m²/m³	Voids Fraction, %
Berl saddles	2	51	8,800	640	110	77
in porcelain	1½	38	22,000	610	150	75
	1	25	80,000	720	250	70
	¾	19	195,000	800	300	67
	½	13	620,000	900	480	65
Intalox saddles	2	51	8,800	600	110	75
in porcelain	1½	38	23,000	600	160	75
	1	25	85,000	600	250	74
	¾	19	210,000	600	300	73
	½	13	630,000	600	480	73

use the results of tower-loading calculations and the number of theoretical trays.

A1.2.2.1 Calculating the Diameter of Packed Towers

Pressure drop, which depends on the type of packing, affects the maximum capacity that can be achieved without flooding the tower. The diameter is calculated for a pressure drop that corresponds to 70–80% of flooding. The loading at flooding is obtained from Fig. A1.7, which shows the relation between

$$\frac{U^2}{g}\,\frac{a}{\epsilon^3}\,\frac{\rho_G}{\rho_L}\,\mu_L^{0.2} \tag{A1.1}$$

and

$$\frac{L}{G}\sqrt{\frac{\rho_G}{\rho_L}} \tag{A1.2}$$

according to one or the other of the various curves (Eg. A1.7), each of which represent a given pressure drop, ΔP, calculated as

$$\Delta P = \alpha(10^{\beta L})\,\frac{G^2}{\rho_G}$$

where G = gas flow per unit of cross section in the column, kg/(m²)(s)
L = liquid flow per unit of cross section in the column, kg/(m²)(s)
ρ_G = density of the gas, kg/m³
ρ_L = density of the liquid, kg/m³
μ_L = viscosity of the liquid, cP
g = acceleration due to gravity, 9.81 m/s²
U = superficial velocity of the gas inside the column, m/s

TABLE A1.6 Coefficients for Calculating Pressure Drop in Packed Towers

Type of Packing	Dimensions: Size, in	Size, mm	Thickness, mm	Voids Fraction, %	Coefficients: α	β
Ceramic	3	76	9.5	75	0.74	0.031
Raschig	2	51	6.5	75	1	0.03
Rings	1	25	3	73	2.2	0.045
	½	13	2.5	64	10	0.1
Steel	2	51	1.2	94	0.95	0.03
Raschig Rings	1½	38	0.9	94	1.2	0.04
	1	25	0.7	93	1.7	0.043
Ceramic	2	51	5	78	0.55	0.025
Pall Rings	1	25	3	73	1.45	0.033
Steel	2	51	1	95	0.25	0.025
Pall Rings	1⅜	35	0.8	95	0.4	0.033
	1	25	0.6	94	0.6	0.03
Berl Saddles	1½	38		75	0.6	0.03
	1	25		70	1.5	0.036
	½	13		65	5	0.04
Intalox Saddles	1½	38		75	0.55	0.03
	1	25		74	1.7	0.033
	¾	19		73	2	0.035
	½	13		73	4.3	0.04

ΔP = pressure drop, mmHg/m of packing height
D' = column diameter, m
a = specific surface of the packing, m^2/m^3
ϵ = voids fraction of the packing, fraction.
α and β = characteristic coefficients of the packing.

First, choose a curve on Fig. A1.7 to represent design conditions. The gas and liquid densities can be obtained from the loading calculations, and the packing characteristics from the tables. Although neither L nor G is known without the tower diameter, the ratio, L/G, is known from the loading calculations, and that ratio plus the choice of a pressure-drop curve in the figure is enough to determine U. Subsequently, G can be approximated by

$$G = 0.8U\rho_G$$

And this value of G, plus the kilogram per second of gas flowing, Q_G, can be used to calculate the diameter from the relation:

$$G = \frac{4Q_G}{\pi D'^2}$$

Finally, the calculated value of G can be used in the equation for pressure drop to check the pressure-drop curve assumed in Fig. A1.7 at the start of the calculation; and the calculation is repeated until the calculated and assumed values check.

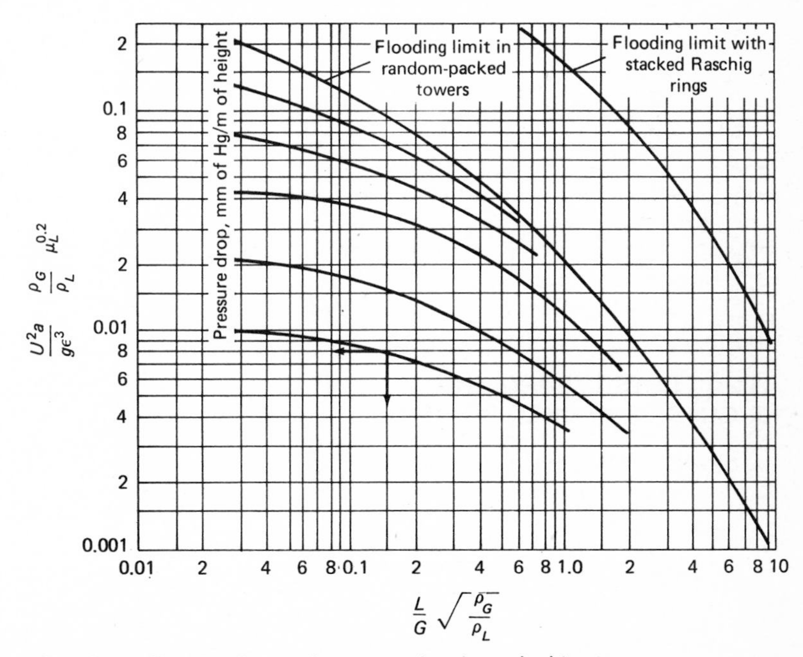

Fig. A1.7 Flooding limit and pressure drop in packed towers.

A1.2.2.2 Determining the Height of Packing

Assume the height equivalent to a theoretical plate (HETP) is one diameter for tower diameters less than 1 meter and two diameters for tower diameters between 1 and 2 meters.

A1.2.3 PACKED EXTRACTION TOWERS

Packing is often used in wash towers with a limited number (4–5) theoretical stages. In such cases, the packing height can be determined from the number of theoretical stages and an HETP assumed as 3 meters. Fabricators should be consulted for solvent extractors with more stages.

The diameters of extraction towers can be determined by means of the following relation, which finds a value for U_c:

$$1 + 0.835\left(\frac{\rho_d}{\rho_c}\right)^{0.25}\left(\frac{Q_d}{Q_c}\right)^{0.5} = 0.6\left[\left(\frac{U_c^2 a}{g\epsilon^3}\right)\left(\frac{\rho_c}{\Delta\rho}\right)\left(\frac{\sigma}{73}\right)^{0.25}\right]^{-0.25}$$

where g = the acceleration due to gravity, 9.81 m/s²
a = the specific surface of the packing, m²/m³
ϵ = the fraction voids in the packing
σ = the surface tension, dyne/cm
ρ = the density, kg/m³
$\Delta\rho$ = $\rho_c - \rho_d$
Q = the flow at the entry to the column, kg/s
U = the superficial velocity in the column, m/s
c,d = subscripts denoting continuous phase (c) and dispersed phase (d)

The ratio Q_d/Q_c should be between 0.1 and 10 for stable operation. The surface tension can be approximated as between 5 (for a liquid that separates slowly) and 40 (for a liquid that separates quickly).

The diameter then can be obtained from

$$D' = 2\sqrt{\frac{Q_c}{0.8\pi\rho_c U_c}}$$

A1.2.4 DETERMINING THE HEIGHTS OF PACKED TOWERS

Assume that the height of a bed of packing is limited to:

1. Three meters for tower diameters less than 1 m
2. Four and one-half meters for diameters greater than 4 m

A support plate is necessary for each bed of packing. These support plates can be priced like 5-mm thick perforated trays, as a first guess. Also, a distributor tray is necessary above the top bed of packing, and if the liquid downflow is less than 0.4–0.5 cm/s, the lower beds will need a distributor plate as well. The spacing between each bed of packing generally runs 0.5–1.0 m.

A1.3 TANK SIZING

First decide on the holdup, and then select the dimensions.

The holdup between maximum and minimum liquid level of a reflux tank should allow for the largest of the following:

- Two minutes of surge time
- Fifteen minutes for drawoff
- Five minutes of reflux

The dimensions are obtained by

1. Sizing for a volume equal to 1.3 times the holdup if the holdup is more than 3 m^3
2. Dimensioning the tank according to Table A1.7 if the holdup is less than 3 m^3
3. Assuming a ratio of length to diameter of
   a. Two to three if the pressure is less than 4 bars
   b. Four to five if the pressure is more than 4 bars

A1.3.1 SEPARATING TANKS

A1.3.1.1 Vapor-Liquid Separators

If the tank is horizontal, it should be sized so that the vapor velocity, *m/s,* is given by

$$V_{\mathrm{m/s}} < 8.15 \times 10^{-2} \sqrt{\frac{d_l}{d_v} - 1}$$

If the tank is vertical, the diameter should be sized so that the vapor velocity, m/s, is given by:

$$V_{\mathrm{m/s}} = 9.6 \times 10^{-2} \sqrt{\frac{d_l}{d_v} - 1}$$

The height of a vertical tank should conform to the equation

$$H = H_1 + H_2$$

TABLE A1.7 Standards for Sizing Reflux Drums

Required Holdup v, m^3	Diameter ϕ m	Length L for Horizontal Tanks, m	Height H for Vertical Tanks, m	Tank Volume V, m^3
0.1	0.6	—	0.8	0.23
0.2	0.6	—	1.2	0.34
0.3	0.9	—	0.9	0.57
0.4	0.9	—	1.1	0.70
0.5	0.9	—	1.25	0.80
0.6	0.9	—	1.4	0.89
0.7	0.9	1.8	—	1.15
0.8	0.9	2.1	—	1.34
0.9	0.9	2.4	—	1.53
1.0	1.0	2.0	—	1.57
1.2	1.0	2.4	—	1.88
1.4	1.0	2.8	—	2.20
1.6	1.0	3.2	—	2.51
1.8	1.2	2.3	—	2.60
2.0	1.2	2.5	—	2.83
2.2	1.2	2.8	—	3.16
2.4	1.2	3.1	—	3.50
2.6	1.2	3.3	—	3.73
2.8	1.2	3.6	—	4.07

where H is the tank height; H_1 is 1.5 m if the diameter is under one meter and 2.0 m if the diameter is over a meter; H_2 corresponds to a liquid holdup of 20 minutes or a minimum of 0.4 m.

A1.3.1.2 Surge Tanks

The diameter should be sized to give a vapor velocity, m/s, as

$$V_{m/s} = 7.5 \times 10^{-2} \sqrt{\frac{d_l}{d_v} - 1}$$

And the height should be determined as for a vertical tank.

A1.3.1.3 Tanks with Liquid-Liquid Separators

Reflux tanks with water drawoff Calculate the holdup required for the reflux. Calculate a holdup of 5 minutes for combined liquids. Calculate a tank volume so that

$$V = \begin{cases} 1.1(v_1 + v_2) & \text{if } v_1 + v_2 \text{ is less than } 2 \text{ m}^3 \\ 1.15(v_1 + v_2) & \text{if } v_1 + v_2 \text{ is less than } 4 \text{ m}^3 \\ 1.20(v_1 + v_2) & \text{if } v_1 + v_2 \text{ is more than } 4 \text{ m}^3 \end{cases}$$

where V = tank volume
v_1 = reflux hold-up
v_2 = 5 minutes holdup on total liquids

Determine the length and diameter of this tank as for a reflux drum. Provide a drawoff pot such that its

	0.30 m	if the tank diameter is under 1.5 m
Diameter =	0.45 m	if the tank diameter is 1.5 to 2.0 m
	0.60 m	if the tank diameter is over 2 m

Height = 0.90 m if the drawoff level is regulated

Liquid-liquid separators Take the larger of the two liquid flows, and calculate a decantation velocity, V, such that:

$$V_{\mathrm{cm/mn}} = 53 \frac{\Delta\rho}{\mu} \times 10^{-6} < 25.0$$

where $\Delta\rho$ = difference in densities of the two liquids, kg/m^3
μ = viscosity of the larger liquid, Pl

Then determine the dimensions, so that the diameter, ϕ_m is

$$\phi_{\mathrm{m}} = 0.6 + 0.634\left(\frac{D}{V}\right)^{0.5} \geqslant 1 \text{ m.}$$

and the length L_m is

$$L_m = \frac{\phi - 0.66}{V \times 10^{-2}}\left(\frac{D}{A}\right) \geqslant 3 \text{ m.}$$

where D = larger of the two liquid flows, m^3/h
A = section corresponding to − 0.66

A1.4 PRICING PRESSURE VESSELS

A1.4.1 CALCULATING WALL THICKNESSES

A1.4.1.1 The Basic Formula

$$e_b = \frac{PR}{\alpha t - 0.6P} \tag{A1.4}$$

where: e_b = required wall thickness, mm
P = operating pressure, bar
R = radius of the vessel, mm
t = maximum allowable stress of the wall material, bar
α = welding coefficient

A1.4.1.2 Values Fixed by the Construction

Welding coefficient α is

$$\alpha = \begin{cases} 1.00 & \text{with a complete X-ray} \\ 0.85 & \text{with spot X-rays} \\ 0.70 & \text{with no X-rays} \end{cases}$$

For a first approximation, assume $\alpha = 1.0$. Maximum stress: Use Table A1.8.

A1.4.1.3 Procedure

Calculate the wall thickness with the above equation A.1.4. Add a corrosion allowance, usually 3 mm. Assume a minimum wall thickness, including corrosion allowance of

- Six millimeters for tanks
- Eight millimeters for towers and reactors

A1.4.2 PRICING TOWERS, TANKS, AND PRESSURE-VESSEL REACTORS

A1.4.2.1 The Estimating Principle

Prices are determined as a function of weight and material of construction for

The externals, including
  Shell and heads
  Skirts, for towers and reactors where applicable
  Accessories, such as nozzles, platforms, ladders, walkways, permanent cranes, and so forth
The internals, including
  Trays
  Packing

Where corrosive materials might require special materials of construction, it has been assumed most economical to use

Solid metals for thickness up to 8 mm
Cladding for thicknesses between 8 and 20 mm
Linings for thicknesses over 20 mm

In all cases, the estimate is made for a base case that assumes mild steel; and correction factors are added for corrosion-resistant materials where necessary.

TABLE A1.8 Maximum Allowable Stress for Steels as a Function of Temperature

ASTM Designation for Steel	Temperature, °C							
	−4 to 343	371	399	427	454	482	510	538
Mild steels								
SA 201 B	1,055	1,009	910	759	608	457	316	176
SA 212 B	1,230	1,167	1,037	844	650	457	316	176
SA 285 A	791	773	721	633	545	457		
SA 285 B	879	851	784	675	566	457		
SA 285 C	967	931	847	717	587	457		
Low-alloy steels								
SA 203 A and B	1,142	1,090	974	801	629	457	316	176
SA 302 B	1,046	1,046	1,046	1,343	1,181	931	703	439
SA 357		942	921	900	872	808	703	513
SA 387 C	1,055	1,055	1,055	1,055	1,012	921	773	548

ASTM-AISI Designation for Steel	Temperature, °C													
	−4 to 38	93	149	204	260	316	343	371	399	427	454	482	510	538
Stainless steels														
SA 240–304	1,318	1,195	1,125	1,086	1,062	1,047	1,044	1,040	1,033	1,023	1,005	984	942	879
304 L	1,230	1,195	1,125	1,055	984	914	879	844	808	773				
310 S	1,318	1,318	1,301	1,279	1,244	1,209	1,188	1,167	1,142	1,104	1,047	970	879	738
316	1,318	1,318	1,258	1,230	1,209	1,202	1,199	1,195	1,188	1,178	1,160	1,125	1,062	984
316 L	1,230	1,230	1,111	1,037	984	956	946	931	914	893	861			
317	1,318	1,318	1,258	1,230	1,209	1,202	1,199	1,195	1,188	1,178	1,160	1,125	1,062	984
321	1,318	1,318	1,195	1,111	1,069	1,047	1,044	1,040	1,033	1,016	1,005	991	974	949
347	1,318	1,318	1,195	1,111	1,069	1,047	1,044	1,040	1,033	1,016	1,005	991	974	949
410	1,142	1,097	1,062	1,026	995	974	963	942	921	896	851	773	619	450

Cladding and lining is treated as a supplement. The skirt is assumed to be in mild steel.

A1.4.2.2 Calculating the Weight

The necessary data are the diameter, the height, and the thickness.

The weight of the shell Use the formula

$$\text{Weight} = 24.7(D')(H)(e) \text{ kilograms}$$

where D' = diameter, m
H = height, m
e = thickness, mm
24.7 = weight of a steel plate 1 m^2 and 1 mm thick.

The weight of the skirt A skirt is usually necessary for distillation columns, and sometimes necessary for reactors. Its characteristics are

Diameter: The same as the tower or shell.

Height: Assume 5 m if there is a bottoms pump requiring NPSH; otherwise, 3 m.

Thickness: According to the diameter and the height of the tower, it will be

Eight millimeters thick for towers with 10 plates or equivalent height.

Ten millimeters for towers with 20 plates or equivalent height.

Twelve millimeters for towers with 35 plates or equivalent height.

With this data, the weight of the skirt can be calculated by the same formula as the shell.

The weight of the heads The two heads close the ends of the shell's cylinder, have a diameter equal to that of the shell, and are assumed to be 2/1 eliptical. Their combined weight is calculated by

$$\text{Weight of heads} = (\text{weight 1 mm thick}) (\text{thickness}) \text{ kilograms}$$

The weight 1 mm thick can be found from Fig. A1.8; and though the heads will be slightly thinner than the shell, they can be assumed to have the same thickness.

A1.4.2.3 Pricing the Shell, Heads, and Skirt

Pricing the shell and heads For a solid material (which will be used when the vessel is in mild steel or its walls are less than 8 mm thick) a base price, $/kg, is obtained from Fig. A1.9, as a function of the shell diameter and

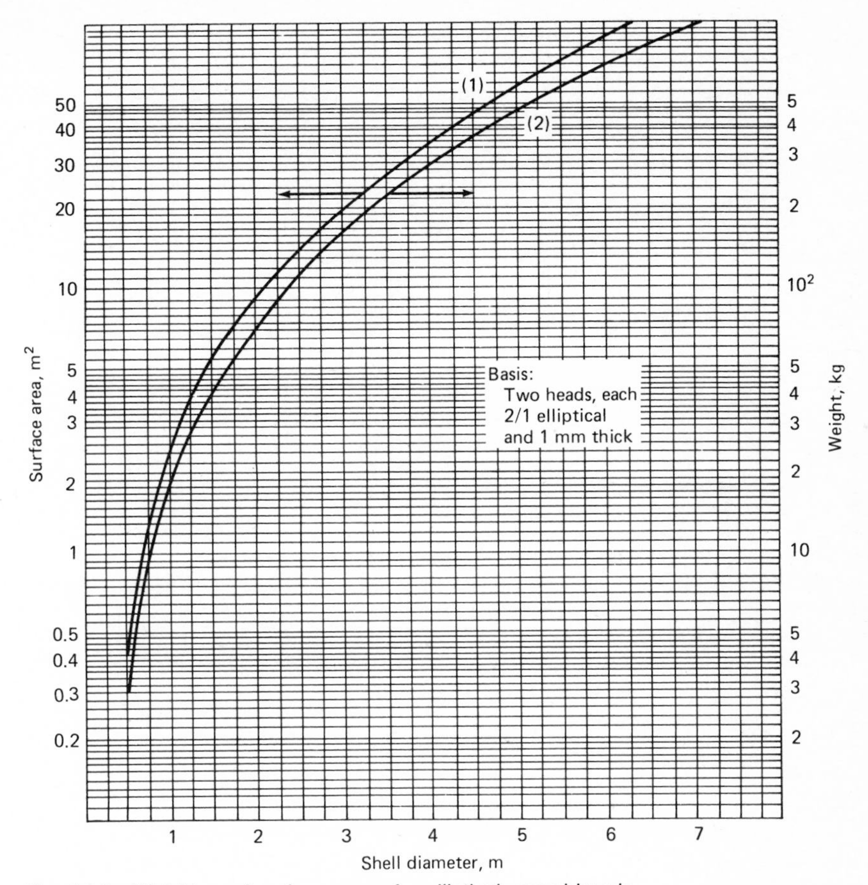

Fig. A1.8 Weights and surface areas for elliptical vessel heads.

assuming 8 mm thickness and mild steel. This base price is modified with the following factors:

1. Factor f_e obtained from Fig. A1.10, for thicknesses other than 8 mm.

2. Factor f_m obtained from Table A1.9, for materials of construction other than mild steel.

The total corrected price is then

$$(\$/\text{kg})(\text{weight in kilograms of shell} + \text{heads})(f_e)(f_m)$$

Cladding is assumed preferred on thicknesses between 8 and 20 mm. The exterior of the vessel will be mild steel of the calculated required thickness, plus an inside layer for corrosion resistance. The cost of the vessel wall thus includes the cost of mild steel plus the cost of cladding. Assuming that the calculated thickness is 8 mm, that the cladding is 10% of the plate thickness,

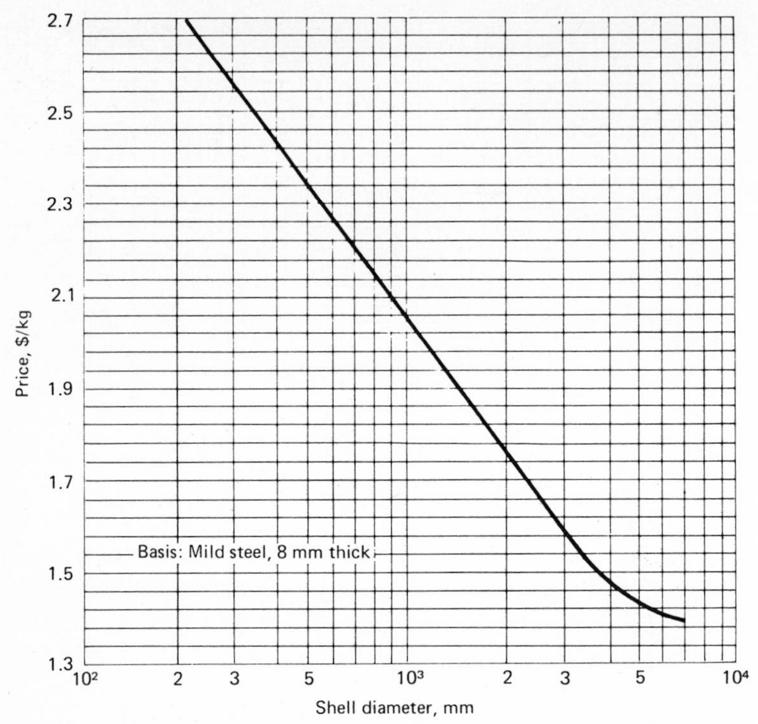

Fig. A1.9 Base price for shell and heads of pressure vessels, mid-1975.

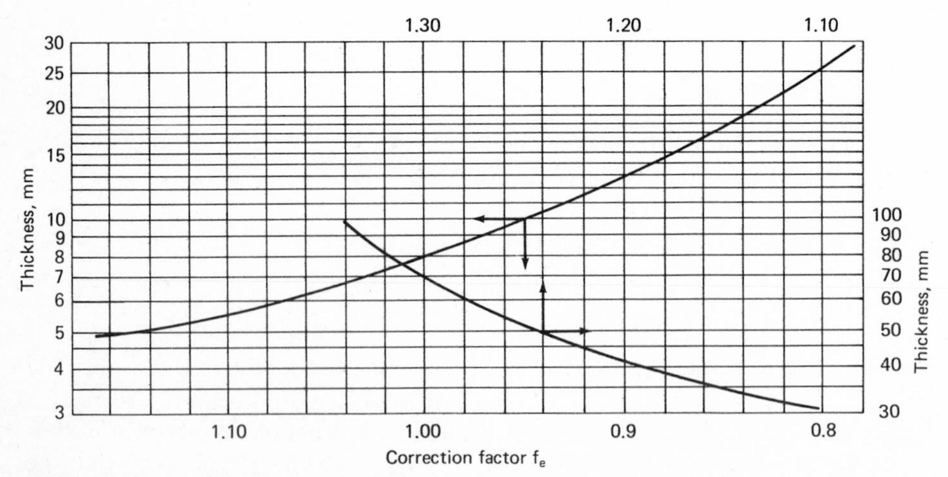

Fig. A1.10 Base-price correction factors for wall thickness.

TABLE A1.9 Base-Price Correction Factors f_m for Materials of Construction

Code Designation			
ASTM	AISI	Miscellaneous	f_m
SA 285 C	—	—	1.0
SA 203 A et D	—	—	1.3
SA 357	—	—	2.0
SA 240	304	—	2.8
SA 240	304 L	—	3.0
SA 240	310 S	—	3.8
SA 240	316	—	2.9
SA 240	316 L	—	3.3
SA 240	316 (Ti)	—	3.1
SA 240	321	—	2.7
SA 240	347	—	2.9
SA 240	410	—	2.4
—	—	Uranus 50	3.6
—	—	Uranus B6	4.8
—	—	Monel 400	8.8
—	—	Inconel 600	10.0
—	—	Inconel 625	13.8
—	—	Hastelloy G	12.5

and that the cladding material is 304 stainless, a base price for cladding in mid-1975 can be calculated from the following formula:

$$\text{Base price} = ((3.14)(D')(H) + S_{\text{heads}}))109 \text{ dollars}$$

where D' = diameter, *m*
H = length of the shell, m
S_{heads} = surface of the heads from Fig. A1.8
This base price is modified with

1. Factor f'_e obtained from Fig. A1.11*a*, for base-metal thickness other than 8 mm.
2. Factor f'_c obtained from Table A1.10, for 20% cladding rather than 10%.
3. Factor f'_m obtained from Table A1.10, for materials other than 304 stainless.

Thus, the corrected price for cladding is

$$\text{Price} = (\text{base price, \$})(f'_e)(f'_c)(f'_m) \text{ dollars}$$

Linings, which are assumed preferred for thicknesses over 20 mm, are calculated analogously to cladding, i.e., the base price is

$$\text{Base price} = [3.14(D')(H) + S_{\text{heads}}]182 \text{ dollars}$$

with the assumptions that the lining is 3 mm thick and of 304 stainless. This base price is modified with

1. Factor f_e'' obtained from Fig. A1.11*b*, for thicknesses other than 3 mm.
2. Factor f_m'' obtained from Table A1.11, for materials other than 304 stainless.

Thus, the corrected price for linings is

$$\text{Price} = (\text{base price, \$})(f_e'')(f_m'') \text{ dollars}$$

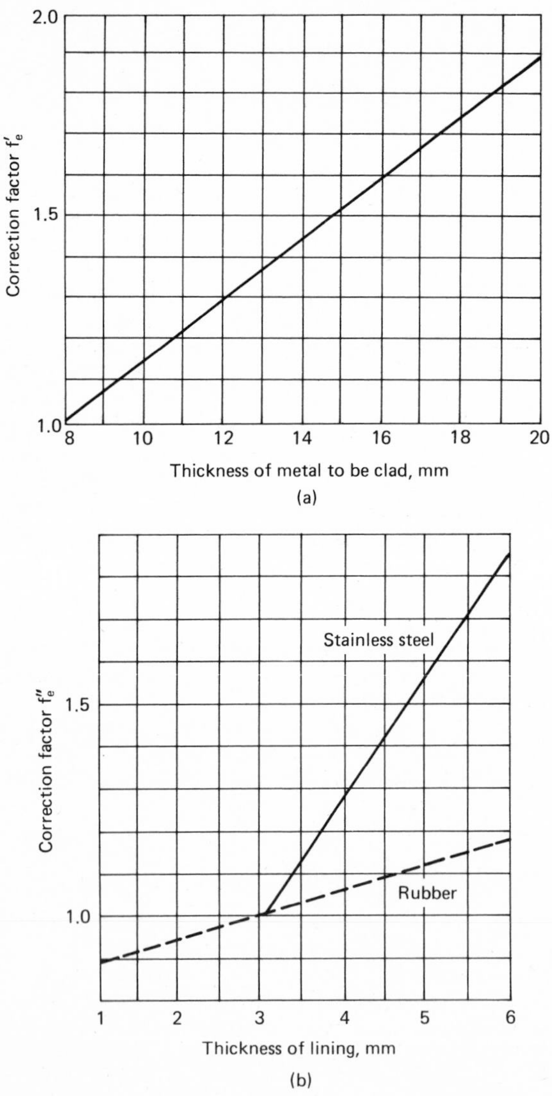

Fig. A1.11 Base-price correction factors for (*a*) cladding and (*b*) lining thickness.

TABLE A1.10 Base-Price Correction Factors for Material f'_m and Thickness f'_c of Cladding

Stainless Steel Designation	f'_m	Cladding Thickness, % of Wall and f'_c	
		10%	20%
304	1.00	1.00	1.15
304 L	1.05		
310 S	1.25	1.00	1.25
316	1.20	1.00	1.15
347	1.05	1.00	1.15
410	0.95	1.00	1.25
Monel	2.40	1.00	1.25
Inconel	2.70	1.00	1.25

TABLE A1.11 Base-Price Correction Factors for Lining Material

Stainless Steel Designation	f''_m
304	1.00
304 L	1.10
310 S	1.35
316	1.25
347	1.10
410	0.95
Monel	2.50
Inconel	2.80
Natural-rubber lined	0.30
Synthetic-rubber lined	0.40

In the event other types of corrosion protection are required, the following costs can be used

Protection	Cost, $/m² mid-1975
Aluminizing	25.00
Bricks	75.00
Gunnite	62.50

Pricing the skirt The skirt is generally of solid mild steel, so that it can be priced from Fig. A1.9, corrected for other thicknesses by a coefficient, f_e, obtained from Fig. A1.10, so that the mid-1975 price is given by

$$\text{Price} = (\text{weight})(\$/\text{kg})(f_e) \text{ dollars}$$

Pricing the accessories The accessories are priced from Fig. A1.12, according to the weight of the empty vessel. In the event special materials of construc-

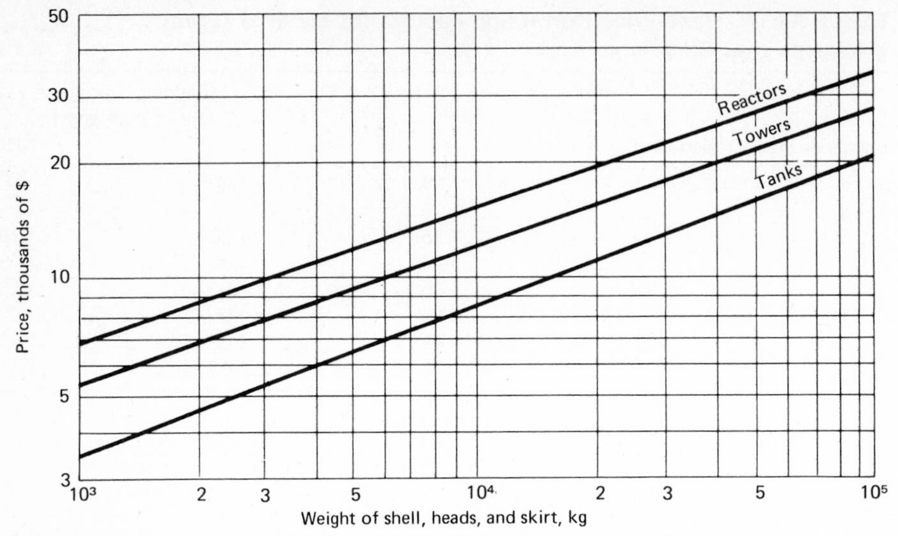

Fig. A1.12 Base price of accessories for reactors, towers, and tanks, mid-1975.

TABLE A1.12 Base-Price Correction Factors for Construction Materials of Accessories to Towers and Tanks

Material Designation		Correction Factor
ASTM	AISI	f_{am}
SA 285 C		1.0
SA 203 A et D		1.2
SA 357		1.8
SA 240	304	3.0
SA 240	310 S	4.1
SA 240	316	3.4
SA 240	321	3.0
SA 240	347	3.3
Monel		9.0
Inconel		11.0

tion are necessary, the base price from Fig. A1.12 is modified by multiplying with correction factor, f_{am}, obtained from Table A1.12.

A1.4.2.4 Calculating the Prices of Vessel Internals

Pricing distillation trays The four principal kinds of distillation trays are

1. Bubble trays, which are common throughout industry in the form of one or another variant

2. Valve trays, which can be weighted according to the vapor flow
3. Sieve trays, which exist in variations and offer low pressure drop, but which are to be avoided with dirty or corrosive fluids
4. Jet trays, which direct the vapor flow usually to assist hydraulic flow

The base price for trays, which is obtained from Fig. A1.13, assumes valve trays in up to 15 trays with single-cross-flow design, and the following tray thicknesses according to the material:

Material	Thickness, mm
Mild steel	3.5
410 Stainless	2
304 Stainless	2
Monel	2

This base price is modified by multiplying with

1. Factor f_{pl} obtained from Table A1.13a, for tray types other than valve.
2. Factor f_e obtained from Table A1.13b, for thicknesses other than those used in Fig. A1.13.
3. Factor f_{pn} obtained from Table A1.13c, for numbers of trays more than 15.
4. Factor f_{pa} for other types of cross flow, as follows:
   a. Double cross flow, $f_{pa} = 1.12$
   b. Quadruple cross flow, $f_{pa} = 1.35$

Also, the effect of labor is sometimes such that the price at large diameters is lower, while the price at low diameters is higher, than indicated in Fig. A1.13.

Pricing packings Table A1.14 gives a first approach for prices of various packings in lots of 20 m^3. This price should be reduced 10% for lots over 50 m^3 and 15% for lots over 100 m^3. It is based on mid-1975 conditions.

A1.4.2.5 Calculating the Final Price

The price of the vessel walls and skirt is added to the price of its accessories and internals, and the sum multiplied by 1.15 for towers and reactors or by 1.10 for tanks.

When the diameter of the shell of a tower or tank is larger than 5 to 6 meters, the shell is often fabricated in the field. In such cases, the cost of field fabrication can be taken as about twice the cost of the bare material.

TABLE A1.13 Base-Price Correction Factors for Distillation Trays

a. Type of Tray

Type	Correction Factor
With bubble caps	1.45
With valves	1.00
Perforated trays	0.70

b. Thickness

Thickness, mm	Correction Factor
2	1.0
3.5	1.25
6	1.60
12	2.50

c. Number of Trays

Number	Correction Factor
0–15	1.0
16–50	0.95
51–75	0.92
Over 76	0.90

TABLE A1.14 Price of packings, $/m³ (mid-1975)

Type		Packing Dimensions		
	in	1	1½	2
	mm	25	38	50
Raschig rings:				
Stoneware		210	160	150
Mild steel		420	290	230
Stainless 304		1,560	1,150	1,040
Pall rings:				
Mild steel		330	240	220
Stainless 304		1,350	960	830
Stainless 316		—	—	1,130
Monel		—	—	2,080
Polypropylene		420	310	250
Intalox saddles:				
Stoneware		270	200	180
Polypropylene		380	—	210

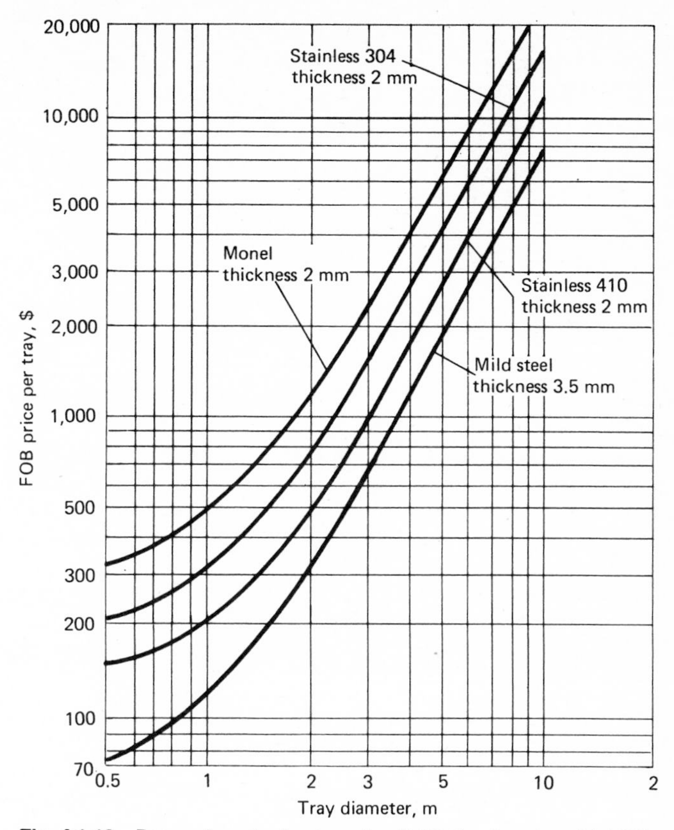

Fig. A1.13 Base price of valve trays for distillation towers, mid-1975.

Appendix 2

Process Design Estimation: Reactors

The main kinds of industrial reactors can be thought of as involving either homogeneous fluids, or multiple phases, or heterogeneous catalysis, or circulating contact.

Reactors involving homogeneous fluids can be tubular, including furnaces, or agitated reactors with a jacketed shell, coil, or external heat exchanger. Reactors involving multiple phases can be extraction or absorption towers or agitated reactors similar to those handling homogeneous fluids. The reactors involving heterogeneous catalysts can be fixed bed (including multitube reactors resembling heat exchangers of supported beds), moving bed with circulation and continuous regeneration, or fluidized bed. The reactors with circulating contact can be agitated reactors with a catalyst in suspension or tubular reactors circulating a mixed phase over a bed of catalyst.

A2.1 SIZING REACTORS

From the point of view of construction, all of the foregoing reactors can be classed as one of three types:

1. Simple vessels with or without trays
2. Multitube vessels
3. Agitated vessels with heat-exchange jacket or coil

A2.1.1 REACTORS CONSISTING OF VESSELS WITH OR WITHOUT TRAYS

There are three possibilities: passing a mixture through the vessel for a predetermined residence time, circulating immiscible liquids or a liquid and a gas either cocurrently or countercurrently over the trays in a tower, and circulating fluids over beds of catalyst.

In the case of simply obtaining the required residence time, the reaction volume is obtained by the expression:

$$v = \frac{Q}{\rho} t$$

where Q = flow, kg/h
ρ = density, kg/m^3
t = residence time, h

Without advice to the contrary, the volume of the actual reactor should be 1.2–1.3 times this reaction volume; and the length-to-diameter ratio of the vessel should be such as to give the minimum internal wall surface, or something over 4 or 5.

When two different fluids are circulated co- or countercurrently over the trays of a tower, each tray becomes a minireactor; and either of the fluids, as well as a homogeneous catalyst, can be added or withdrawn from any tray. It is possible to establish for each tray the relations between residence time, reaction constants, conversions, and mole ratio of the fluids. Generally, there will be either a number of equal-volume stages with a decreasing conversion per stage, or equal-conversion stages with decreasing volume per stage. In the first case, the economic optimum will be achieved when a few added percentage points of conversion will require an inordinant number of stages; and in the latter, the economic optimum will be achieved when the volume of successive stages becomes too small to be practical for construction.

In sizing and estimating such reactors, the actual volume per stage should be 1.2–1.3 times the calculated reaction volume; and the diameter should be that required for the desired liquid level, allowing for heat-exchange apparatus.

When fluids are circulated over beds of catalyst or packing supported on trays, the reaction volume will be predicated on the space velocity, volume of feed per unit time, per volume of catalyst. With the space velocity fixed by the experimental data, a critical feature of tray sizing is often the permissible pressure drop. Generally, the pressure drop can be calculated by the following equation:

$$\Delta P = \frac{f'}{1.271 \times 10^{12}} \left(\frac{1 - \epsilon}{\phi_s}\right)^{(1 + d)} \frac{G^2 L}{\epsilon^3 \rho D_p}$$

where ΔP = pressure drop, bar
G = mass velocity in the free area under the bed, kh/(h)(m^2)

L = height of the solids bed, m
D_p = average diameter of a grain of solid, m
ρ = density of the fluid, kg/m^3
f' = coefficient of friction determined from Fig. A2.1, when the fluid viscosity, kg/h/m, is 3.6 μ, cP
d = exponent obtained from Fig. A2.1 as a function of R_e'
ϵ = voids fraction obtained from Fig. A2.2
D_t = internal diameter of the shell, m
ϕ_s = form factor for particles with a simple geometric form, according to the formula

$$\phi_s = 4.84 \frac{V_p^{0.67}}{A_p}$$

V_p = volume of the particle, m^3
A_p = surface of a particle, m^2

Particles with complicated shapes exhibit form factors as follows:

Particle	Form-factor
Berl saddles	0.3
Raschig rings	0.3
Cork	0.69
Sand (average)	0.75
Powdered carbon	0.73

Given the allowable pressure drop and the space velocity, it is possible to calculate the overall height of a catalyst bed and hence the diameter of the reactor. The overall volume of the reactor is usually determined by the number of beds, which is often a function of the heat-transfer requirements. For example, an exothermic reaction may be carried out through beds whose heights are varied according to the completion of the reaction and the heat to be removed, i.e., as a function of the allowable temperature difference.

The actual volume of the reactor will be 1.2–1.3 times the required volume of catalyst plus the volume of any necessary heat-transfer apparatus.

A2.1.2 MULTITUBE REACTORS

These reactors are similar to shell-and-tube exchangers, including both those handling homogeneous liquids and those with fixed catalytic beds.

A2.1.2.1 Multitube Reactors Handling Homogeneous Fluids

A typical example of this group might be that of a continuous tube arranged in successive passes inside a shell so as to achieve an isothermic reaction with

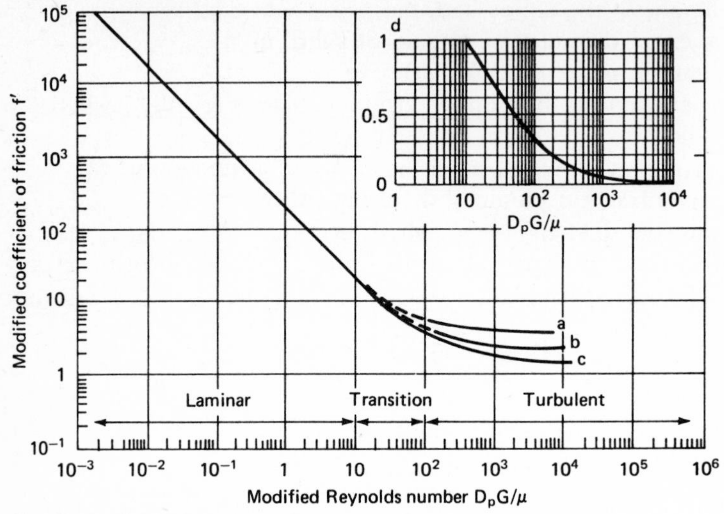

Fig. A2.1 Calculation for coefficient of friction.

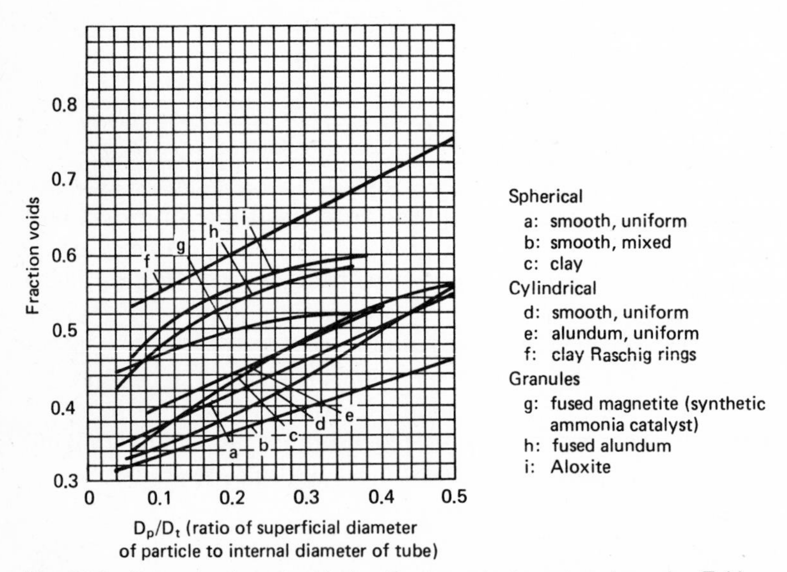

Fig. A2.2 Determination of voids fraction in packed reactors (see also Tables A1.3 to A1.5).

a graduated concentration. The reactants might be injected at the entrance of the tube or at several stages along it. The reaction volume is obtained as a function of the residence time or the kinetics leading to the desired conversion level. A safety factor of 20–30% is assumed. The length and diameter of such tubes is based on the three following factors:

1. A Reynolds number compatible with the reaction conditions in the turbulent region.

$$\text{Re} = \rho \frac{Ud}{\mu}$$

where d = the tube diameter, m
U = the linear velocity of the fluid, m/s
μ = the viscosity, poisseuilles (1.0 poisseuille = 10 P = 1,000 cP)
ρ = the fluid density, kg/m^3

Since the linear velocity U is related to the flow Q, kg/h, by the expression

$$\rho U = \frac{Q}{3{,}600} \frac{4}{\pi d^2} = \frac{Q}{900\pi d^2}$$

the Reynolds number can be expressed as

$$\text{Re} = \frac{Q}{900\pi d \mu}$$

2. The assumed diameter, which would normally be that of a standard heat-exchanger tube.
3. The heat-transfer surface required by the reaction.

It is possible either to treat the reactor as a heat exchanger and select a standard exchanger with the required surface or to calculate the tubes, shell, and then the weight of the entire reactor. In the latter case, the following exchanger standards should be used for the reactor design:

- The pitch should be square or triangular.
- The tube length should be 8, 12, 16, or 20 ft.
- The diameter of the shell should conform to standard exchanger shell diameters.
- The tube count should be standard for the shell diameter.
- The tube weights should be standard.
- The number of baffles should be 4 or 5.

A2.1.2.2 Reactors with Catalyst-Filled Tubes

The volume of catalyst and tube length depend on the space velocity and allowable pressure drop, as described in Sec. A2.1.1. The pressure drop is related to the heat-transfer surface, for any given space velocity and its corresponding volume, with small-diameter tubes producing

More surface than longer tubes in direct proportion to the diameters squared

More pressure drop than larger tubes in direct proportion to the diameters to the fifth power

Consequently, this type of reactor is sized in a trial-and-error calculation in which (1) a tube diameter is assumed, (2) the tube length is calculated for the allowable pressure drop, and (3) the number of sized tubes required for the given space velocity is checked for the heat-transfer surface afforded.

A2.1.3 AGITATED REACTORS

Extraction towers that involve chemical reactions, but which are essentially extraction towers, will not be included in this section. Rather, agitated reactors will be treated as those for Grignard-type reactions and those for staged or stepwise reactions.

A2.1.3.1 Grignard-Type Reactors

These consist of jacketed vessels with or without a supplementary coil for heat transfer. Pricing them requires (1) determining the volume, (2) determining the required heat-exchange surface, and (3) determining the characteristics of the agitator.

A2.1.3.1a Sizing Grignard-Type Reactors

The volume is the required reaction volume, based on holdup time, plus an added 10% for safety. The length-to-diameter ratio usually varies between 0.8 and 1.5.

A2.1.3.1b Calculating a Heat-Transfer Surface

The heat-transfer surface is calculated as for exchangers in App. 3, with the required jacket surface equal to the duty divided by the overall transfer coefficient times the log-mean temperature difference. Approximate overall transfer coefficients can be obtained from Table A2.1 The surface required for a coil is also calculated according to the discussion in App. 3.

TABLE A2.1 Overall Heat-Transfer Coefficients for Agitated Reactors kcal/(h)(m²)(°C)

		Reactants		
Type of Exchanger	Heat-transfer Fluid	Viscous Liquid	Liquid	Boiling Liquid
Jacket	Condensation	150–250	400–1,200	600–1,500
	Vaporization liquid	50–120	150–300	300–500
Coil	Condensation	200–300	600–2,000	1,000–3,000
	Vaporization liquid	50–150	400–1,000	600–1,200

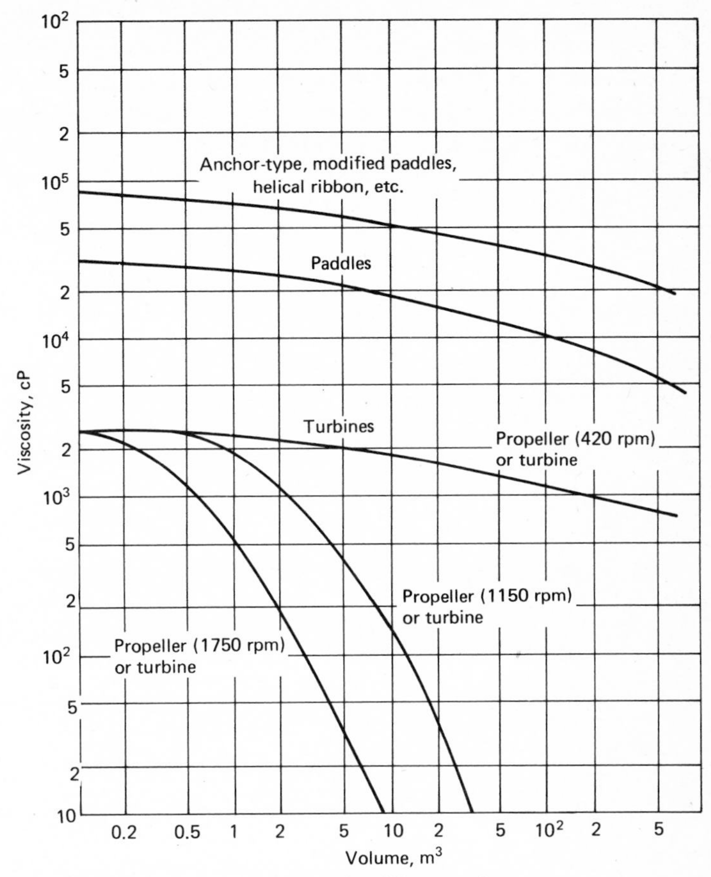

Fig. A2.3*a* Range of application for different agitators.

A2.1.3.1c Selecting and Sizing the Agitator

The type of agitator should be selected from Fig. A2.3*a*. Low-speed agitators, with the ends of the arm or blade moving at about 2 m/s, include paddles, modified paddles, anchor type, helical ribbon, etc., and are frequently used in reactors with diameters of 0.5–1.0 m. High-speed agitators, with blade tips moving at about 3 m/s, include propellers and turbines and are used in liquids with viscosities less than 20–30 P. The power consumed by an agitator can be obtained from Fig. A2.3*b*.

A2.1.3.2 Agitated Reactors Handling Stepwise Reactions

These consist of a single shell containing five to six compartments with mechanical agitation, each of which is sized according to Sec. A2.1.1. If heat

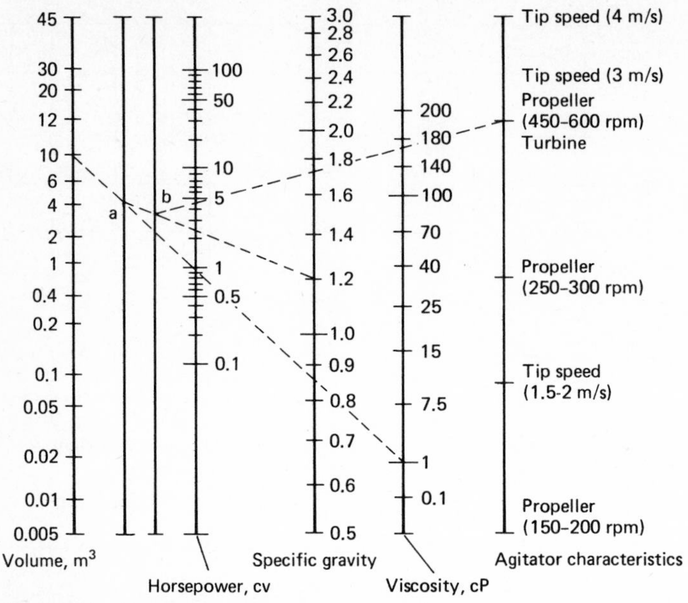

Fig. A2.3*b* Horsepower consumption for agitators.

transfer is required, a jacket is avoided by using either a coil or by circulating liquid from each compartment through an external heat exchanger.

A2.2 PRICING REACTORS

In most cases, reactors are essentially towers or tanks, and the price is obtained as for towers and tanks in App. 1. For jacketed reactors, use Fig. A2.4, which gives total cost, including mixer and motor. For the other reactors, calculate the pressure vessel, then use Fig. A2.5 for agitators, which are based on assumed 304 stainless and a viscosity of 1,000 cP. Horsepowers for various sizes of reactor are given in Table A2.2. Finally, the agitator price is corrected with Table A2.3.

TABLE A2.2 Typical Horsepower Ratings for Agitator Motors

Tank Volume, m³	Motor Horsepower, cv
0.150	4–8
0.300	10–15
1.0	3–10
4.0	5–15
6.0	7–20
10.0	10–30

TABLE A2.3 Price-Correction Factors for Materials of Agitated Reactors

Material	f_m
Stainless 304	1.00
Stainless 316	1.05
Rubber-lined	1.25
Monel	1.30

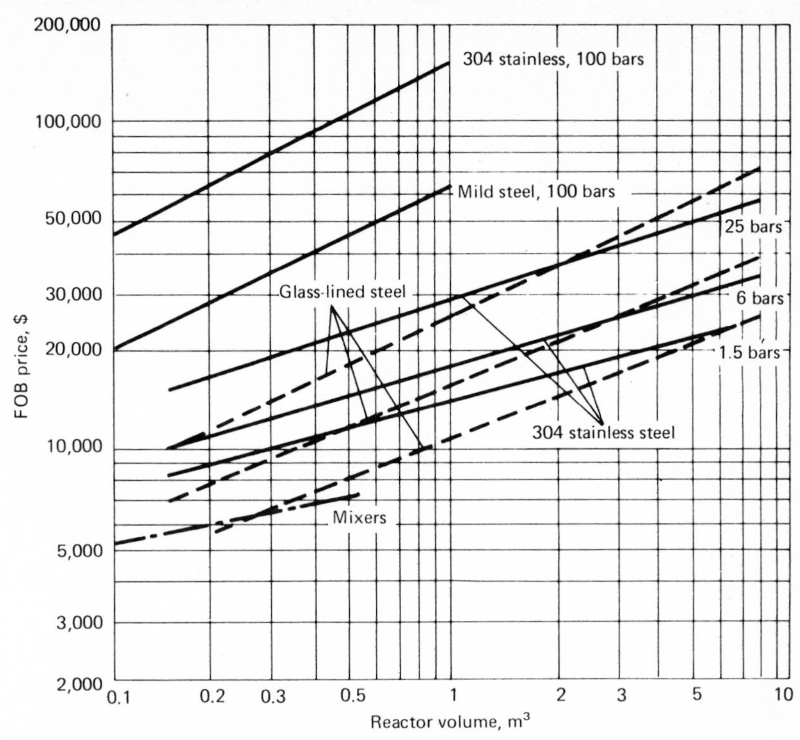

Fig. A2.4 Price of agitated jacketed reactors, including mounting seal, explosion-proof motor, speed reducer, agitator, as well as simple mixers including motor and mount, mid-1975.

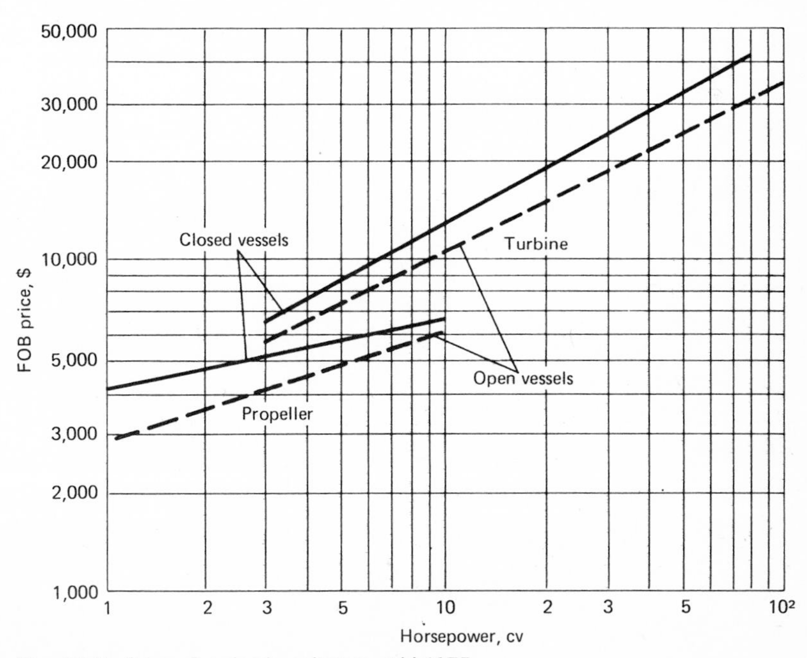

Fig. A2.5 Prices for simple agitators, mid-1975.

Appendix 3

Process Design Estimation: Heat Exchangers

Two types are included: shell-and-tube exchangers and air coolers.

A3.1 SHELL-AND-TUBE HEAT EXCHANGERS

The following sections present procedures for selecting, sizing, and pricing shell-and-tube exchangers.

A3.1.1 SIZING SHELL-AND-TUBE EXCHANGERS

Sizing exchangers here includes calculation of the required surface, and of the consumption of coolant or heating fluid, then the selection of exchanger type. Calculating the transfer surface requires prior calculation of the heat-exchange duty, the effective temperature difference, and of the overall transfer coefficient; and these all impinge on the coolants or heating fluids employed. Choice of exchanger types involves consideration of certain peculiarities of heat exchanger construction.

A3.1.1.1 Quick Calculation for Transfer Surface

Two categories are generally distinguished for process evaluation:

1. Condensers and exchangers in which the hot-side and cold-side temperatures can cross

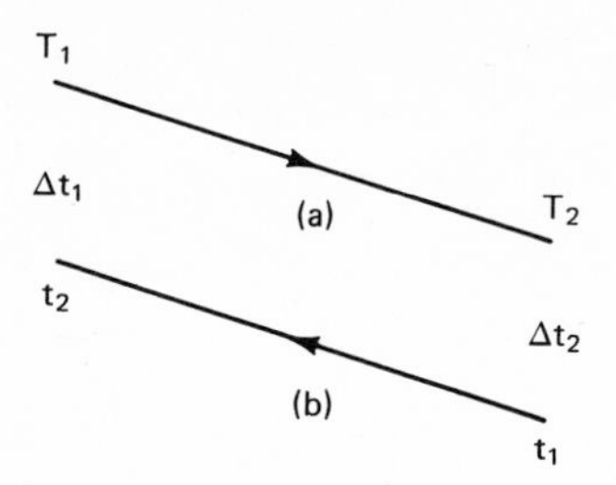

2. Reboilers in which hot-side and cold-side temperatures are usually constant

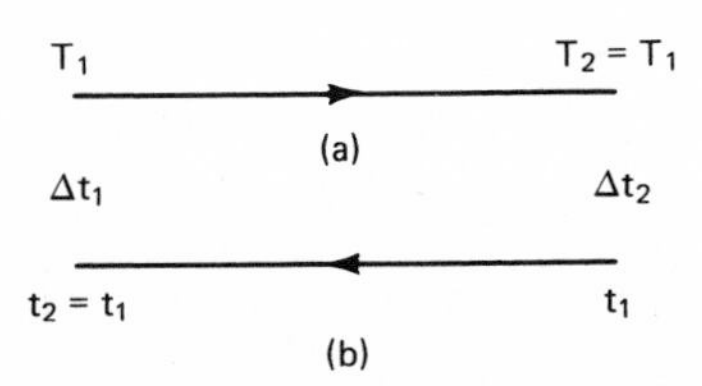

The duty is calculated differently, depending on which category an exchanger falls into. The basic equation for heat-exchange duty is

$$S = \frac{Q}{U\Delta t}$$

where S = surface, m^2
Q = duty, kcal/h
U = overall transfer coefficient, kcal/(h)(m^2)(°C)
Δt = log-mean temperature difference, °C

When developing the parts of this equation, it is first necessary to decide whether the hot side is to be the tubes or the shell. Generally, it is more economical to put that fluid with the greatest flow (e.g., cooling water) through the tubes. Also, the tube-side fluid is usually more corrosive, dirtier, at higher pressure, at higher temperature, less viscuous, and better able to take a low-pressure drop.

Accordingly, in the first of the above two categories, with respect to condensers, the condensing fluid will generally be on the shell side, while the cooling water or refrigerant will be on the tube side. The four temperatures are known. (The inlet and outlet temperatures of the water depend on the ambient temperature and the maximum allowable temperature with the water quality—about 10°C rise, for 25 to 35°C typically.) The duty is calculated on the process fluid for T_1 and T_2 as

$$Q = D \cdot \Delta H$$

where Q = duty, kcal/h
D = flow of process fluid, kg/h
ΔH = difference in enthalpy of the process fluid between inlet and outlet, kcal/kg

When the crossing temperatures involve direct heat exchange, the inlet temperatures are generally the only ones known; and it is necessary to assume one of the outlet temperatures and calculate the other by means of the equations

$$Q_a = Q_b = Q$$

From which

$$Q_a = D_a \, \Delta H_a \qquad \text{and} \qquad Q_b = D_b \, \Delta H_b$$

And for purposes of approximation,

$$Q_a = D_a C_{pa} \, \Delta t_a \qquad \text{and} \qquad Q_b = D_b C_{pb} \, \Delta t_b$$

assuming an average value of C_p over the temperature range.

In the reboiler category, the process fluid is usually in the shell side, while steam of other heating fluid is in the tubes. Thus, of the four temperatures in the diagram, t_1 is known and assumed equal to t_2. T_1 is selected so that $T_1 - t_1 = \Delta t_1$ is more than about 20–25°C. Barring exceptions, the duty is the reboiling duty, which can be assumed equal to 1.1 times the overhead condensing duty.

Calculating the effective temperature difference: First calculate the log-mean temperature difference by means of Fig. A3.1. Then determine

$$E = \frac{t_2 - t_1}{T_1 - t_1} \qquad \text{and} \qquad R = \frac{T_1 - T_2}{t_2 - t_1}$$

Then, by means of Fig. A3.2, determine the correction factor f for the log-mean temperature difference (LMTD). f should be larger than about 0.8, and the number of shells should be increased to obtain such a factor. The required number of shells will have an effect on the price of heat-exchange surface for the given duty. As a first approximation, a factor of 0.8 can be assumed, along with one shell. Note that when $T_2 = t_2$, $f = 0.8$.

Calculating the overall transfer coefficient The precise calculation of transfer coefficients lies beyond the scope of project evaluation. Also, many overall transfer coefficients have been published, so that it is possible to use a table like Table A3.1 or a chart like Fig. A3.3. In reviewing such data, it is important to remember that 1 kcal/(h)(m^2)(°C) is equal to 4.882 Btu/(h)(ft^2)(°F). The fouling factor, though appreciable, can be assumed included in the published

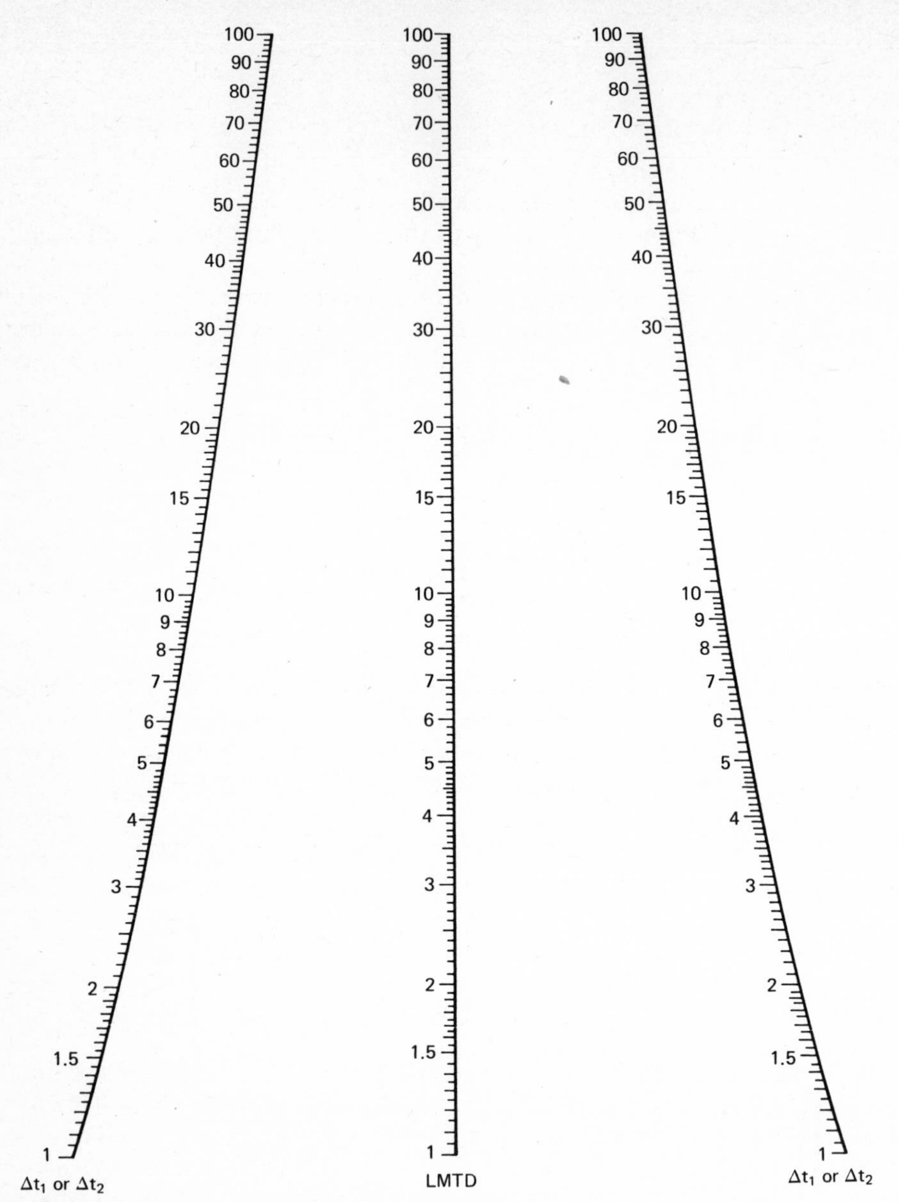

Fig. A3.1 Calculation for LMTD.

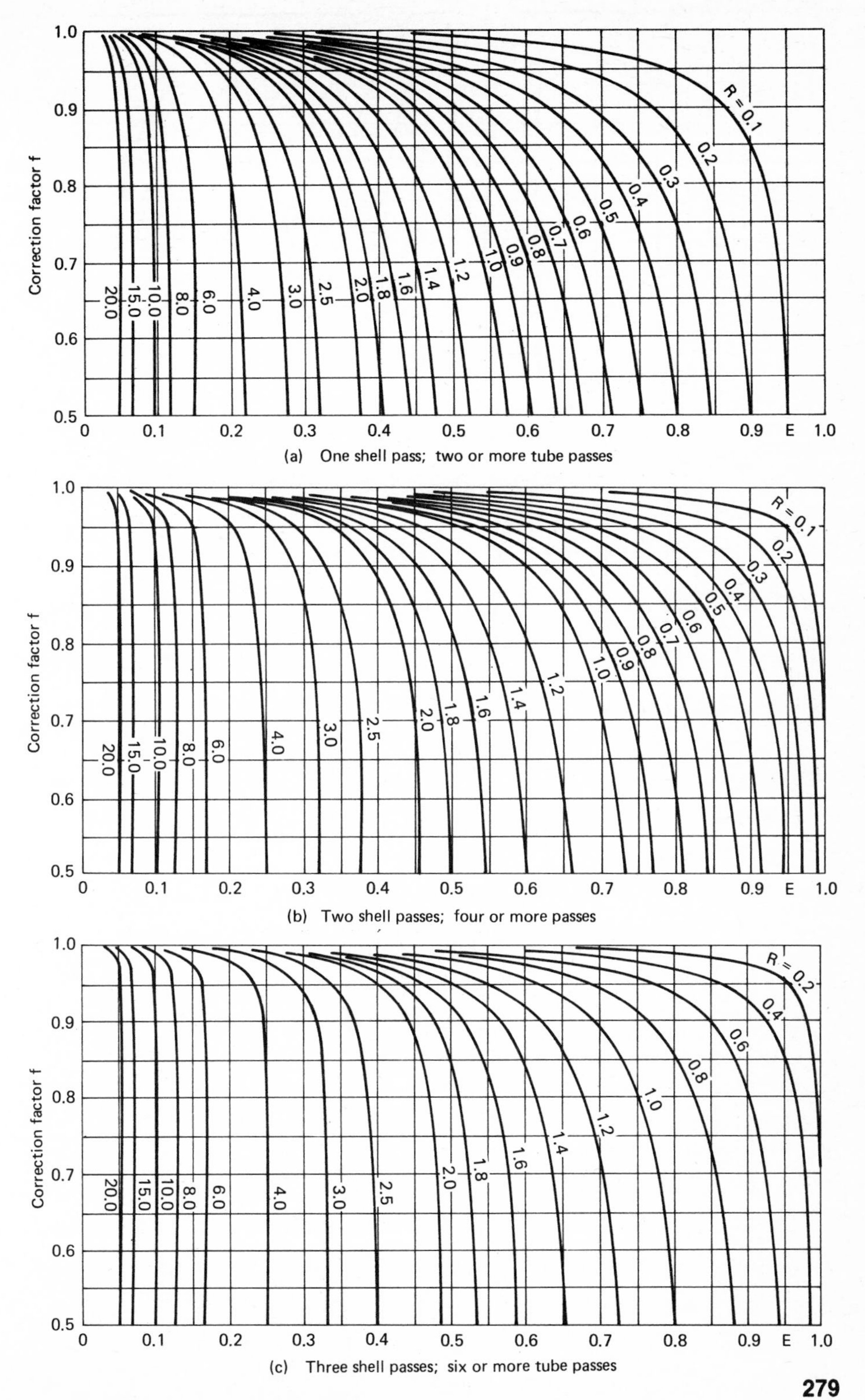

Fig. A3.2 LMTD correction factors.

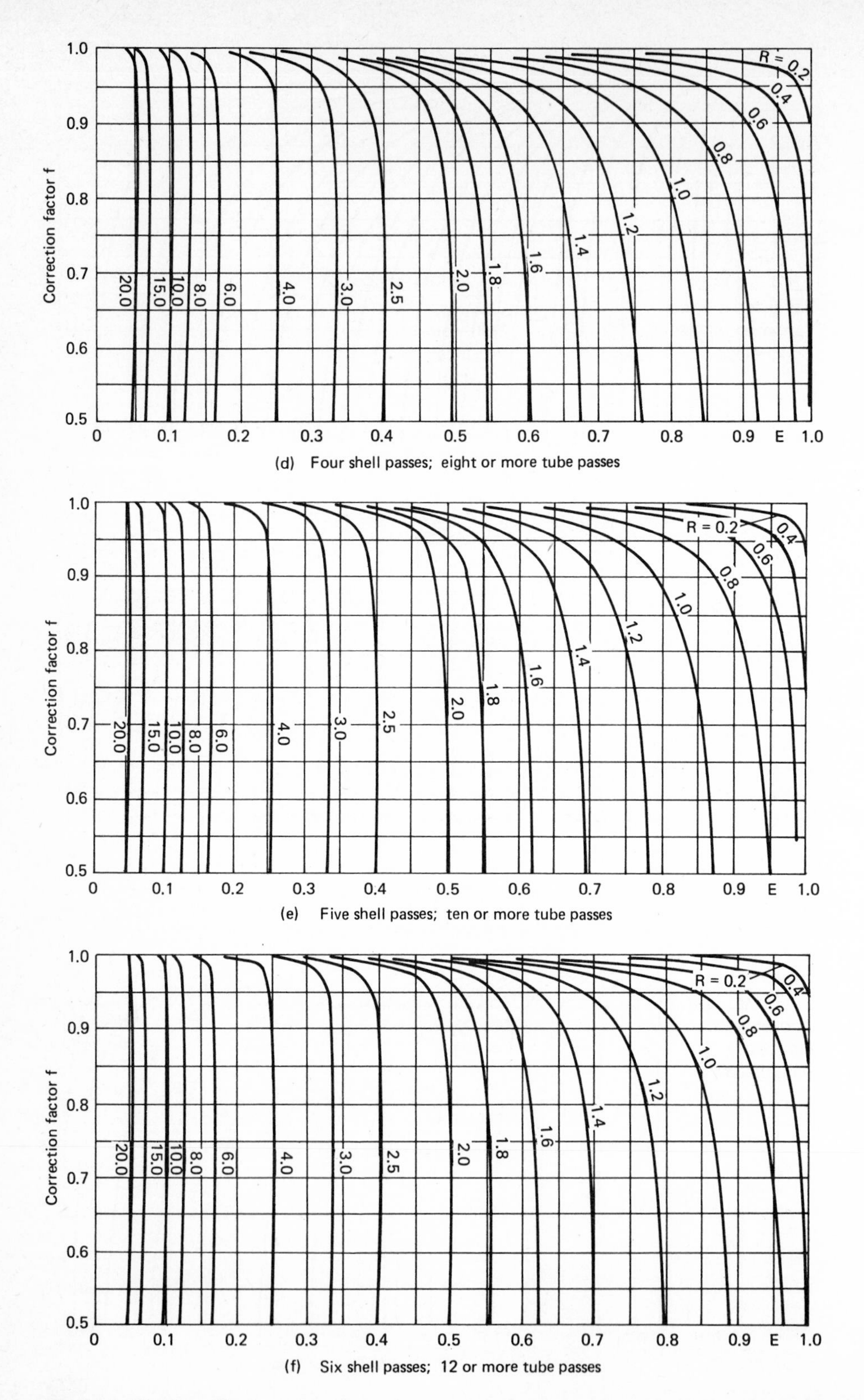

(d) Four shell passes; eight or more tube passes

(e) Five shell passes; ten or more tube passes

(f) Six shell passes; 12 or more tube passes

Fig. A3.2 ***(Continued)***

TABLE A3.1 Typical Overall Heat-Transfer Coefficients U[kcal/(h)(m^2)(°C)], for Industrial Applications

Tube Side	Shell Side	U
General Exchange Services		
Butadiene	Steam	60
C_4 olefins	Propylene (vaporizing)	65–90
Vaporized ethylene	Steam and condensate	450–600
Vaporized ethylene	Cold water	250–400
Liquid ethylene	Vaporized ethylene	50–100
Propane vapor	Liquid propane	30–75
Light olefinic hydrocarbon vapors containing CO, CO_2, and H_2	Steam	50–100
Light chlorinated hydrocarbons	Steam	60–150
Ethanolamine	Steam	75–125
Solvent	Propylene (vaporizing)	150–200
Solvent	Solvent	170–200
Solvent	Cold water	170–350
Oil	Oil	300–400
Steam condensate	Propylene	300–600
Calcium chloride 25%	Chlorinated methane	200–300
Steam	Air mixture	50–100
Steam	Styrene and tar	250–300
Cooling water	Freon 12	500–600
Cooling water	Recycle oil	200–350
Water	Treated water (35–45°C)	500–600
Water	Treated water (100–35°C)	800–1,100
Water	Light chlorinated C_2 hydrocarbons	30–50
Water	Heavy chlorinated C_2 hydrocarbons	150–220
Water	Perchloroethylene	170–270
Water	Air plus chlorine	40–90
Water	HCl	35–75
Water	Air and steam	100–170
Water	Absorption oil	400–560
Condensers		
Butadiene	Propylene (refrigerant)	330–390
C_4 olefins	Propylene	290–330
Ethylene	Propylene	300–450
Light olefinic hydrocarbons	Propylene	250–300
HCl	Propylene	300–550
Chloroethanes and light ends	Propylene	75–125
Chlorinated olefinic hydrocarbons	Water	450–600
Solvent and noncondensables	Water	75–125
Water	Propylene vapor	650–750
Chlorinated hydrocarbons	Water	100–150

TABLE A3.1 Typical Overall Heat-Transfer Coefficients U[kcal/(h)(m²)(°C)], for Industrial Applications *(Continued)*

Tube Side	Shell Side	U
Condensers *(Continued)*		
Water	Propylene	300–500
Water	Steam	600–1,100
Treated water	Steam	100–150
Oil	Steam	350–550
Cold water	Air + Cl_2 (partial condensation)	40–75
Water	Propylene refrigerant	125–750
Water	Light hydrocarbon refrigerant	200–450
Water	Ammonia	700–800
Air and steam	Freon	50–250
Reboilers		
Olefinic C_4s	Steam	450–550
Chlorinated hydrocarbons	Steam	170–120
Chlorinated olefinic hydrocarbons	Steam	500–700
Dichloroethane	Steam	350–450
Heavy solvent	Steam	350–550
Mono- and diethanolamine	Steam	750–1,000
Water and organic acids	Steam	300–500
Amine and water	Steam	600–700
Steam	Naphtha	75–100
Propylene	Ethane and ethylene	600–700
Propylene and butadiene	Butadiene and olefins	75–90

data, unless the data are precise. For recommended fouling factors, publications like the Standards of the Tubular Exchanger Manufacturers Association (TEMA) should be consulted.

A3.1.1.2 Utilities Consumption

Generally, the consumption of cooling water or refrigerant is given by the expression

$$m = \frac{Q}{C_p(t_2 - t_1)} = \frac{Q}{\Delta H}$$

and in particular when the cooling-water temperature rise is 10°C, its consumption is

$$m = \frac{Q}{10}$$

where m = consumption, kg/h
Q = duty, kcal/h
ΔH = change in enthalpy, kcal/kg

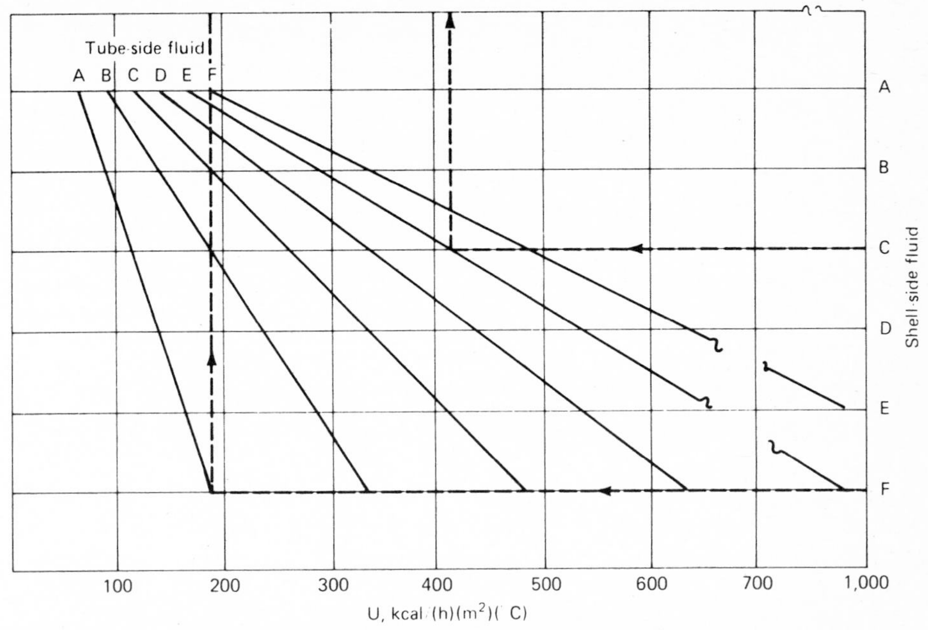

Fig. A3.3 Overall heat-transfer coefficient, *U*. *A* heavy hydrocarbon residue, no change of phase; *B* middle-distillate hydrocarbons, no change of phase; *C* gasoline and light hydrocarbons, no change of phase; *D* light hydrocarbons with change of phase; *E* cooling water; *F* condensing or vaporizing steam.

C_p = specific heat, kcal/(kg)(°C)
t = temperature, °C

For reboilers, the consumption of condensing heating fluid is given by

$$m = \frac{Q}{\Delta H}$$

When the heating fluid is condensing steam, ΔH can be obtained from Table A3.2.

A3.1.1.3 Selecting the Construction for Tubular Exchangers

First determine the preferred construction type for the service by studying Table A3.3 on page 286. The choice will fall between double-pipe exchangers, for which the maximum surface is about 10 m², and exchangers with tube bundles, for which the minimum surface is on the order of 3–5 m². The latter will generally fall into one of three categories:

1. Condensers: simple shell-pass arrangements with floating head and removable tube bundle.

TABLE A3.2 Latent Heat of Vaporization of Water ΔH, kcal/kg

Pressure bar abs.	Temperature, °C	ΔH	Pressure, bar abs.	Temperature, °C	ΔH	Pressure, bar abs.	Temperature, °C	ΔH
0.1	45.58	574.75	5.1	151.73	499.54	10.25	179.96	479.03
0.2	59.76	564.84	5.2	152.47	499.01	10.50	181.01	478.27
0.3	68.74	558.53	5.3	153.19	498.49	10.75	182.04	477.51
0.4	75.47	553.81	5.4	153.90	497.98	11.00	183.05	476.77
0.5	80.90	549.99	5.5	154.59	497.47			
0.6	85.48	546.75	5.6	155.28	497.07	11.25	184.05	476.04
0.7	89.47	543.94	5.7	155.96	496.48	11.50	185.03	475.32
0.8	93.00	541.44	5.8	156.63	496.00	11.75	185.99	474.62
0.9	96.19	539.20	5.9	157.29	495.52	12.00	186.99	473.92
1.0	99.09	537.15	6.0	157.94	495.05			
						12.25	187.87	473.24
1.1	101.76	535.26	6.1	158.59	494.58	12.50	188.78	472.57
1.2	104.24	533.50	6.2	159.22	494.12	12.75	189.69	471.90
1.3	106.55	531.86	6.3	159.85	493.67	13.00	190.57	471.25
1.4	108.72	530.33	6.4	160.47	493.22			
1.5	110.76	528.87	6.5	161.08	492.78	13.25	191.45	470.61
1.6	112.70	527.49	6.6	161.68	492.34	13.50	192.31	469.97
1.7	114.54	526.18	6.7	162.28	491.91	13.75	193.16	469.34
1.8	116.29	524.84	6.8	162.87	491.48	14.00	194.00	468.73
1.9	117.97	523.74	6.9	163.45	491.06			
2.0	119.57	522.60	7.0	164.03	490.64	14.25	194.83	468.12
						14.50	195.64	467.52
2.1	121.11	521.50	7.1	164.60	490.22	14.75	196.45	466.92
2.2	122.59	520.46	7.2	165.16	489.82	15.00	197.24	466.34
2.3	124.02	519.43	7.3	165.72	489.41			
2.4	125.40	518.44	7.4	166.27	489.01	16.00	200.32	464.07
2.5	126.73	517.49	7.5	166.82	488.62	17.00	203.20	461.83

2.6	128.02	516.57	7.6	167.36	488.22	18.00	206.07	459.81
2.7	129.26	515.68	7.7	167.89	487.83	19.00	208.75	457.82
2.8	130.48	514.81	7.8	168.42	487.45			
2.9	131.65	513.97	7.9	168.94	487.07	20.00	211.34	455.89
3.0	132.80	513.15	8.0	169.46	486.69			
3.1	133.91	512.35	8.1	169.97	486.32			
3.2	135.00	511.57	8.2	170.48	485.95			
3.3	136.06	510.81	8.3	170.98	485.58			
3.4	137.09	510.07	8.4	171.48	485.22			
3.5	138.10	509.35	8.5	171.98	484.86			
3.6	139.09	508.67	8.6	172.47	484.50			
3.7	140.05	507.95	8.7	172.95	484.15			
3.8	141.00	507.27	8.8	173.43	483.80			
3.9	141.92	506.61	8.9	173.91	483.45			
4.0	142.82	505.96	9.0	174.38	483.11			
4.1	143.71	505.32	9.1	174.85	482.77			
4.2	144.58	504.70	9.2	175.31	482.43			
4.3	145.43	504.08	9.3	175.77	482.09			
4.4	146.27	503.48	9.4	176.23	481.76			
4.5	147.09	502.89	9.5	176.68	481.43			
4.6	147.90	502.31	9.6	177.13	481.10			
4.7	148.69	501.73	9.7	177.57	480.78			
4.8	149.47	501.17	9.8	178.01	480.45			
4.9	150.24	500.62	9.9	178.45	480.14			
5.0	150.99	500.07	10.0	178.89	479.82			

TABLE A3.3 Selection Criteria for Heat-Exchanger Types

Type	Characteristics	Where Used	Limitations
Fixed tube sheet	The two tube sheets are larger than the internal diameter of the shell and cannot be pulled through. They are bolted to the shell or integral with it.	Wherever this less expensive design can be used. See limitations.	Stresses result from temperature differences in shell and tubes. Shell-side baffles, etc., cannot be replaced. Maximum Δ T is 100°C.
Floating head	The rear-end tube sheet is smaller in diameter than its cover and free to move with expansion.	Either this construction or an expansion joint in the shell is required for Δ T over 100°C.	Possibility for leakage between floating head and cover or shell.
Pull through	Tube-sheet diameter is smaller than shell for pulling through.		
Split ring	Construction used to seal tube side from shell side at floating head.		
U-tubes	One tube sheet with tubes in U shape.	On clean fluids at higher temperature differences.	U-tubes cannot be reamed out like straight tubes. The bends are susceptible to erosion and corrosion by dirty or vaporizing liquids.
Kettle	A horizontal, removable tube bundle in a shell enlarged to permit vapor disengaging above and liquid holdup behind an overflow weir at the end of the bundle.	Eliminates bottom section and skirt of a distillation tower. Is generally less expensive wherever net positive suction head (NPSH) for a bottoms pump does not require elevation.	More expensive than thermosiphon reboilers when elevation is necessary.
Thermosiphon	Vertical with boiling liquid in tubes, or horizontal with boiling liquid in shell. Suitable shell-and-tube construction.	Distillation towers, strippers.	Vertical construction makes tube pulling difficult. Requires 35–40% vaporization, or a pump must be added.
Double pipe	Concentric tubes	Limited exchange surface. Elevated pressures.	Limited surface only partly offset by finned tubes.

2. Coolers and exchangers: simple U-tube exchangers, simple pass with removable tube bundle, or simple pass with fixed tube sheet.

3. Reboilers: thermosiphon, forced circulation with submerged tube bundle, or kettle type.

The many possible variations for these types can be identified according to TEMA designations for channel end, shell, and tube-sheet arrangement, as shown in Fig. A3.4, which is adapted from the TEMA Standards. The preferred construction can be identified by the letters of Fig. A3.4, and this designation used to establish the price by means of Fig. A3.5 (type AES) and Table A3.6.

A3.1.1.4 Important Features of Heat Exchangers

Sometimes it is helpful to have the standard dimensions of exchangers, if not for the exchangers, then for similar equipment like the reactors (App. 2), and where more careful studies are required. The critical dimensions involve the tubes and the tube bundles, as shown in Tables A3.4 and A3.5, respectively.

A3.1.2 PRICING TUBULAR EXCHANGERS

Two types are considered: those with tube bundles and double-pipe types.

A3.1.2.1 Pricing Heat Exchangers with Tube Bundles

The pricing method is based on the formula

$$\text{Corrected price} = (\text{base price})\, f_d f_\phi f_l f_{np} f_p f_t f_m$$

where f_d = correction factor according to exchanger type
f_ϕ = correction factor according to tube pitch and diameter
f_l = correction factor according to tube length
f_{np} = correction factor according to number of tube passes
f_p = correction factor according to the pressures in shell and tubes
f_t = correction factor according to the temperatures
f_m = correction factor according to the materials of construction

The base price is obtained from Fig. A3.5, which relates surface, m^2, to dollars in mid-1975 for a simple pass, floating heat exchanger that has a removable tube bundle with 16-ft-long tubes on $\frac{3}{4}$-in^2 pitch and having wall thicknesses of 14 BWG; the design pressure is 10 bars, and the design temperature 350°C; materials of construction are mild steel throughout; and installation is included in this price.

In order to obtain this base price, it is only necessary to know the surface. Corrections by the above factors require knowing, in order of importance, the exchanger type, the materials of construction, and the maximum

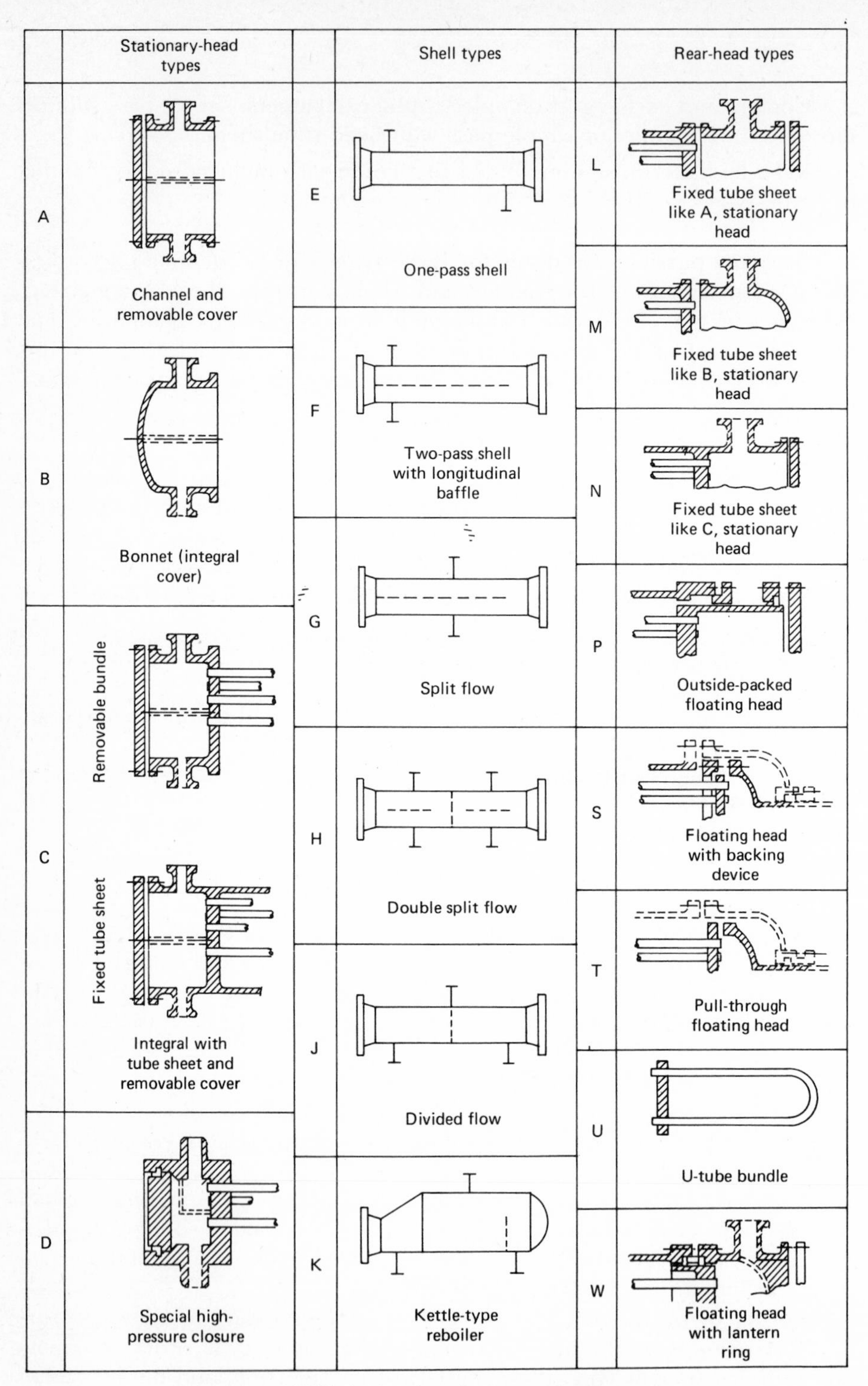

Fig. A3.4 TEMA designations for types of exchangers.

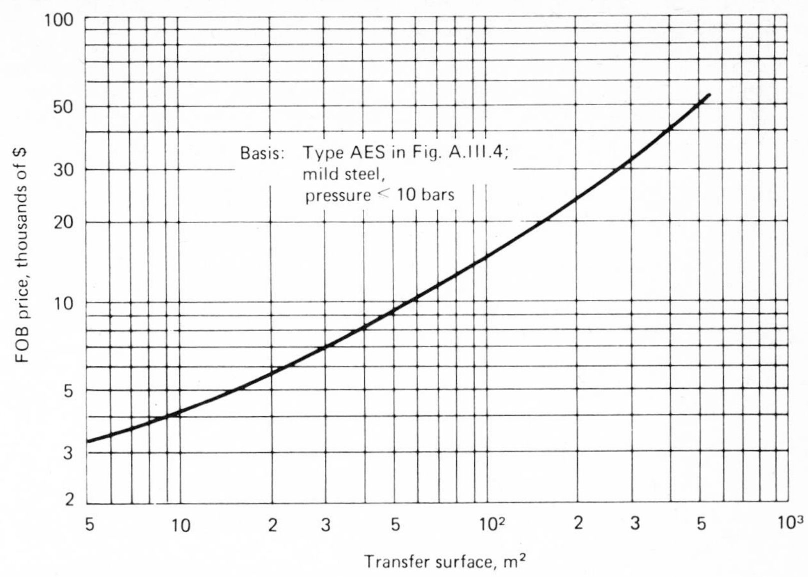

Fig. A3.5 Base price for shell-and-tube exchangers, mid-1975.

TABLE A3.4 Characteristics of Heat-Exchanger Tubes

Outside Diameter, in (mm)	BWG*	Thickness, mm	Inside Diameter, cm	Area cm^2	Surface, m^2/m		Weight, kg/m
					Outside	Inside	
½ in (12.7 mm)	14	2.10	0.848	0.565	0.0399	0.0266	0.600
	16	1.65	0.940	0.694		0.0295	0.490
	18	1.24	1.021	0.819		0.0321	0.384
¾ in (19.05 mm)	10	3.40	1.224	1.177	0.0598	0.0384	1.436
	12	2.77	1.351	1.434		0.0424	1.216
	14	2.10	1.483	1.727		0.0466	0.963
	16	1.65	1.575	1.948		0.0495	0.774
	18	1.24	1.656	2.154		0.0520	0.597
1 in (25.4 mm)	10	3.40	1.859	2.714	0.0798	0.0584	2.024
	12	2.77	1.986	3.098		0.0624	1.696
	14	2.10	2.118	3.523		0.0665	1.324
	16	1.65	2.210	3.836		0.0694	1.057
	18	1.24	2.291	4.122		0.0720	0.811
1¼ in (31.75 mm)	10	3.40	2.494	4.885	0.0997	0.0783	2.604
	12	2.77	2.616	5.375		0.0822	2.158
	14	2.10	2.743	5.909		0.0862	1.682
	16	1.65	2.845	6.357		0.0894	1.340
	18	1.24	2.291	6.701		0.0918	1.024
1½ in (38.1 mm)	10	3.40	3.124	7.665	0.1197	0.0981	3.185
	12	2.77	3.251	8.300		0.1021	2.634
	14	2.10	3.378	8.962		0.1061	2.039
	16	1.65	3.480	9.512		0.1093	1.622
	18	1.24	3.556	9.931		0.1171	1.237

*BWG: Birmingham Wire Gauge.

TABLE A3.5 Number of Tubes in a Tube Bundle

Shell Diameter		Square Pitch: Tube Diameter on Center-to-Center Pitch											
		¾ in on 1 in Tube Passes				1 in on 1¼ in Tube Passes				1¼ in on 19/16 in Tube Passes			
in	cm	2	4	6	8	2	4	6	8	2	4	6	8
8	20.3	26	20	20		16	14						
10	25.4	52	40	36		32	26	24		12	10		
12	30.5	76	68	68	60	45	40	38	36	24	22	16	16
13¼	33.7	90	82	76	70	56	52	48	44	30	30	22	22
15¼	38.7	124	116	108	108	76	68	68	64	40	37	35	31
17¼	43.9	166	158	150	142	112	96	90	82	53	51	48	44
19¼	48.9	220	204	192	188	132	128	122	116	73	71	64	56
21¼	54.0	270	246	240	234	166	158	152	148	90	86	82	78
23¼	59.1	324	308	302	292	208	192	184	184	112	106	102	96
25	63.5	394	370	358	346	252	238	226	222	135	127	123	115
27	68.6	460	432	420	408	288	278	268	260	160	151	146	140
29	73.7	526	480	468	456	326	300	294	286	188	178	174	166
31	78.8	640	600	580	560	398	380	368	358	220	209	202	193
33	83.8	718	688	676	648	460	432	420	414	252	244	238	226
35	88.9	824	780	766	748	518	488	484	472	287	275	268	258
37	94.0	914	886	866	838	574	562	544	532	322	311	304	293
39	99.0	1024	982	968	948	644	624	612	600	362	348	342	336

Shell Diameter		Triangular Pitch: Tube Diameter on Center-to-Center Pitch											
		¾ in on 15/16 in tube passes				¾ in on 1 in tube passes				1 in on 1¼ in tube passes			
in	cm	2	4	6	8	2	4	6	8	2	4	6	8
8	20.3	32	26	24	18	30	24	24		16	16	14	
10	25.4	56	47	42	36	52	40	36		32	26	24	
12	30.5	98	86	82	78	82	76	74	70	52	48	46	44
13¼	33.7	114	96	90	86	106	86	82	74	66	58	54	50
15¼	38.7	160	140	136	128	138	122	118	110	86	80	74	72
17¼	43.9	224	194	188	178	196	178	172	166	118	106	104	94
19¼	48.9	282	252	244	234	250	226	216	210	152	140	136	128
21¼	54.0	342	314	306	290	302	278	272	260	188	170	164	160
23¼	59.1	420	386	378	364	376	352	342	328	232	212	212	202
25	63.5	506	468	446	434	452	422	394	382	282	256	252	242
27	68.6	602	550	536	524	534	488	474	464	334	302	296	286
29	73.7	692	640	620	594	604	556	538	508	376	338	334	316
31	78.8	822	766	722	720	728	678	666	640	454	430	424	400
33	83.8	938	878	852	826	830	774	760	732	522	486	470	454
35	88.9	1 068	1 004	988	958	938	882	864	848	592	562	546	532
37	94.0	1 200	1 144	1 104	1 072	1 044	1 012	986	870	664	632	614	598
39	99.0	1 330	1 258	1 248	1 212	1 176	1 128	1 100	1 078	736	700	688	672

pressure. The other considerations have a smaller effect on the corrected price. These correction factors are all assembled in Table A3.6.a through A.III.6.g. An additional 5% should be added to the final price for contingencies.

A3.1.2.2 Pricing Double-Pipe Exchangers

The pricing method is the same as for exchangers with tube bundles, except that since only the corrections for pressure and materials have much meaning for double-pipe exchangers, these two alone are used. The base price is obtained from Fig. A3.6 according to surface, and the correction factors are obtained from Table A3.7 on page 294. The final price should include an additional 5% for contingencies.

A3.1.2.3 Prices for Heat-Exchanger Tubes

Since exchanger tubes are sometimes used separately, an estimate of their cost can be of interest. Figure A3.7 gives the cost of exchanger tubes in mild steel. As a first approximation, this cost can be modified with the correction factors for double-pipe exchangers. The cost of welded and extruded tubes is heavily dependent on supply and demand. In 1974 prices were particularly high, as shown by the Nelson index. Currently, they have come down. Figure A3.7 furnishes average prices, such as were prevalent in mid-1975. Incidentally, it seems that the Nelson index for exchangers does not completely take into account such price fluctuations, whereas other indexes like the Los Alamos Scientific Laboratory (LASL) index do. Figure A3.7 is established for orders of 20,000 meters and for tubes in mild steel. Corrections should be made when the conditions are different. For seamless tubes, the following factors can be used to correct for materials: low-alloy steel (ASTM A199), 1.7; 304L stainless, 4.5; 316L stainless, 7.8.

A3.2 AIR COOLERS

Air coolers are used for condensing and cooling the same as shell-and-tube exchangers; and the project evaluator is faced with the problem of choosing between the two, first, on the question of availability of utilities, and second, on the question of economics. Since it is difficult to figure the optimization of equipment or even units during project evaluation, only the availability of cooling water is generally considered.

Air coolers are made up of horizontal beds or rows of long fin tubes supported side by side. A fan forces air through these rows. Although there are many interdependent variables, the sizing and cost calculation of this equipment can be carried out in a detailed fashion through trial-and-error calculations normally done by the fabricator.

TABLE A3.6 Base-Price Correction Factors for Shell-and-Tube Exchangers

a. Type of Exchanger in Fig. A3.4

Type	f_d	Type	f_d
AES	1.00	BEM	0.80
AEM	0.87	BEU	0.75
AEU	0.85	BKT	1.10
AKT	1.20	Thermosiphon	1.35
BES	0.92		

b. Influence of the Tube Diameter and Pitch

Diameter			
in	mm	pitch, in	f_Φ
¾	19.2	1-in. square	1.00
¾	19.2	15/16-in triangular	0.95
¾	19.2	1-in triangular	0.97
1	25.4	1¼-in square	1.07
1	25.4	1¼-in triangular	0.97

c. Tube Length

Length		
ft	m	f_l
8	2.4	1.35
12	3.7	1.13
16	4.9	1.00
20	6.1	0.92
24	7.3	0.90

d. Number of Tube Passes

No. of Passes	f_{np}
2	1.00
4	1.02
6	1.04
8	1.06
12	1.08

e. Maximum pressure

Pressure, bars	f_p		
	50 m^2	100 m^2	500 m^2
$\leqslant$ 10	1.00	1.00	1.00
10–20	1.03	1.08	1.18
20–30	1.15	1.20	1.32
30–40	1.28	1.35	1.50
40–65	1.67	1.75	1.93
65–85	1.80	1.90	2.10
85–130	2.35	2.45	2.70
130–180	3.00	3.15	3.45

f. Temperature

Temperature, °C	f_t
$t < 350$	1.00
$350 < t < 550$	1.08

g. Materials of construction

Surface, m^2	Materials of construction,* shell/tubes									
	CS/CS	CS/Cu	CS/Mo	CS/304	304/304	CS/316	CS/Monel	Monel/Monel	CS/Ti	Ti/Ti
$\leqslant$ 10	1.00	1.05	1.40	1.55	2.30	1.95	2.80	4.50	4.70	11.00
10–50	1.00	1.10	1.55	1.75	2.55	2.15	3.05	4.80	5.80	12.20
50–100	1.00	1.15	1.75	2.15	2.90	2.55	3.55	5.30	7.30	13.90
100–500	1.00	1.30	2.05	2.60	3.40	3.00	4.35	6.10	9.40	16.30
500–1 000	1.00	1.50	2.35	3.20	4.15	3.65	5.25	7.10	12.00	19.10

*CS = carbon steel; Cu = admiralty; 304 = 304-type stainless, 316 = 316-type stainless; Mo = molybdenum; Monel = Monel; and Ti = titanium.

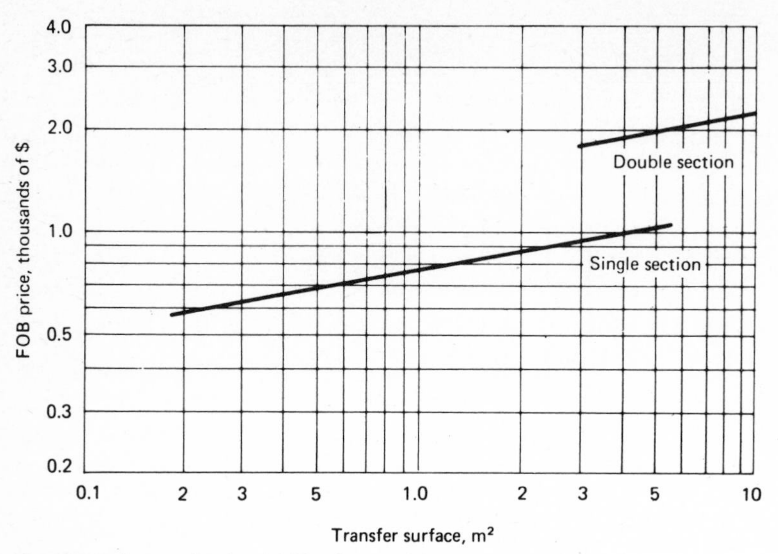

Fig. A3.6 Base price for double-pipe exchangers, early 1975.

TABLE A3.7 Base-Price Correction Factors for Double-Pipe Exchangers

a. Pressure

Pressure, bar	f_p
⩽ 4	1.00
4–6	1.10
6–7	1.25

b. Material of Construction

Material: Shell/Tube	f_m
cs/cs	1.0
cs/304L stainless	1.9
cs/316 stainless	2.2

A3.2.1 SIMPLIFIED METHOD FOR SIZING AND PRICING AIR COOLERS

A3.2.1.1 The Principle

The simplified method consists of determining the characteristics of the tube rows and the power absorbed by the fan, and then the cost, by means of coefficients K and C, which are functions of

Heat duty

Inlet and outlet temperatures, T_1 and T_2, of the process fluid

Inlet temperature of the air, t_a

Heat-transfer and fouling resistance of the process fluid

A3.2.1.2 Preliminary Calculations

The extensive trial-and-error assumptions of a detailed calculation are avoided by using the empirical relations in Table A3.9 and Figs. A3.9 and A3.10 to calculate the following five parameters:

1. The temperature correlating factor R as $R = \dfrac{T_1 - T_2}{T_1 - t_a}$

2. The reduced duty S as $S = \dfrac{Q \times 10^{-3}}{T_1 - t_a}$

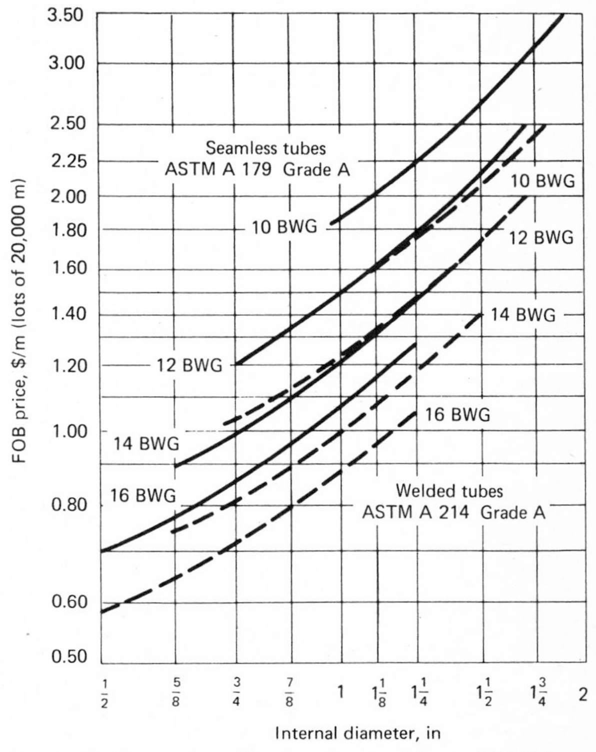

Fig. A3.7 Base price of heat-exchanger tubes, mid-1975.

3. The resistance to heat transfer of film plus fouling on the process side, which can be obtained from Fig. A3.8.

4. The overall heat-transfer coefficient U, kcal/(h)(m^2)(°C), as

$$\frac{1}{U} = r_i + r_d + r_m + r_a$$

where r_m = resistance of the fin-tube wall, which is often equal to about 0.00015 (h)(m^2)(°C)/kcal

r_a = resistance of the air film, as obtained from Table A3.9. Note that since r_a is dependent on $(T_1 - t_a)/U$, its determination involves a simplified trial and error.

5. The corrected LMTD, as obtained from Fig. A3.8. This requires the calculation of P as

$$P = \frac{t_h - t_a}{T_1 - T_2}$$

And that in turn requires determination of the air exit temperature t_h, which is calculated from the equation

$$t_h = \frac{Q}{1{,}061 V_f P_{cv}} + t_a$$

where Q = duty, kcal/h

V_f = air velocity (superficial) through the tube rows, m/s

P_{cv} = fan horsepower, CV

t = air-side temperatures, °C

h,a = subscripts denoting out and inlet, respectively

Solution of this equation now requires knowing two more variables, V_f and P_{cv}. V_f is identified in table A3.9 with r_a, which was determined in step 4. P_{cv} is determined as

$$P_{cv} = KS$$

and the value of S was determined in step 2, while the value of K is determined from Fig. A3.9 from R (step 1) and r (Table A3.8).

A3.2.1.3 Sizing Air Coolers

The tubes are assumed to be 10 m long with this method; and for a 10-m length, the approximate width is given by the relation

$$l = 0.112\, P_{cv}$$

where l is the width in meters.

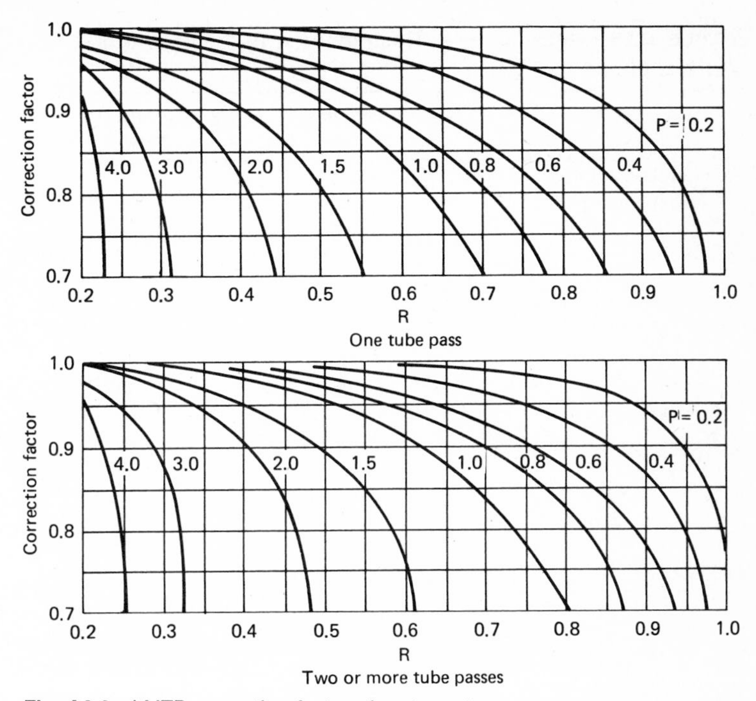

Fig. A3.8 LMTD correction factors for air coolers.

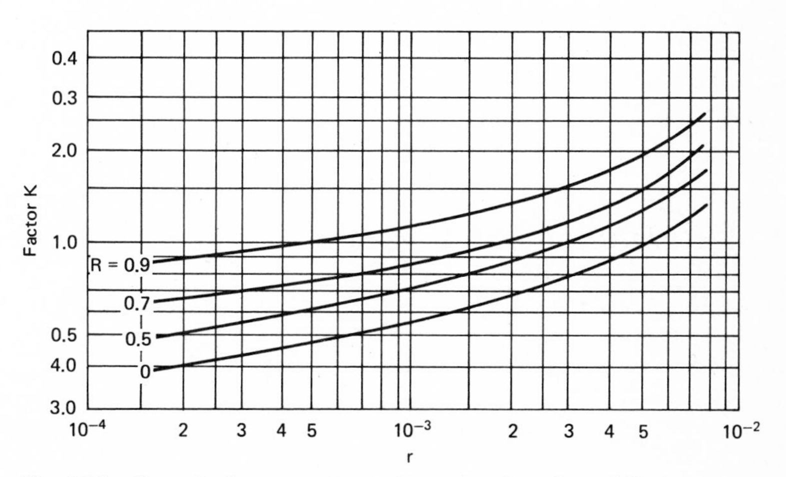

Fig. A3.9 Air-cooler fan power factor K as a function of r and R.

TABLE A3.8 Heat-Transfer Resistance r, (h)(m²)(°C)/kcal, Offered by Various Fluids

Fluid	Film Resistance r_i	Fouling Resistance r_d	Overall Resistance r
Refrigeration service			
Cooling water	0.00016	0.0002	0.00036
Aqueous solutions	0.0002	0.0004	0.0006
LPG	0.0004	0.0002	0.0006
Light hydrocarbons	0.0005	0.0003	0.0008
Medium hydrocarbons:			
average viscosity 1 cP	0.0010	0.0004	0.0014
average viscosity 5 cP	0.0032	0.0006	0.0038
average viscosity 10 cP	0.0050	0.0008	0.0058
Condensation			
Steam	0.0002	0.0001	0.0003
Ammonia	0.00028	0.0002	0.00048
LPG	0.0006	0.0002	0.0008
Light hydrocarbons	0.0008	0.0003	0.0011
Light naphtha	0.0010	0.0004	0.0014
Heavy naphtha	0.0014	0.0004	0.0018
Gasoline	0.0008	0.0002	0.0010
Gas and oil	0.0014	0.0004	0.0018

The number of rows of tubes, N, is obtained from Table A3.9.
The heat-transfer surface is given by

$$A = \frac{Q}{U(LMTD \text{ corrected})}$$

where A = surface, m²
Q = duty, kcal/h
U = overall heat-transfer coefficient from step 4, kcal/(h)(m²)(°C)
$LMTD$ = corrected log-mean temperature difference from step 5

A3.2.1.4 Pricing Air Coolers

The base price assumes

Tubes: 10 m long, 1-in diameter, mild steel, 12 BWG wall thickness, and with type G aluminum fins inserted into grooves in the tube surface
Six rows of tubes
Ten bars design pressure
Bolted return-type end cover in mild steel
Manually regulated pitch on the fan blades
Totally enclosed fan-cooled (TEFC) motor as fan driver

The base price for these assumptions is a product of coefficient C, which is obtained from Fig. A3.10, and coefficient S determined in step 2 above.

The corrected price is obtained as

$$\text{Corrected price} = (\text{base price})\, f_e f_p f_l f_N f_m$$

and the correction coefficients are obtained from Table A3.10.

This method is not applicable for vacuum condensers.

TABLE A3.9 Relations for Quickly Sizing Air Coolers

$\frac{T_1 - t_a}{U}$	Tube Rows N	External Film Resistance r_a, (h)(m²)(°C)/kcal	Superficial Air Velocity V_f, m/s
$(T_1 - t_a)/U \leqslant 0.13$	3	0.00096	3.20
$0.13 < (T_1 - t_a)/U \leqslant 0.17$	4	0.00102	3.02
$0.17 < (T_1 - t_a)/U \leqslant 0.22$	5	0.00107	2.87
$0.22 < (T_1 - t_a)/U \leqslant 0.28$	6	0.00112	2.74
$0.28 < (T_1 - t_a)/U \leqslant 0.36$	7	0.00118	2.58
$0.36 < (T_1 - t_a)/U \leqslant 0.46$	8	0.00121	2.48
$0.46 < (T_1 - t_a)/U \leqslant 0.58$	9	0.00125	2.38
$0.58 < (T_1 - t_a)/U \leqslant 0.73$	10	0.00128	2.26
$0.73 < (T_1 - t_a)/U$	11	0.00132	2.16

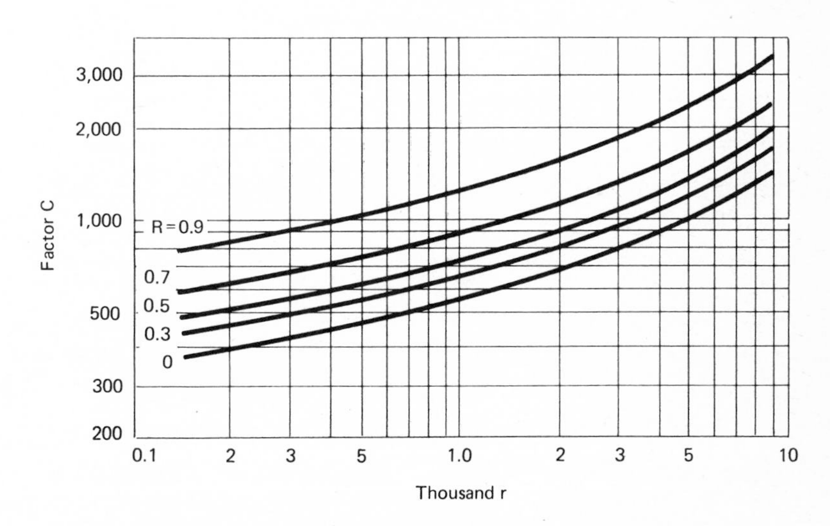

Fig. A3.10 Air-cooler price-factor C as a function of r and R.

TABLE A3.10 Base-Price Correction Factors for Air Coolers

a. Tube thickness, mm	BWG*	f_e	b. Pressure, bar	f_p
2.77	12	1.0	⩽ 10	1.00
2.11	14	0.9	10–20	1.03
1.65	16	0.8	20–30	1.06
1.24	18	0.7	30–50	1.10
			50–75	1.13
			75–100	1.15
			100–150	1.20

c. Tube length, m	f_l	d. Number of tube rows	f_N
		3	1.25
12	0.90	4	1.15
10	1.00	5	1.05
8	1.05	6	1.00
6	1.12	8	0.90
5	1.15	10	0.85

e. Tube material (aluminum fins)	f_m
Mild steel	1.00
Aluminum 3S	1.30
Aluminum alloy	1.50
Stainless 304	2.20
Stainless 321	2.50
Stainless 316	3.00
Monel	3.20

*See Table A3.4.

Appendix 4

Process Design Estimation: Pumps and Compressors

A4.1 PUMPS

A4.1.1 SELECTING PUMPS

There is no simple basis for classifying pumps, particularly centrifugal pumps, which differ not only by their application but also in their performance (head, capacity, NPSH, etc.) and in their design (type of casing joint, impeller configuration, nozzles, number of stages, etc.), so that different kinds of pumps fall together with each one of these criteria. However, it is possible to make a broad separation into volumetric pumps and centrifugal pumps.

A4.1.1.1 Volumetric Pumps

These might be generally thought of as pumps that trap portions of a liquid so that there is no possibility of backward flow as those portions are moved from a lower suction pressure to a region of higher pressure at the discharge. Such pumps can be further divided into reciprocating pumps and rotating pumps.

Reciprocating pumps These can be further divided into the following catagories:

1. Steam pumps, which are obsolete in most applications
2. Motor-driven pumps, including

a. Plunger pumps (metering or nonmetering)
b. Diaphragm pumps (metering or nonmetering)

Rotating pumps These include

1. Pumps with sliding or rotating vanes
2. Cam pumps
3. Gear pumps
4. Screw pumps
5. Lobe pumps

A4.1.1.2 Centrifugal Pumps

Centrifugal pumps are here classified according to their application and according to their design.

Application categories

1. Industrial pumps, for near-ambient pressures and temperatures.
2. Process pumps, for continuous operation at high temperatures and pressures up to a maximum of about 50 bars, which pumps conform to the standards of the American Petroleum Institute (API).
3. Multistage pumps for high pressures.
4. Chemical pumps, for continuous operation at pressures and temperatures slightly less than process pumps—these pumps are frequently made of cast iron.
5. Slurry pumps.
6. Vacuum pumps.

Design categories

1. Process pumps
   a. Single stage
   b. Two stage
   c. Multistage
2. Hot-oil pumps
   a. Multistage (volute or barrel)
3. Impeller configurations
   a. Axial flow
   b. Radial flow
   c. Mixed axial and radial flow

4. Francis screw type
5. Vertical pumps
6. In-line pumps
7. Pumps with horizontally split casings
8. Pumps with verticaly split casings
9. Pumps with liquid seal
10. Self-priming pumps

A guide to selecting among the many types of pumps can be obtained from Fig. A4.1. It can be seen from this figure that centrifugal pumps have the widest range of application.

A4.1.2 DETERMINING PUMP HORSEPOWER

The hydraulic power is first determined from the liquid's operating conditions, and this is then converted to brake horsepower demanded by the pump from its driver.

A4.1.2.1 Hydraulic Horsepower

This is given by the expression

$$PH = 3.8 \times 10^{-2}(Q_c)\Delta P$$

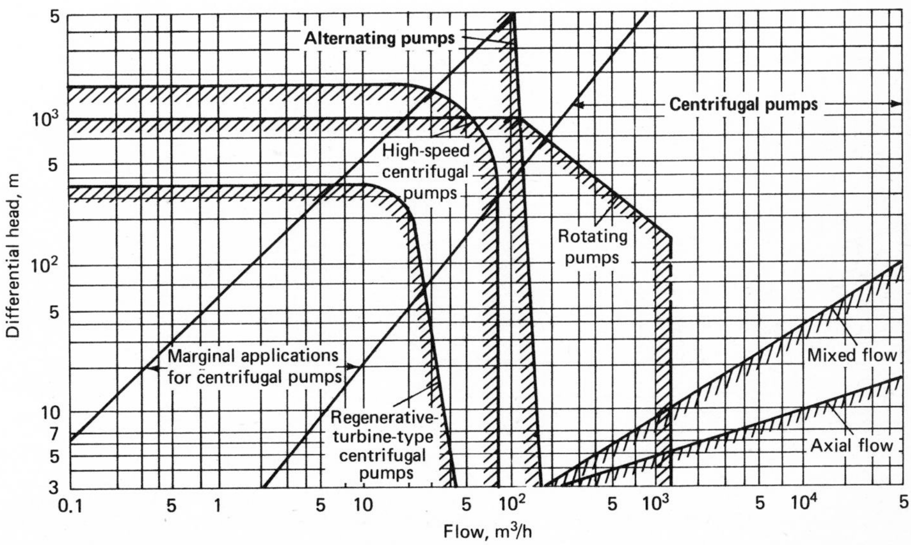

Fig. A4.1 Selection diagram for process pumps.

where PH = hydraulic horsepower, cv
Q_c = design flow, m^3/h
= $Q_n K$
Q_n = operating flow, m^3/hr.

K = safety factor, being $\begin{cases} 1.25 & \text{for feed pumps} \\ 1.20 & \text{for reflux pumps} \\ 1.10 & \text{for other pumps} \end{cases}$

ΔP = total pressure differential
= $P_2 - P_1 + \Delta H + \delta P$

where P_2 = pressure in the downstream receiver, bars
P_1 = pressure in the upstream supply vessel, bars
ΔH = difference in elevation, bars
δP = friction loss, bar (allow 0.30 bar for an exchanger, 1.0 bar for a valve, and neglegible loss for the piping)

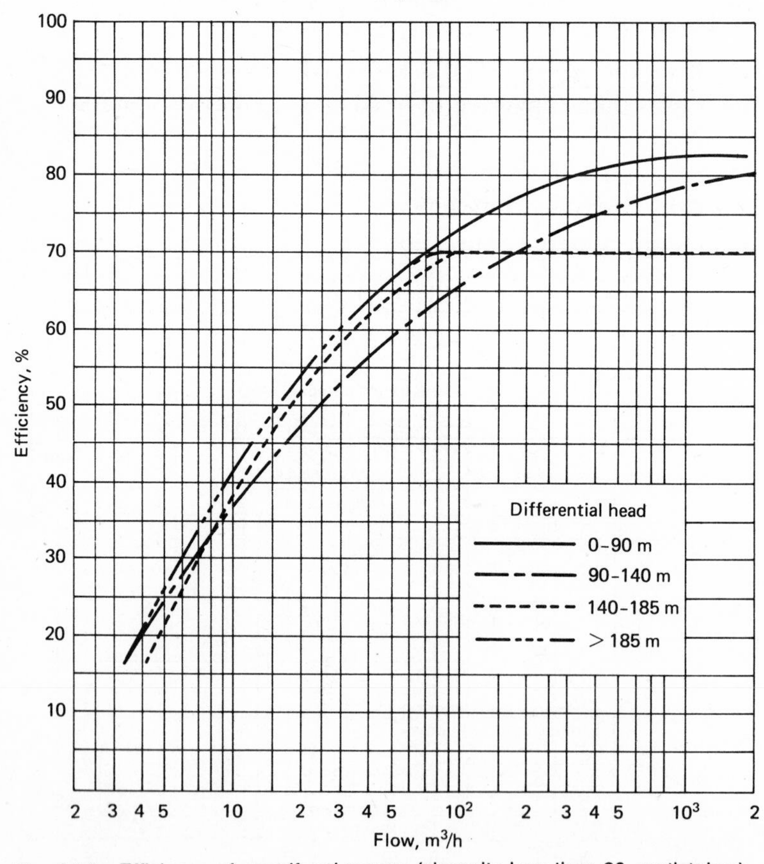

Fig. A4.2 Efficiency of centrifugal pumps (viscosity less than 20 centistokes).

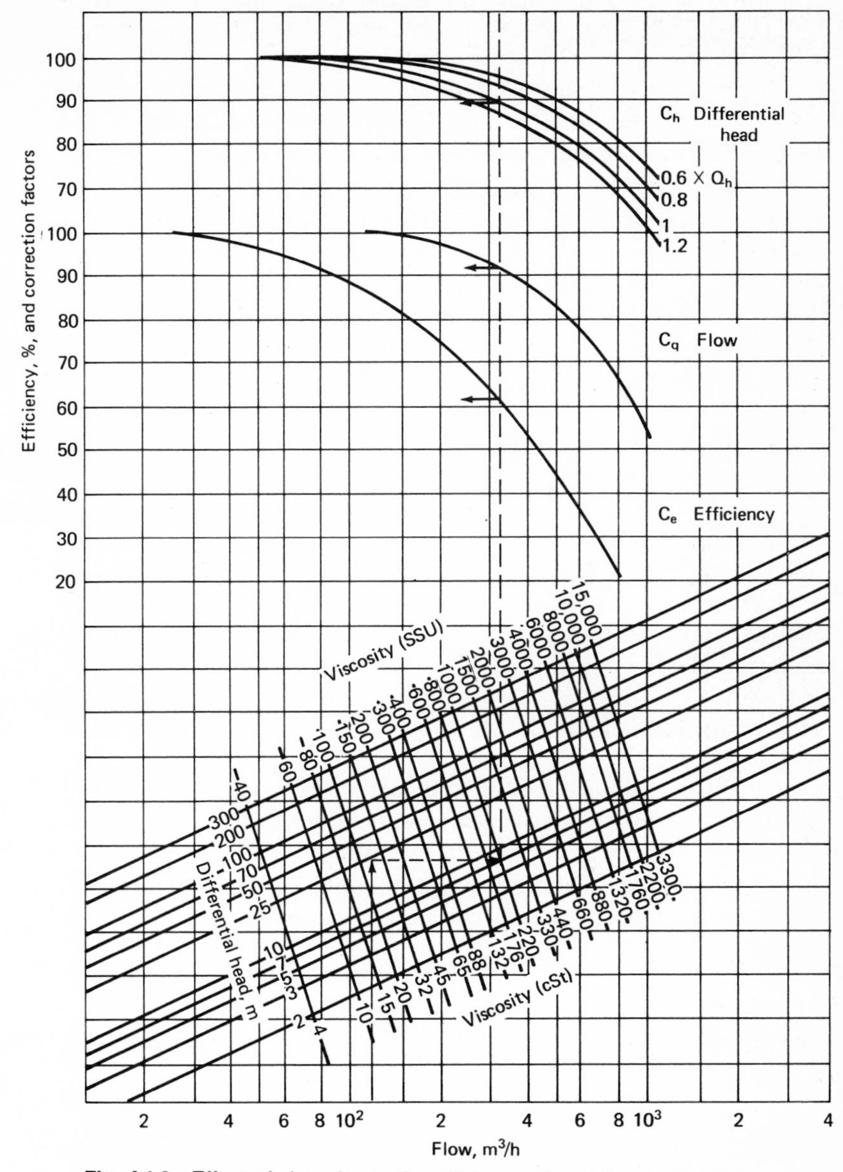

Fig. A4.3 Effect of viscosity on the efficiency of centrifugal pumps.

A4.1.2.2 Determining the Pump Efficiency

When the liquid viscosity is less than 20 cSt, centrifugal pump efficiency is obtained from Fig. A4.2. When the pump flow is less than 3.5 m^3/h (the minimum flow in Fig. A4.2), a centrifugal pump should be equipped with a recirculating bypass.

When the liquid viscosity is more than 20 cSt, the efficiency is determined from Fig. A4.3, and the brake horsepower with the expression

$$\text{Brake horsepower} = \frac{\text{PH}}{C_e C_q C_h}$$

where C_e = base efficiency, fraction
C_q = efficiency-correction factor for flow, fraction
C_h = efficiency-correction factor for head, fraction

The efficiency of reciprocating pumps depends on the volumetric efficiency, the hydraulic efficiency, and the mechanical efficiency. The volumetric efficiency, which relates to the effectiveness of the pump's piston displacement, will be about 95% for a new pump and 50% for an old pump. The hydraulic efficiency, which relates to pressure losses within the pump, can be assumed equal to 100% for process estimation. The mechanical efficiency, which relates to mechanical losses, can be taken as 80–90% for nonmetering pumps, and 90% for metering pumps.

The efficiency of rotating pumps can be obtained from Fig. A4.4, with viscosity corrections from Table A4.1.

It is conventional for plants to have a standby pump on line for all essential services, sometimes with one standby serving two or more pumps. Consequently, consideration should be given to pump sparing arrangements in a project analysis.

A4.1.3 PRICING PUMPS

Centrifugal pumps are sold in standard casing sizes, for which the impellers are varied to meet specific requirements. Since the price depends primarily on the casing, the prices for centrifugal pumps will cover ranges of operating requirements corresponding to the capability of the casings, as shown in Fig. A4.5*a*. Other types of pump can be priced from curves resembling those of other equipment.

A4.1.3.1 Pricing Centrifugal Pumps

The method consists of establishing a base price and then correcting that for differences of type, of material, of temperature, and of pressure, according to the formula

$$\text{Corrected price} = (\text{base price})\, f_d f_m f_t f_p$$

where f_d = base-price modification for pump type, from Table A4.2a
f_m = base-price modification for materials, from Table A4.2b
f_t = base-price modification for temperature, from Table A4.2c
f_p = base-price modification for suction pressure, from Table A4.2d

The base price, as obtained from Fig. A4.5*a* assumes the following:

Type: vertically split process pump

Materials: casing in cast steel, internals in cast iron

Operating temperature: 0–150°C

Maximum flow: 790 m^3/h at 3,500 rpm; 420 m^3/h at 1,750 rpm

Maximum head: 450 m at 3,500 rpm; 90 m at 1,750 rpm

Suction pressure: 20 to 40 bars

This base price does not include driver. Vertical submerged pumps with large capacity can be priced, with driver, from Fig. A4.5*b*. When sizing and pricing centrifugal pumps, it is often useful to keep in mind the following general correlation:

$$\left(\frac{Q_1}{Q_2}\right)^2 = \left(\frac{N_1}{N_2}\right)^2 = \frac{\Delta P_1}{\Delta P_2}$$

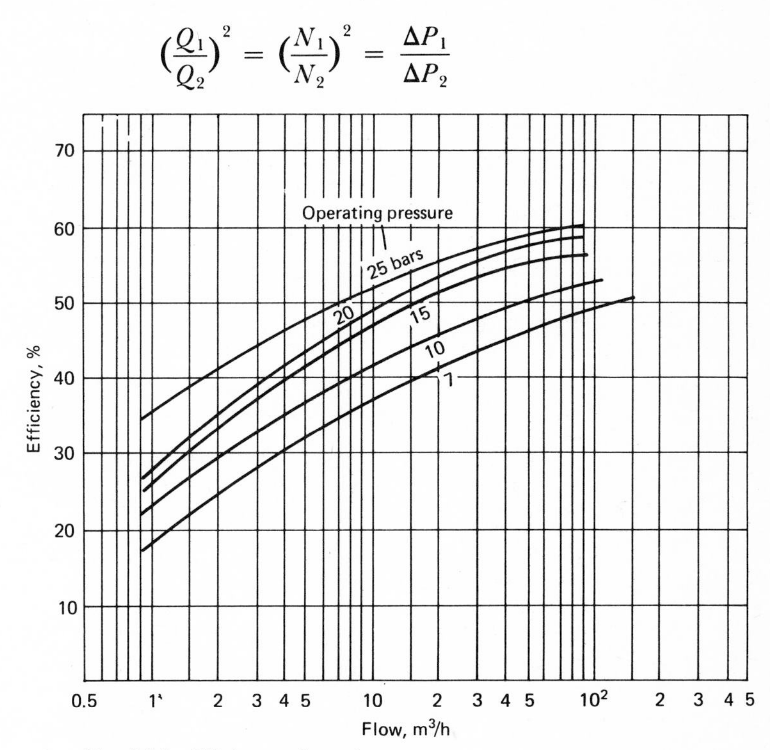

Fig. A4.4 Efficiency of rotating pumps.

TABLE A4.1 Viscosity Correction Factor for Rotating Pumps

Viscosity, SSU	f_v
10,000–30,000	0.95
30,000–50,000	0.85
50,000 (11,000 cSt)	0.35

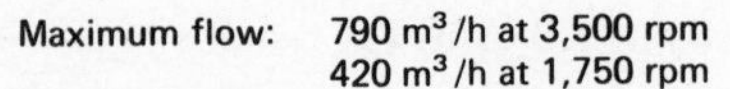

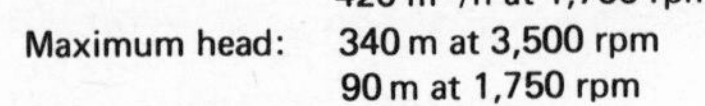

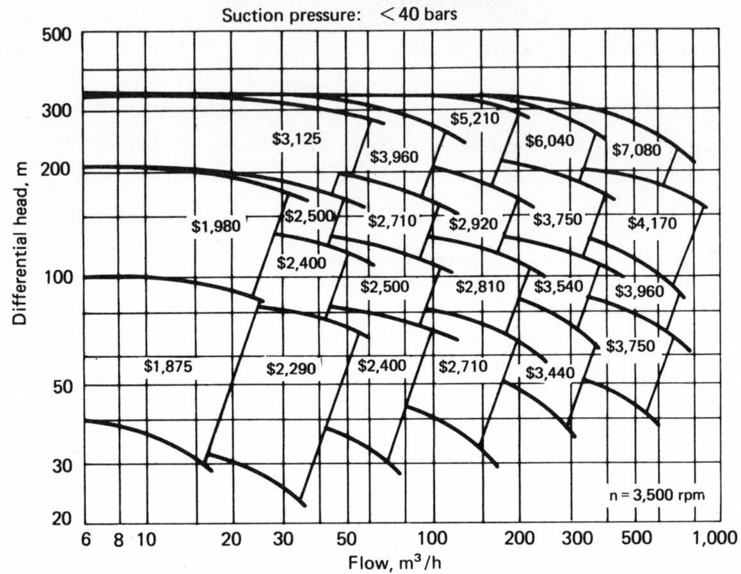

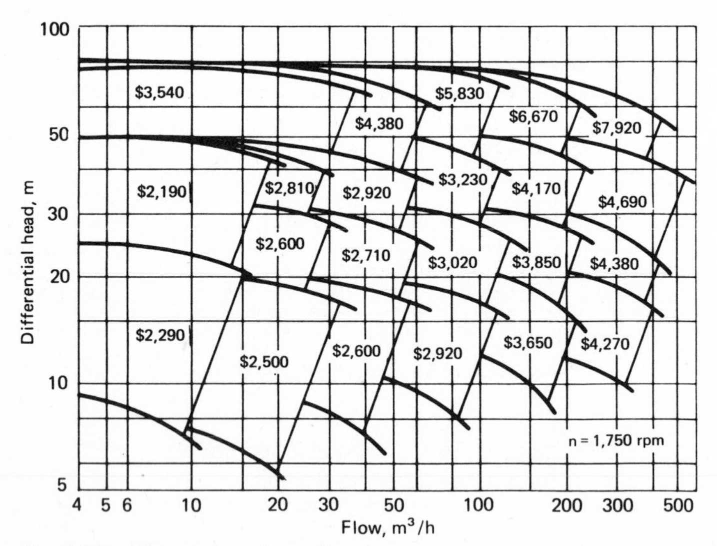

Fig. A4.5*a* Base price of centrifugal process pumps, $ mid-1975.

TABLE A4.2 Base-Price Correction Factors for Centrifugal Pumps

a. Type of Pump

Type of pump	f_d
Process, with vertically split casing	1.0
Process with double seal and fluid recycled between seals	1.5
Vertical, in-line	0.5
Turbine type	0.8
Chemical type	0.5–0.7

b. Material of Construction

Material of Construction	f_m
Cast steel	1.00
Bronze	1.25
Cast steel with 316 stainless internals	1.50
Stainless type 316	1.80
Uranus	2.00
Hastelloy C	2.80

c. Operating Temperature

Temperature, °C	f_t
< 150	1.00
150–250	1.15
> 250	1.30

d. Suction Pressure

Suction Pressure, bars	f_p
< 20	0.7
20–40	1.0
40–70	1.3

where Q = flow
N = speed of rotation
ΔP = differential head

A4.1.3.2 Reciprocating Pumps

Use Fig. A4.6 to get the cost of reciprocating pumps for two pressure ranges, as a function of the hydraulic horsepower. These curves correspond to duplex pumps in cast iron, with manual regulation, and for a suction pressure less than 20 bars. These curves can be modified according to the following:

- Number of cylinders, as compared to two for duplex
- Type of regulation

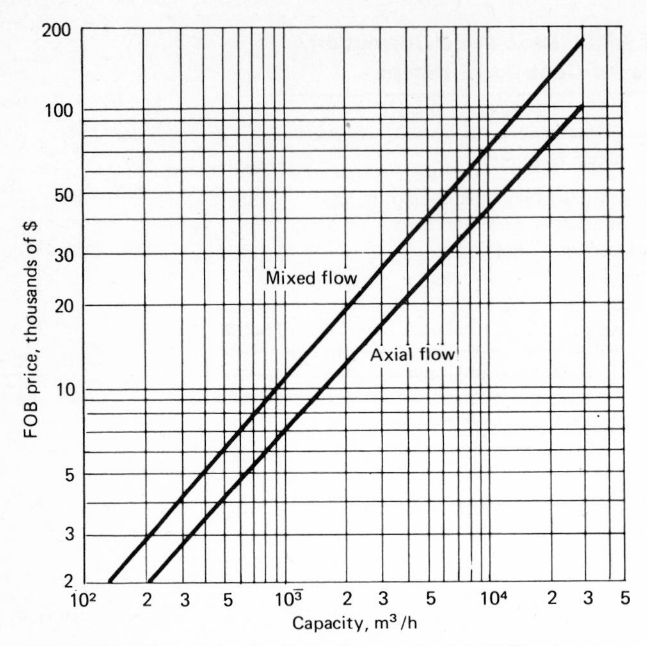

Fig. A4.5*b* Base price for large-flow vertical centrifugal pumps with drive, mid-1975.

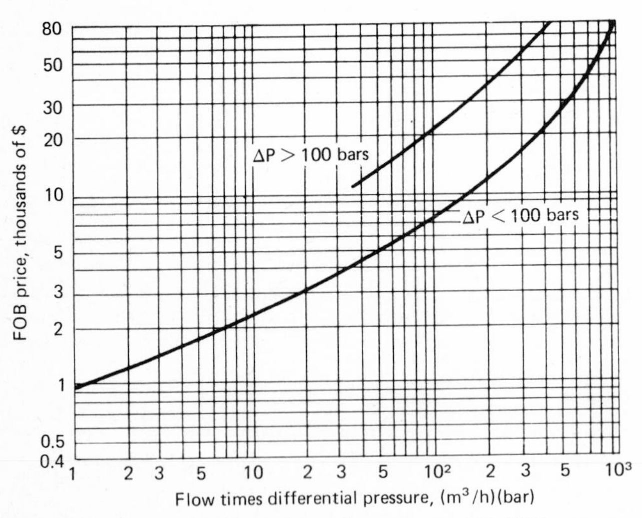

Fig. A4.6 Base price for reciprocating pumps, mid-1975.

- Lubrication system
- Material of construction
- Discharge pressure

Correction factors for materials of construction and the suction pressure are shown in Table A4.3a and b, respectively.

TABLE A4.3 Base-Price Correction Factor for Reciprocating Pumps

a. Material of Construction

Material	f_m
Cast iron	1.00
Rubber-lined cast iron	1.15
Cast steel	1.10
Stainless 304	1.30
Stainless 316	1.50

b. Suction Pressure

Suction Pressure, bar	f_p
< 20	1.0
20–40	1.3
40–70	1.5

A4.1.3.3 Pricing Rotating and Metering Pumps

Figure A4.7 furnishes, as a function of flow, the following:

- The price of lobe pumps, without driver, in 316 stainless.
- The price of screw pumps, without driver, with a cast iron casing and stainless rotor.
- The price of metering pumps, both piston and diaphragm, in 316-type stainless, except for coupling and sleeve in aluminum alloy, and driven by an explosion-proof electric motor. The head, which can be adapted to specific flows, is also included.

A4.2 COMPRESSORS

The costs of compressors are best correlated with the brake horsepower, or with the suction volume, depending on compressor type. The type of compressor best suited for any given service is most conveniently correlated with suction volume and discharge pressure. Consequently, the process design estimation of compressors requires determination of the suction volume, choice of compressor type, and calculation of the brake horsepower for those compressors whose costs are not correlated with suction volume.

Determining the suction volume requires knowledge of the flow rate, the inlet pressure, the inlet temperature, and calculation of the compressibility factor of the gas to be compressed. Determining the brake horsepower requires (1) calculating the number of stages, (2) calculating the compression ratio per stage, (3) calculating the suction volume at each stage, (4) calculating the adiabatic horsepower, (5) choosing the type of compressor, and (6) calculating the brake horsepower for the chosen compressor type.

A4.2.1 COMPRESSOR TYPES

The various compressor types, their range of application and their base costs, are shown in Figs. A4.9, A4.13, A4.14, and A4.15. These include the followings in approximate order of importance:

1. Centrifugal
2. Reciprocating
3. Fans and blowers
4. Rotary screw
5. Rotary sliding vane

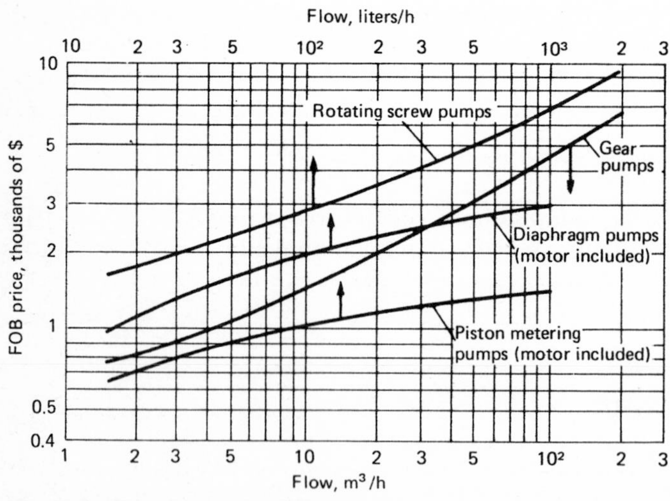

Fig. A4.7 Base price for rotating and metering pumps, mid-1975.

A4.2.2 SIZING COMPRESSORS

A4.2.2.1 Calculating the Average Properties of the Gas to Be Compressed

The calculation for the average physical properties of the gaseous mixture at the compressor inlet is made according to Table A4.4. The ratio of specific heats, $C_p/C_v = \gamma$, is calculated empirically as

$$\gamma = \frac{\text{Average } MC_p}{\text{Average } MC_p - 1.99}$$

A4.2.2.2 Calculating the Number of Stages and the Average Compression Ratio

As a rule, the compression ratio of discharge-to-suction pressure should never be more than 5. Consequently, the number of stages can be calculated as

$$s = 1.43 \log \frac{P_2}{P_1}$$

where s = number of stages, rounded off to the next highest whole number
P_1 = suction pressure, bars
P_2 = discharge pressure, bars

When there is more than one compression stage, the gas must be cooled between stages, although not necessarily to the temperature of the first-stage inlet. These intercoolers impose a pressure drop on the gas; and this pressure drop is assumed as 0.350 bars. Consequently, the overall compression ratio is

$$R = \frac{P_2 + (s - 1)0.350}{P_1}$$

where R is the overall compression ratio, and the average compression ratio per stage, ρ, is given by

$$\rho = R^{1/s}$$

A4.2.2.3 Calculating the Suction Volume at Each Stage

The basic formula is

$$V = 0.08315D\left(\frac{T}{P}\right)Z$$

where V = inlet volume, m^3/h
D = flow rate in moles, kmol/h
T = inlet temperature, K

TABLE A4.4 Calculation for the Physical Properties of a Gaseous Mixture

	(1)	(2)	(3)	(4)	(5)	(6)	(7)	(8)
Component	Mole %	Flow kmole/hr	T_c	P_c	Average T_c (1) × (3)	Average P_c (1) × (4)	MC_p	Average MC_p (1) × (7)
TOTAL								

Note: For T_c, P_c, and MC_p, use Table A4.5.

P = inlet pressure, bars
Z = compressibility factor

All terms correspond to the inlet conditions at any given compression stage, after the gas has been cooled in the intercooler or at the first-stage inlet.

The compressibility factor is next obtained from Fig. A4.8, as a function of the reduced temperature ($T_R = T/T_c$) and the reduced pressure ($P_R = P/P_c$). The critical temperature (T_c) and pressure (P_c) for a mixture of gases is obtained from the calculation in Table A4.4.

A4.2.2.4 Calculating the Adiabatic Horsepower Requirements

The basic equation is

$$\text{PTA} = \left(\frac{P_1V_1}{26.5}\right)\left(\frac{\gamma}{\gamma - 1}\right)\left[\rho^{\frac{\gamma - 1}{\gamma}} - 1\right]$$

where PTA = theoretical adiabatic horsepower per stage, cv
P_1 = suction pressure for that stage, bar
V_1 = suction volume for that stage, m^3/h.
γ = ratio of specific heats for the gas, C_p/C_v, as calculated from Table A4.4 and obtained from Table A4.5
ρ = average compression ratio

Since adiabatic mandates isentropic, the adiabatic horsepower can also be found by reading along a constant entropy line on a Mollier diagram, where

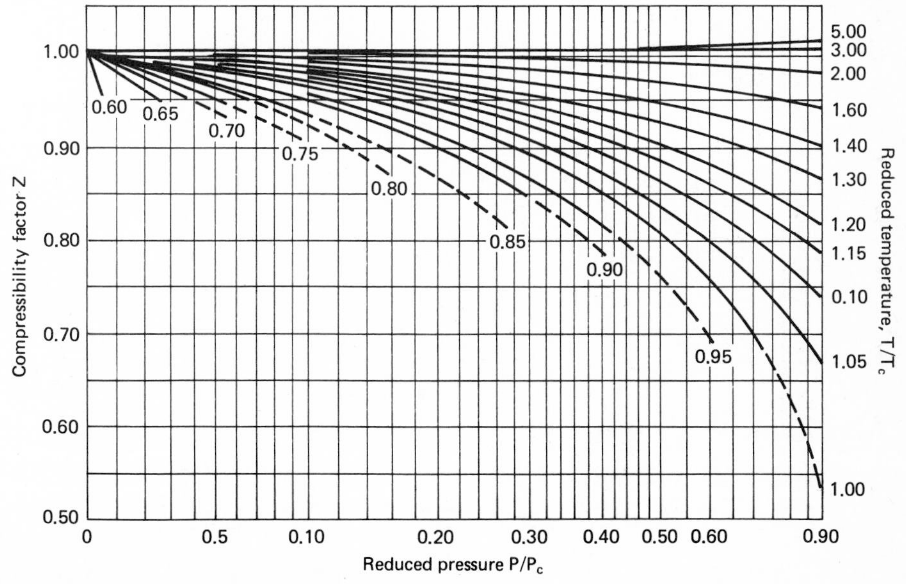

Fig. A4.8 Compressibility factor Z.

TABLE A4.5 Thermodynamic Constants of Various Compounds

Compound	Chemical Formula	Molecular Weight	$\gamma = C_p/C_v$ at 15.6°C	Critical Temperature T_c, °C	Critical Pressure P_c, bar	Molar Specific Heat MC_p* at 10°C	Molar Specific Heat MC_p* at 150°C
Acetylene	C_2H_2	26.04	1.24	36	62.8	10.22	12.21
Air	$O_2 + N_2$	28.97	1.40	−141	37.	6.95	7.04
Ammonia	NH_3	17.03	1.31	132	113.0	8.36	9.45
Argon	A	39.94	1.66	−122	48.7	4.97	4.97
Benzene	C_6H_6	78.11	1.12	288	48.4	18.43	28.17
Blast furnace gas	—	30.00	1.39	—	—	7.18	7.40
i-Butane	C_4H_{10}	58.12	1.10	134	37.5	22.10	31.11
n-Butane	C_4H_{10}	58.12	1.09	153	36.5	22.83	31.09
i-Butene	C_4H_8	56.10	1.10	145	40.0	20.44	27.61
n-Butenes	C_4H_8	56.10	1.11	147	40.2	20.45	27.64
Carbon dioxide	CO_2	44.01	1.30	31	74.0	8.71	10.05
Carbon monoxide	CO	28.01	1.40	−139	35.5	6.96	7.03
Cat-cracking off-gas	—	29.00	1.20	13	46.5	11.3	15.00
Chlorine	Cl_2	70.91	1.36	144	77.1	8.44	8.52
Coke-oven gas	—	11.00	1.35	−164	28.1	7.69	8.44
n-Decane	$C_{10}H_{22}$	142.28	1.03	346	22.1	53.67	74.27
Ethane	C_2H_6	30.07	1.19	32	49.5	12.13	16.33
Ethanol	C_2H_5OH	46.07	1.24	243	64.0	17	21
Ethyl chloride	C_2H_5Cl	64.52	1.19	290	52.7	14.5	18
Ethylene	C_2H_4	28.05	1.24	10	51.2	10.02	13.41
Freon	CCl_2F_2	120.91	1.13	112	40.1	18	22.
Helium	He	4.00	1.66	−268	2.3	4.97	4.97
n-Heptane	C_7H_{16}	100.20	1.05	267	27.2	39.52	53.31
n-Hexane	C_6H_{14}	86.17	1.06	235	29.9	33.87	45.88

TABLE A4.5 Thermodynamic Constants of Various Compounds *(Continued)*

Compound	Chemical Formula	Molecular Weight	$\gamma = C_p/C_v$ at 15.6°C	Critical Temperature T_c, °C	Critical Pressure P_c, bar	Molar Specific Heat MC_p*	
						at 10°C	at 150°C
Hydrogen	H_2	2.02	1.41	−240	13.0	6.86	6.98
Hydrogen chloride	HCl	36.46	1.48	51	82.7	6.93	7.00
Hydrogen sulfide	H_2S	34.08	1.32	100	90.1	8.09	8.54
Methane	CH_4	16.04	1.31	−82	46.4	8.38	10.25
Methanol	CH_3OH	32.04	1.20	240	79.8	10.5	14.7
Methyl chloride	CH_3Cl	50.69	1.20	143	66.7	11.0	12.4
Natural gas	—	19.00	1.27	−63	46.6	8.40	10.02
Neon	Ne	20.18	1.64	−229	26.3	4.97	4.97
Nitric oxide	NO	30.01	1.40	−94	65.9	6.9	7.2
Nitrogen	N_2	28.02	1.40	−147	34.0	6.96	7.03
n-Nonane	C_9H_{20}	128.25	1.04	323	23.8	48.44	67.04
n-Octane	C_8H_{18}	114.22	1.05	296	24.9	43.3	59.90
Oxygen	O_2	32.00	1.40	−119	50.4	6.99	7.24
i-Pentane	C_5H_{12}	72.15	1.08	188	33.2	27.59	38.70
n-Pentane	C_5H_{12}	72.15	1.07	197	33.5	28.27	38.47
Pentenes	C_5H_{10}	70.13	1.08	201	40.4	25.08	34.46
Propane	C_3H_8	44.09	1.13	97	42.6	16.82	23.57
Propylene	C_3H_6	42.08	1.15	92	45.6	14.75	19.91
Sulfur dioxide	SO_2	64.06	1.24	157	90.1	9.74	10.63
Steam	H_2O	18.02	1.33	374	221.4	7.98	8.23
Vent gas	—	30.00	1.38	−126	38.8	7.23	7.50
Water gas	—	19.48	1.35	−143	31.3	7.60	8.33

*Use linear extrapolation or interpolation for temperatures at compressor inlet.

such diagrams are available. Either way, the theoretical horsepower thus determined applies to all types of compressor, whether centrifugal or reciprocating.

Theoretically, the calculation for horsepower should be applied to each stage of compression and the stage horsepowers added for the overall power consumption. Practically, however, sufficiently accurate results can be obtained by multiplying the calculated horsepower for the first stage by the number of stages.

A4.2.2.5 Selecting the Compressor Type

With the suction volume established in Sec. A4.2.2.3 and the discharge pressure given in the basic data, Fig. A4.9 can be used to obtain the preferred type of compressor for the service. In using Fig. A4.9, the conventional sparing arrangements for reciprocating compressors should be kept in mind, i.e., either 50% sparing by dividing the total volume into two parts and providing three machines each with 50% of the total capacity or 33% sparing by dividing the total volume into three parts and providing four machines each with 33% of the total capacity. Centrifugal compressors are not conventionally provided with spares.

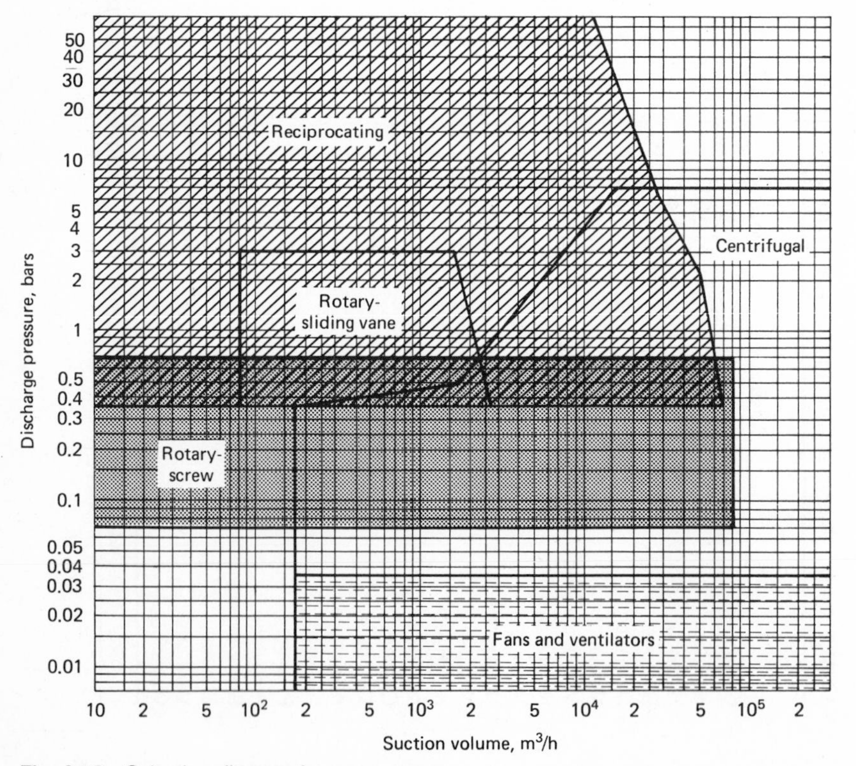

Fig. A4.9 Selection diagram for compressors.

A4.2.2.6 Determining the Efficiency and Brake Horsepower

The brake horsepower of centrifugal compressors equals the adiabatic horsepower (PTA) divided by the isentropic efficiency E_i plus losses p, or

$$\text{Actual horsepower} = \frac{\text{PTA}}{E_i} + p$$

The isentropic efficiency E_i is determined from the polytropic efficiency E_p with the aid of Fig. A4.11. E_p is in turn obtained from Fig. A4.10 and the known suction volume. The losses p can be estimated as follows:

Losses, *p*, cv	Suction Volume, m^3/h
5–20	Less than 30,000
30	30,000–50,000
50	50,000–100,000
60	Over 100,000

The action of reciprocating compressors approaches adiabatic compression. However, reciprocating compressors require that the adiabatic horsepower be corrected for piston clearances and inlet-outlet valve losses, which are here summarized as the mechanical efficiency—a loss that is often confused with the isentropic efficiency determined for centrifugal compressors above.

This mechanical efficiency of reciprocating compressors is determined from the average compression ratio by means of Fig. A4.12. From this,

$$\text{Actual horsepower} = \frac{\text{PTA}}{E}$$

where E is the mechanical efficiency for the reciprocating compressor.

A4.2.2.7 Calculating Discharge Temperatures

The discharge temperatures for the stages are frequently required for sizing intercoolers and aftercoolers. The isentropic discharge temperature, T_2, is calculated from the inlet temperature for that stage, T_1, by

$$T_2 = T_1\left(\frac{P_2}{P_1}\right)^{\frac{(\gamma-1)}{\gamma}}$$

The actual temperature rise will also include the effects of isentropic efficiency, so that

$$\Delta T_a = \frac{T_2 - T_1}{E_i}$$

where ΔT_a is the actual temperature rise, which can then be used to calculate the actual discharge temperature for the stage as

$$T_{2a} = T_1 + \Delta T_a$$

The actual discharge temperature for a stage of reciprocating compression can be assumed equal to the isentropic discharge temperature.

A4.2.3 PRICING COMPRESSORS

A4.2.3.1 Base Prices

The base prices for centrifugal, reciprocating, and rotating compressors are obtained from Figs. A4.13 to A4.15. Figure A4.13 applies to general centrifugal compressors and reciprocating compressors. When the estimated horsepower goes beyond the ranges of this chart, either the total duty is divided

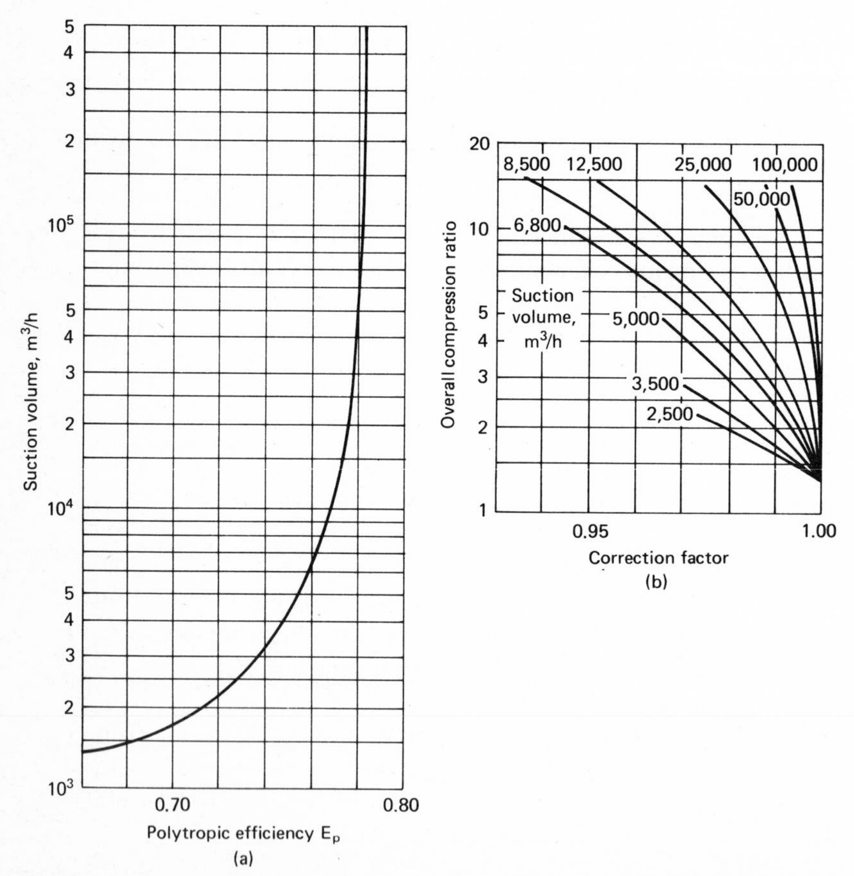

Fig. A4.10 Chart for polytropic efficiency. (*a*) Rough polytropic efficiency; (*b*) efficiency-correction factor.

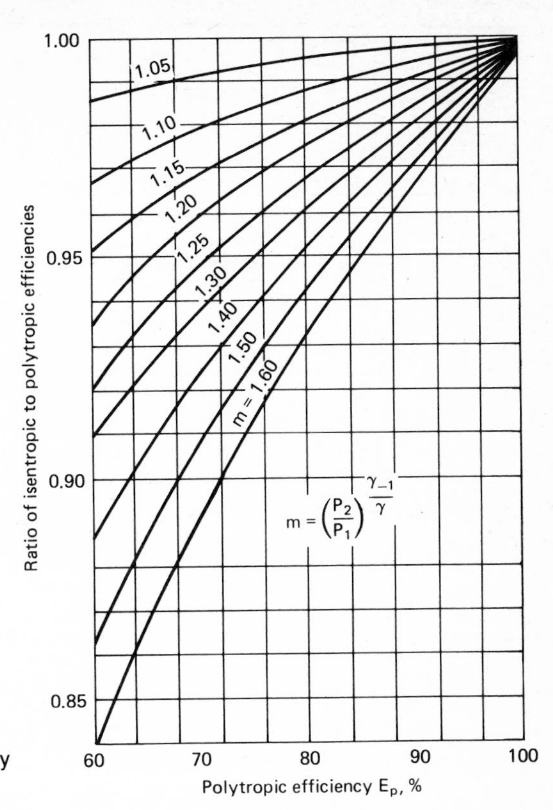

Fig. A4.11 Isentropic efficiency from polytropic efficiency.

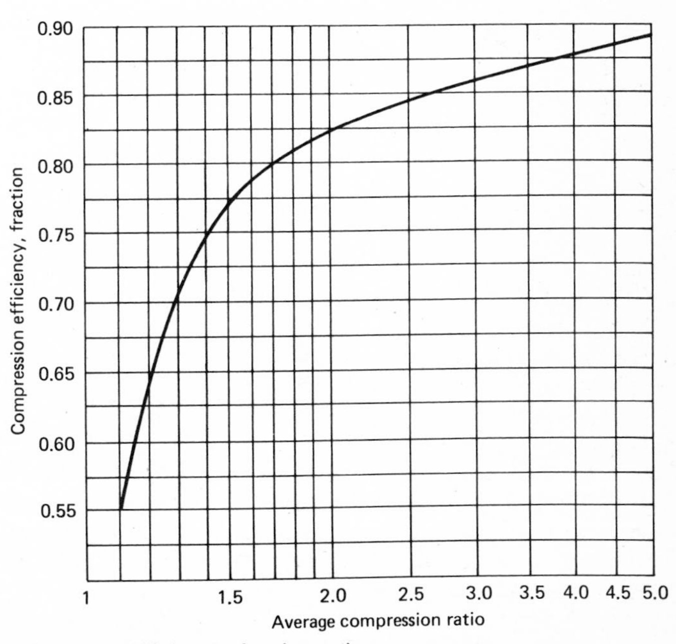

Fig. A4.12 Efficiency of reciprocating compressors.

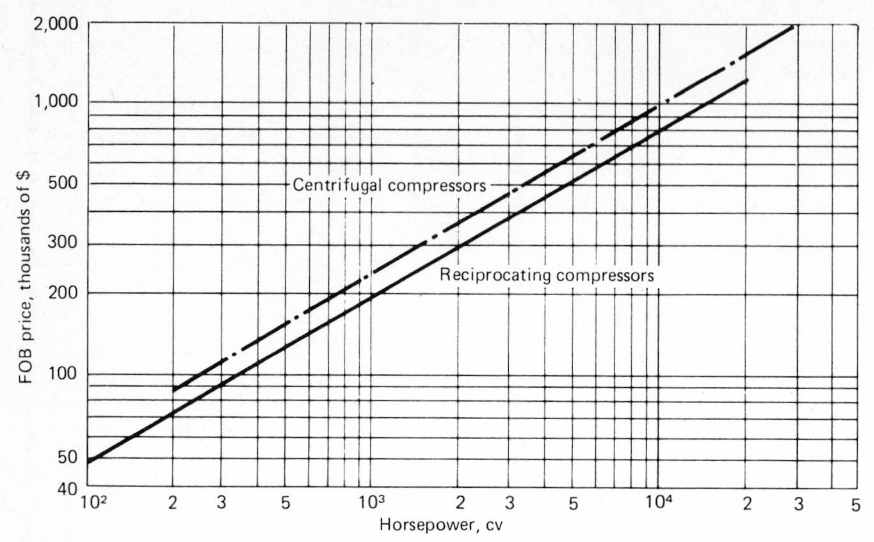

Fig. A4.13 Base price for reciprocating and centrifugal compressors, including speed changer, coupling, and auxiliaries, but not drivers, mid-1975.

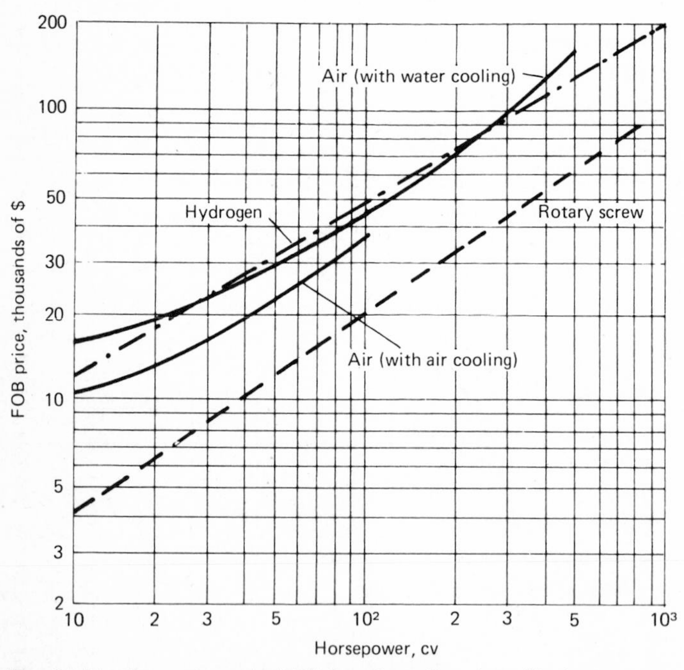

Fig. A4.14 Base price, mid-1975, for alternating air and hydrogen compressors and rotary-screw compressors, including speed reducer, coupling, and accessories, but not drivers.

among several machines or the compressors are specially constructed. Figure A4.14 applies to air and hydrogen compressors, which are generally smaller and include special design features. Figure A4.15 applies to various types of blowers and ventilators.

A4.2.3.2 Assumptions Included in the Base Prices

The curves of Fig. A4.13 do not include installation costs, which should be obtained from Table 4.16, page 175. Once the installation costs are added, it will be seen that the reciprocating compressors, which are less expensive than centrifuging in Fig. A4.13, become more expensive for the same brake horsepower. The prices of Fig. A4.13 assume a cast steel compressor, plus a

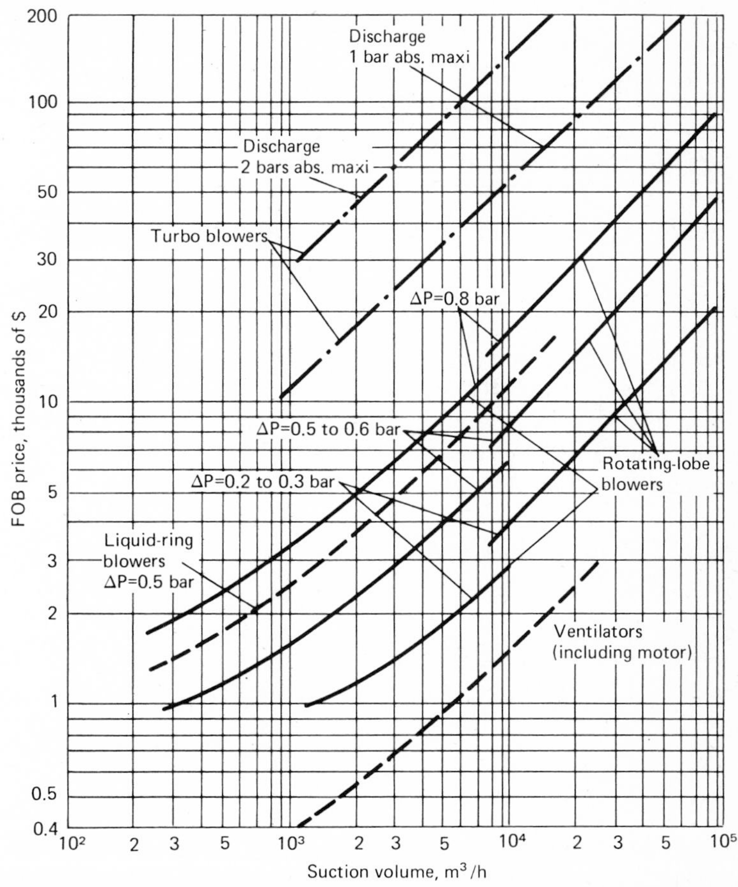

Fig. A4.15 Base price mid-1975 for blowers (complete to shaft only) and for ventilators (including motors).

TABLE A4.6 Base-Price Correction Factors for Compressors

a. Effect of Discharge pressure

Maximum pressure, bars	f_{pmax} Centrifugal	Reciprocating
⩽75	1.00	1.00
75–100	1.08	1.05
100–150	1.16	1.11
150–200	1.25	1.18
200–300	1.38	1.30
300–400	1.52	1.43
400–500	1.68	1.57
500–600	1.86	1.72

b. Effect of materials of construction

Material	f_m Compressors	Blowers
Cast steel	1.0	1.0
For the moving parts:		
Stainless type 304	1.4	2
Stainless type 316	1.7	2.5

reducer, coupling and auxiliary equipment, but not the driver, which must be obtained from App. 5.

The curves of Fig. A4.14, also, do not include installation costs. The prices of Fig. A4.14 include the compressor, reducer, coupling, auxiliary equipment, but not the motor. The material is cast steel.

The curves for blowers in Fig. A4.15 assume the bare machine without accessories or driver. On the average those prices should be multiplied by 1.5 to account for the accessories. The curves for ventilators in Fig. A4.15 assume the driver included.

A4.2.3.3 Correction Factors

Base-price correction factors for Figs. A4.13 through A4.15 are shown in Table A4.6a and b.

Appendix 5

Process Design Estimation: Drivers

Although drivers in general include diesel engines and gas turbines, as well as electric motors and steam turbines, this appendix will cover only electric motors and gas turbines.

A5.1 ELECTRIC MOTORS

Motors may be explosion-proof, or totally enclosed and fan-cooled, or open. Depending on their horsepower, they are built for different voltage requirements; thus motors up to 150 cv may be considered 480 V, and motors over 150 horsepower may be considered 4,160 V.

A5.1.1 SIZING MOTORS

Motors are sized from Table A5.1, by, first, selecting a motor horsepower to suit the corresponding pump horsepower, then determining the percent demand (as ratio of required to supplied horsepower) for the motor, and finally (from the table) the motor efficiency. The electrical consumption is calculated as

$$\text{kWh/h} = 0.735P_{cv}/\rho$$

where P_{cv} is the demand horsepower and ρ is the efficiency.

A5.1.2 PRICING ELECTRIC MOTORS

Explosion-proof motors or totally enclosed motors are used almost exclusively in the petrochemical industries. The base price for these and open motors is given as a function of motor horsepower in Fig. A5.1. This price corresponds to a motor speed of 1,750 rpm. For other motor speeds, the base price should be corrected from Fig. A5.2.

TABLE A5.1 Efficiency of Electric-Motor Pump Drivers

Pump Horsepower, cv	Motor Horsepower, cv	Efficiency of motor, % — % Utilization of motor		
		50	75	100
0–0.5	1	81	82	82.5
0.51–0.75	1.5	67	73	75
0.76–1.00	2	75	78	80
1.01–2.00	3	75	79	80
2.01–4.00	5	81	83	84
4.01–6.00	7.5	75	80	81.5
6.01–8.00	10	80	84	85
8.01–12.0	15	81	85	86.5
12.1–16.0	20	80	83	86
16.1–20.0	25	83	86.5	88
20.1–26.1	30	83	86.5	88.5
26.2–34.8	40	85	88	88.5
34.9–43.5	50	80	85	87.5
43.6–52.2	60	84	88	89.5
52.3–65.2	75	87	89.5	90.5
65.3–87.0	100	84	89	91
87.1–114	125	85	89.5	91.5
115–136	150	86	89	91
137–182	200	88	91	92.5
183–227	250	90	92.5	93.5
228–273	300	90.5	93	94
274–318	350	91	93	94
319–364	400	91	93	93
365–409	450	91	93	93
410–455	500	91.5	93	93.5
456–545	600	93	94	94.5

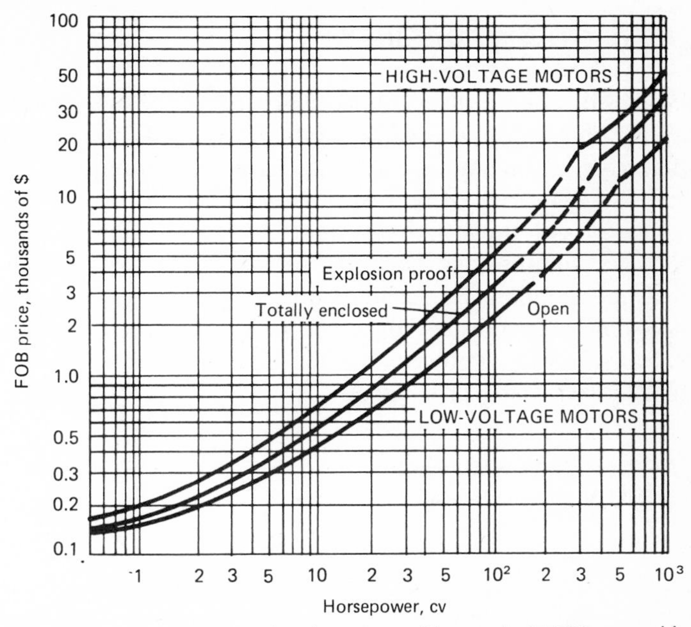

Fig. A5.1 Base price for electric motors with speed of 1,750 rpm, mid-1975.

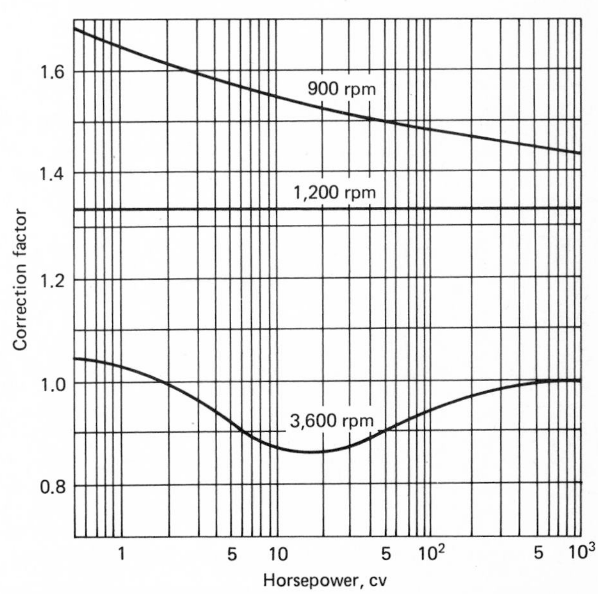

Fig. A5.2 Base price correction factor for motor speeds other than 1,750 rpm.

A5.2 STEAM TURBINES

Steam turbines with a single wheel are called *single stage;* those with two or more wheels on the same shaft are called *multistage.* Steam turbines exhausting steam under vacuum to a condenser are called *condensing turbines;* those exhausting to a steam system under pressure are called *back-pressure turbines;* and those exhausting to the atmosphere are called *noncondensing turbines.* Single-stage back-pressure turbines are frequently used for pump drivers. Condensing turbines are frequently used as drivers for large electric-power generators. Back-pressure turbines require careful analyses of a plant's energy requirements in order to benefit from the higher thermal efficiencies made possible.

A5.2.1 DETERMINING TURBINE PERFORMANCE

The efficiency of steam turbines can be obtained from Fig. A5.3 as a function of horsepower requirements and turbine type, and this base efficiency, which assumes steam inlet and outlet pressures of 7.6 and 1.25 bars absolute, respectively, is corrected for other inlet and outlet pressures by means of the factor obtained from Fig. A5.4. The horsepower to allow for the turbine is then the required brake horsepower of the pump divided by the corrected efficiency.

The steam consumption, kilograms per horsepower-hour, for turbines is obtained from Figs. A5.5 and A6.6. Figure A5.5 covers the range of back-pressure turbines operating on medium pressure steam; Fig. A5.6 covers turbines operating on high-pressure steam and discharging both against back pressure and to a vacuum condenser. The intermediate value, heat available in kilocalories per kilogram, shown in these figures can be increased 8% for each 50°C of superheat.

A5.2.2 PRICING TURBINES

Use Fig. A5.7 on page 332.

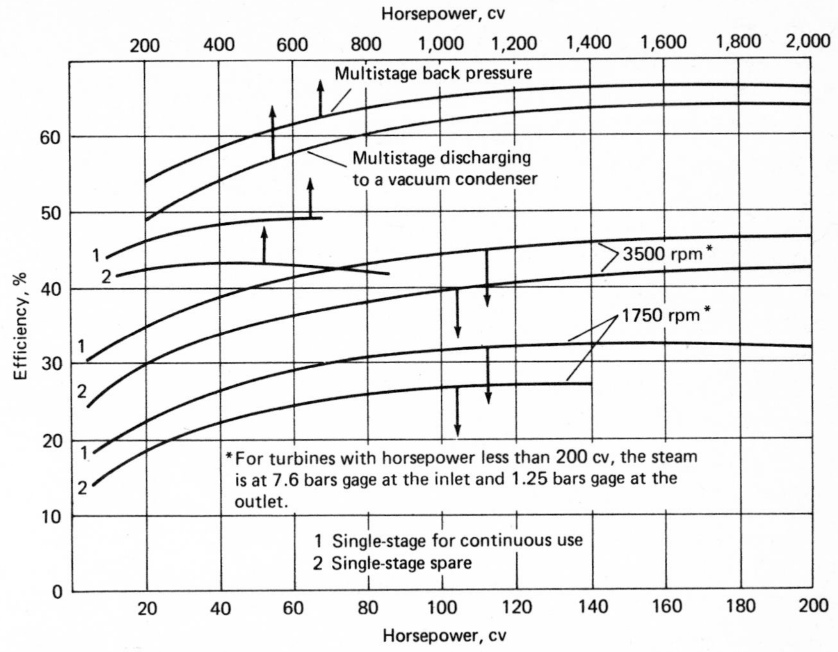

Fig. A5.3 Turbine efficiencies. For turbines with horsepower less than 200 cv. The steam is at 7.6 bars gage at the inlet and 1.25 bars gage at the outlet.

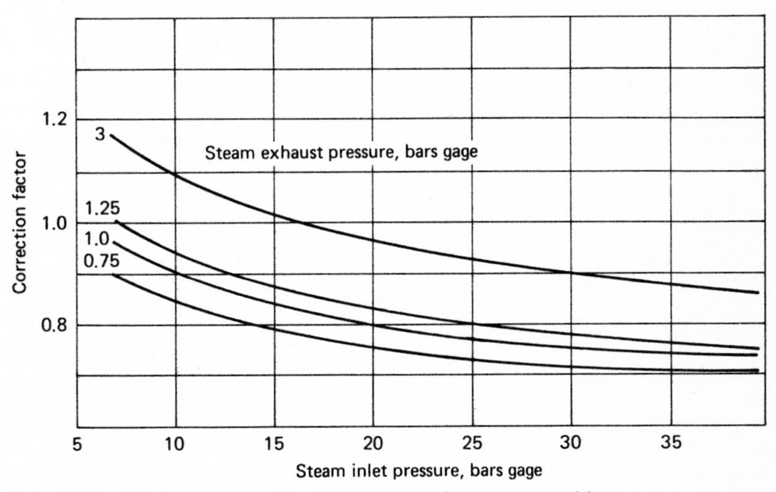

Fig. A5.4 Efficiency correction factor for back-pressure turbines.

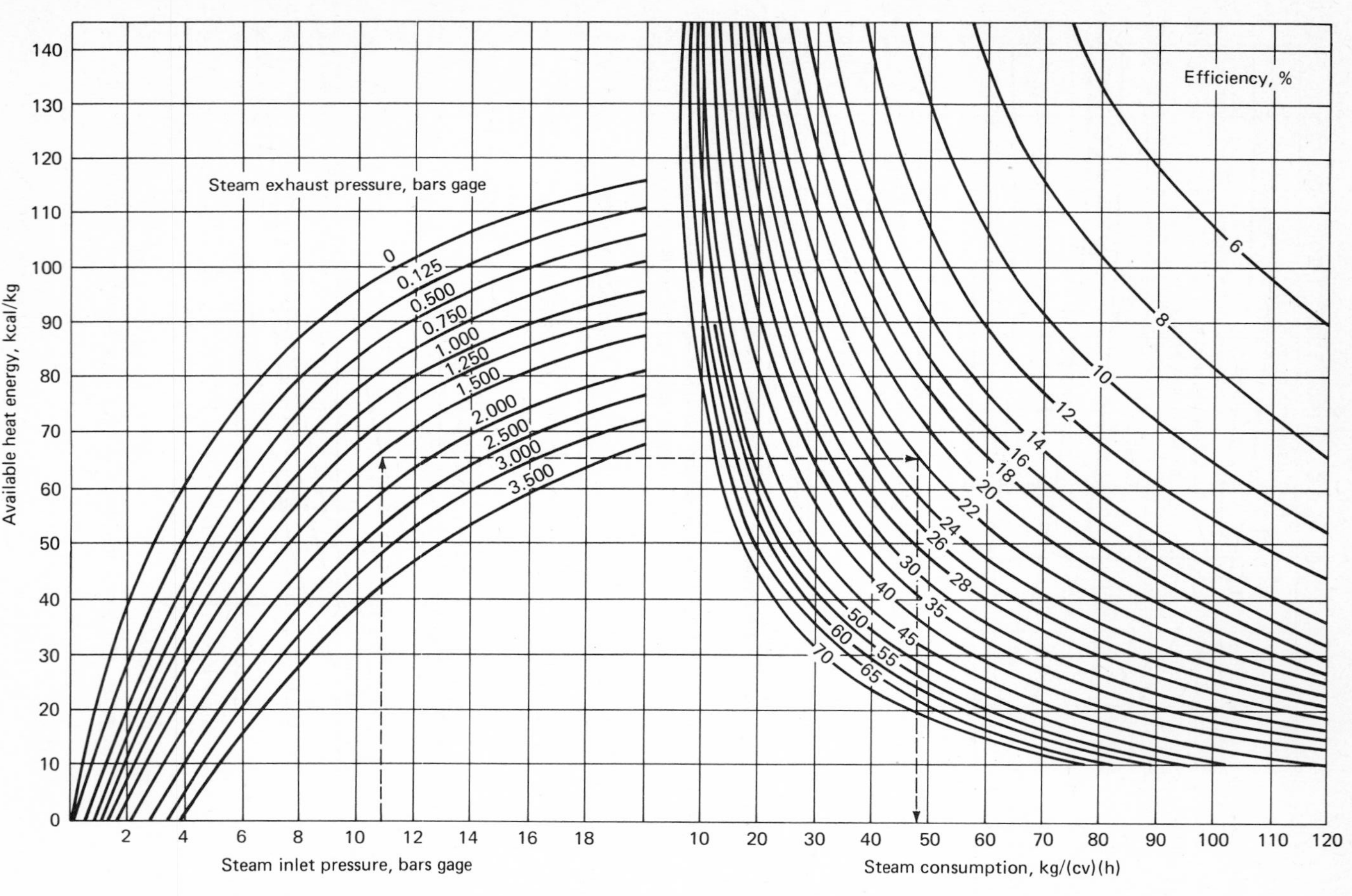

Fig. A5.5 Calculation for steam consumption in back-pressure turbines using medium-pressure steam.

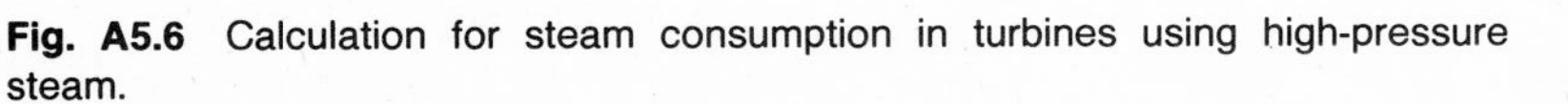

Fig. A5.6 Calculation for steam consumption in turbines using high-pressure steam.

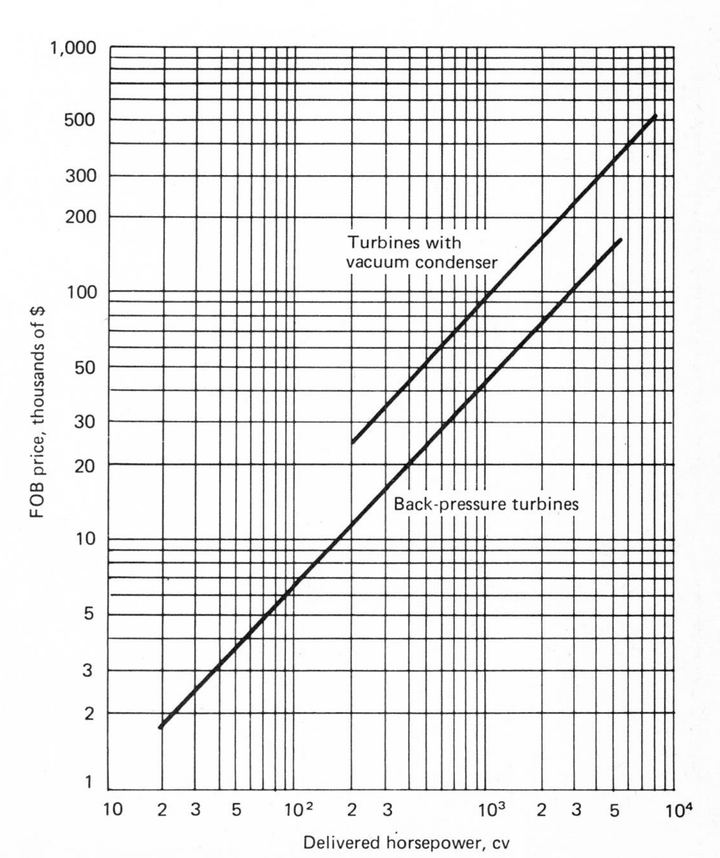

Fig. A5.7 Prices in mid-1975 for turbines.

Appendix 6

Process Design Estimation: Furnaces

During the late 1970s, engineering companies and furnace designers have expanded the practice of firing supplementary fuel in furnace convection sections. Such additional firing, which can considerably extend the size of a furance, is not considered here.

A6.1 GENERAL CHARACTERISTICS OF FURNACES

Two types of furnace are distinguished: fired heaters and reaction furnaces. Fired heaters are used in services where the temperatures are so high or the duties so large as to give them an advantage over heat exchangers. They can be used to heat a heat-transfer fluid, to heat and vaporize crude oil feeding an atmospheric distillation tower, or even to reboil a large distillation tower. Fired reactors, by contrast, are usually required to maintain a given high temperature once the furance feed has been heated to that point. Because of this required residence time, the design of fired reactors is usually more complex.

In either case, the design of a furnace is too complex to be reduced to a short-cut method for sizing furnaces. On the other hand, the duty of a furnace is a dominating primary variable that is easily determined and that can be used to price the furnace. In this case, however, the relatively small packaged furnaces should be distinguished from larger ones that are field erected.

A6.2 PRICING FURNACES

A6.2.1 OBTAINING THE BASE PRICE

Figure A6.1 gives approximate prices for fired heaters and reaction furnaces, not installed, as a function of their duty. The price obtained from this figure is based on the following assumptions:

Tubes in mild steel

Maximum pressure = 30 bars

Conventional application

A6.2.2 CORRECTION FACTORS

The base price of Fig. A6.1. is corrected for other conditions by the following expression:

$$\text{Corrected price} = (\text{base price})\ (1 + f_d + f_m + f_p)$$

where f_d = correction for furnace type
f_m = correction for materials of construction
f_p = correction for pressure

These correction factors are found in Table A6.1a to c for type, materials used, and pressure, respectively.

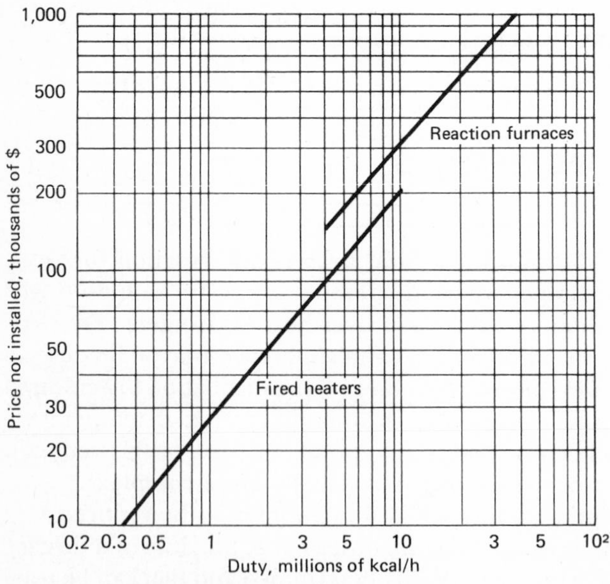

Fig. A6.1 Base price in mid-1975 for furnaces.

TABLE A6.1 Base-Price Correction Factors for Furnaces

a. Application

	Type	f_d
Fired heaters	Cylindrical furnaces	0.00
	Furnaces heating a heat-transfer fluid	0.15
Reaction furnaces	Preheaters	0.0
	Pyrolysis furnaces	0.1
	Others	0.3

b. Materials of Construction

Material	f_m
Carbon steel	0.00
Chrome-molybdenum	0.50
Stainless steel 18-8	0.75

c. Pressure

Pressure, bar	f_p
30	0.00
30–70	0.10
70–100	0.15
100–150	0.25
150–200	0.45
200–250	0.60

TABLE A6.2 Heat of Combustion of Various Fuels

Fuel	Low heating value, th/m³
Synthesis gas	4.5
Butane	30
Propane	23
Natural gas	10.6

A6.2.3 FUEL REQUIREMENTS FOR FURNACES

Fuel requirements can be calculated from the furnace duty by assuming an overall efficiency of 80% to determine the total heat liberated. This is then divided by the heating value of the fuel. Thus,

For heavy fuel oil, such as is burned in refineries, the heating value may be taken as 10,000 kcal/kg.

The heating value for fuel gases may be taken from Table A6.2.

Appendix 7 Process Design Estimation: Steam Ejectors

Vacuum steam ejectors do not require a precise analysis and estimate of installed cost, because their costs are small compared to the costs of other process equipment. However, it is important to determine the consumption of steam and water that such ejectors require. The vacuum that can be attained by one or more stages of steam ejectors is shown in Table A7.1.

Industrial plants generally use one or two stages. When the gas to the ejector contains condensable components, the vacuum system will include a precondenser and interstage condensors to reduce the flow of vapors and thus ejector-steam consumption. Also, an after-condenser is usually employed to prevent the steam from escaping to the atmosphere. The cooling fluid in these condensers will be usually cooling water or brine.

A7.1 CALCULATIONS FOR STEAM EJECTORS

A7.1.1 DETERMINING THE MAXIMUM SUCTION CAPACITY

The procedure is to (1) determine the vacuum, mm Hg, at the inlet, (2) determine the overall volume of the equipment to be evacuated, and (3) determine the maximum suction capacity. These three values have been related in a chart by the Heat Exchange Institute, which is redrawn in Fig. A7.1. The

TABLE A7.1 Suction Pressures Possible with Ejectors

Number of stages	1	2	3	4	5
Suction pressure, mmHg	$\geqslant 76$	10–100	1–25	0.15–4	0.05–0.20

suction flow obtained from Fig. A7.1 should be doubled for process design estimation.

A7.1.2 DETERMINING THE FLOW EQUIVALENT OF DRY AIR AT 20°C

A7.1.2.1 Concentration of Condensable Components

This calculation consists of (1) determining the partial pressure of the condensable components at the operating temperature, (2) calculating the partial pressure of noncondensables (usually air), and (3) calculating the moles of noncondensables; thus,

$$P_v = \Sigma P_{vi} \qquad P_{nc} = \pi - P_v$$

$$N_{nc} = \frac{2\text{ (maximum flow at inlet from Fig. A7.1)}}{\text{molecular weight of noncondensables}}$$

where P = partial pressure, bars
N = number of moles
π = vacuum pressure, bars
v = subscript denoting condensable vapor
vi = subscript denoting condensable component i
nc = subscript denoting noncondensables

The molecular weight of air is taken as 29; that of the condensables may be taken as the average, or the components may be each calculated separately by the ratio of partial pressures as follows:

$$N_v = N_{nc}\frac{P_v}{P_{nc}}$$

The mass flow, kg/h, of condensables, D_v, is then obtained from the molar flow:

$$D_{\text{v}} = N_v(\text{molecular weight})$$

If the average molecular weight of the condensables is used, the corresponding mass flow is calculated from N_v as above; if the molecular weights of individual components are used, the moles of each component is calculated from its partial pressure, the mass flow of each is calculated from its molar flow and molecular weight, and the mass flows for all the condensable components added.

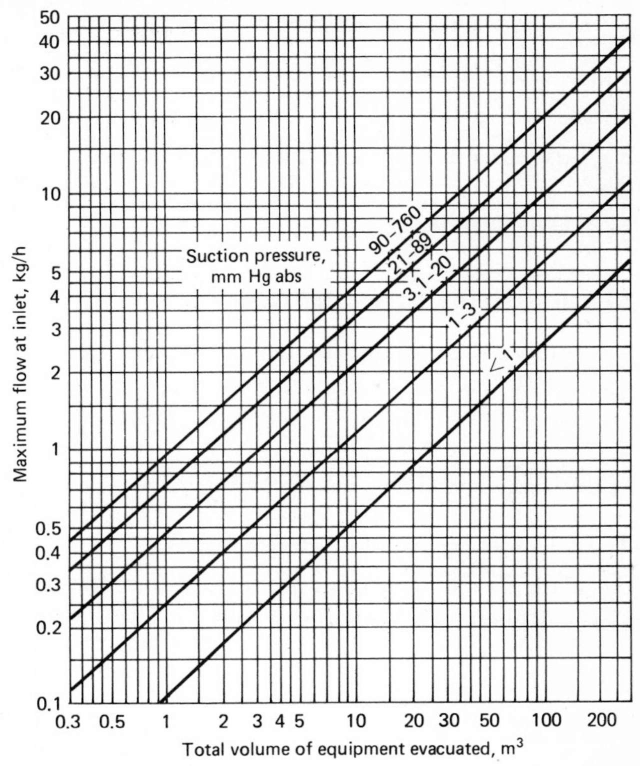

Fig. A7.1 Flow requirements for ejectors.

A7.1.2.2 Calculating the Flow of Dry-Air Equivalent at 20°C

The mass flows calculated above are now converted to their equivalent in dry air at 20°C by means of Fig. A7.2. Although this calculation is simple, it easily leads to errors, unless the specific calculation is identified with one of the following five cases and treated accordingly:

1. Air only: Obtain the entrainment coefficient corresponding to the operating air temperature and divide that coefficient into the calculated air flow.

2. Water vapor only: Obtain the entrainment coefficient corresponding to the operating temperature for water vapor, obtain the entrainment coefficient (0.80) corresponding to the molecular weight of water, multiply the two coefficients, and divide the multiple into the calculated flow of water vapor.

3. Mixtures of air and water vapor: Calculate the corrected air flow as in case 1.; calculated the corrected water-vapor flow as in case 2.; add the results.

4. Gaseous mixtures without water vapor: Calculate the average molecular weight; obtain the entrainment correction coefficient corresponding to this molecular weight; obtain the temperature coefficient as for air in case 1; multiply the two coefficients; and divide the multiple into the calculated mass flow.

5. Gaseous mixtures with water vapor: Calculate the air equivalent for the gases, as in case 4; calculate the air equivalent for the water vapor as in case 2; add the results.

A7.1.3 UTILITY CONSUMPTIONS

The utilities depend on whether or not the ejector system contains intercondensers. (Intercondensers are assumed part of the ejector system; precondensers and aftercondensers are calculated separately.) Without intercondensers, only steam is consumed; with intercondensers, a reduced amount of steam plus cooling water (or brine) is consumed.

A7.1.3.1 Ejectors without Intercondensers

Calculate the steam consumption as a function of the dry-air equivalent at 20°C calculated above, using Table A7.2 and interpolating when necessary.

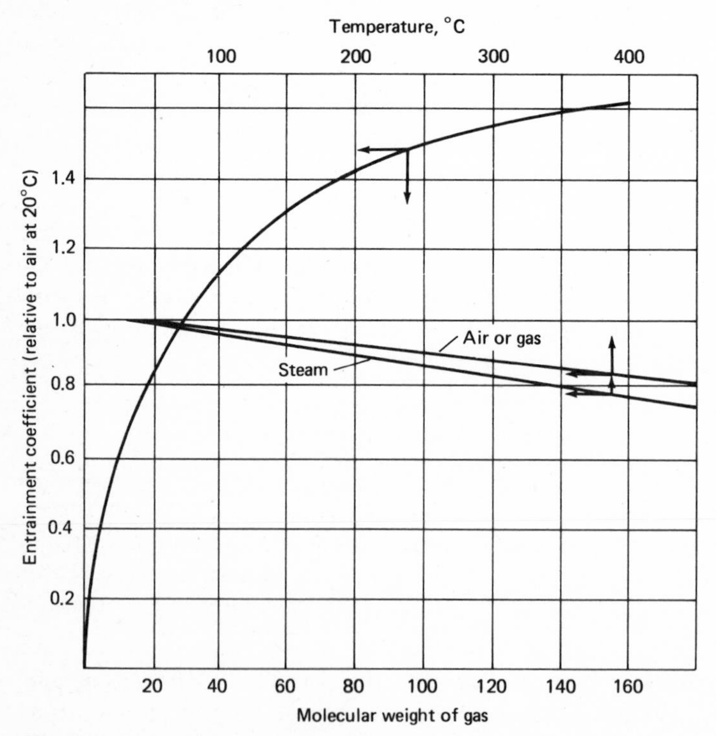

Fig. A7.2 Determining flow equivalent of dry air at 20°C.

TABLE A7.2 Base Steam Requirements for Ejectors Without Intermediate Condensers

Number of States	Suction Pressure, mm Hg	Kilograms of 7-bar Steam per Kilogram of Air Equivalent at 20°C
1	100	6.90
	125	4.90
	150	3.70
	175	3.00
	200	2.60
	225	2.30
	250	2.05
2	12.5	34.5
	25.0	16.5
	50.0	9.30
	75.0	6.75
	100.0	5.30
	125.0	4.60

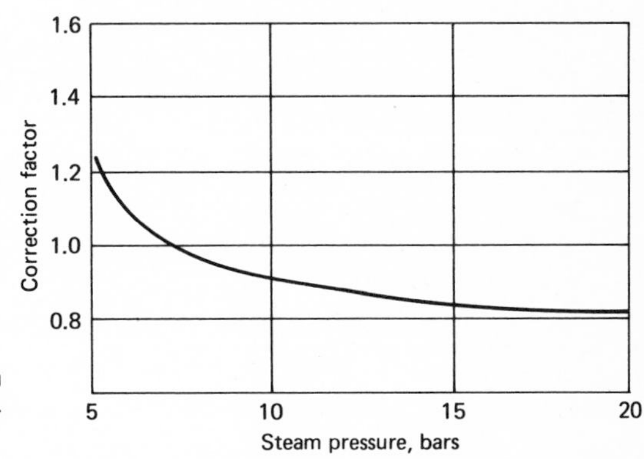

Fig. A7.3 Pressure-correction factor for ejector steam requirements.

The steam pressure, assumed as 7 bars in Table A7.2, can be corrected for in Fig. A7.3. Also, Table A7.2 does not provide data for more than two stages, since it is rare that an industrial ejector will use more than that number of stages.

A7.1.3.2 With Intercondensers

Calculate the weight percent of noncondensables in the inlet to the ejector, and use Table A7.3, interpolating where necessary, to obtain both steam and cooling water requirements. Correct the steam consumption, if its pressure is different from 7 bars, by means of Fig. A7.3.

The intercondenser may be a barometric condenser using direct contact with a water jet, but will most likely be a surface condenser, which can be sized and priced as prescribed in App. 3.

TABLE A7.3 Compressors with Intermediate Condensers: Base Requirements for Steam (kg/kg air) and Water (m^3/kg air)

Noncondensables, % wt.	Suction Pressure, mmHg	Kilograms of 7-bar Steam per Kilograms of Dry-Air Equivalent at 20°C	Cubic Meters of 20°C Water per Kilograms of Dry-Air Equivalent at 20°C
100	12.5	10.0	0.200
	25.0	6.30	0.125
	50.0	4.45	0.075
	75.0	3.50	0.057
	100.0	3.00	0.047
75	12.5	8.70	0.172
	25.0	5.45	0.122
	50.0	3.80	0.080
	75.0	3.00	0.065
	100.0	2.55	0.057
50	12.5	7.05	0.155
	25.0	4.45	0.112
	50.0	3.05	0.085
	75.0	2.40	0.070
	100.0	2.05	0.062
25	12.5	5.20	0.165
	25.0	3.20	0.110
	50.0	2.10	0.077
	75.0	1.65	0.067
	100.0	1.45	0.062

A7.2 PRICING EJECTORS

The mid-1975 prices for ejectors can be obtained from Figs. A7.4*a* through A7.4*c*, according to whether or not there are one or two stages, or with or without an intercondenser. These prices assume stainless steel ejector nozzles in cast iron bodies. Also, the intercondenser is assumed to be a barometric condenser. To obtain the price with a surface condenser between stages, use Fig. A7.4*b*, and calculate the intercondenser as prescribed in App. 3.

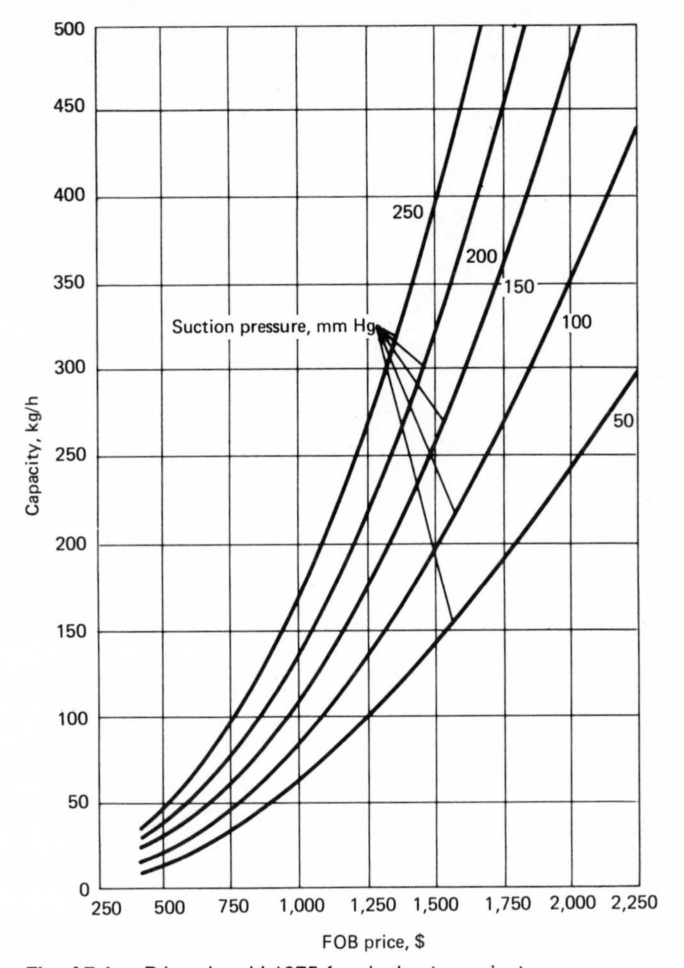

Fig. A7.4a Prices in mid-1975 for single-stage ejectors.

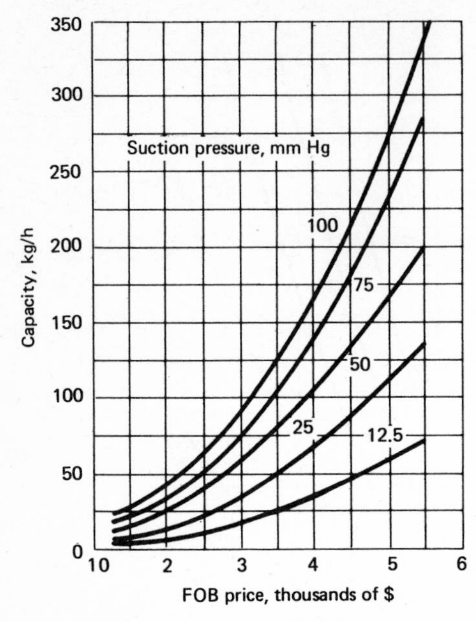

Fig. A7.4*b* Price in mid-1975 for two-stage ejectors without intermediate condensers.

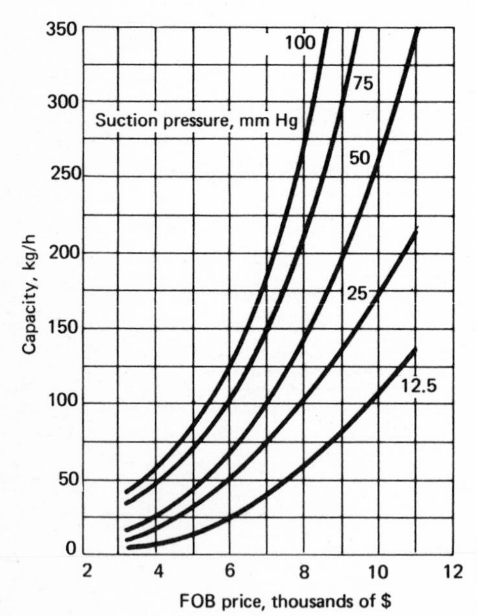

Fig. A7.4*c* Price in mid-1975 for two-stage ejectors with intermediate condensers.

Appendix 8

Process Design Estimation: Special Equipment

In a general way, process equipment can be divided into two broad categories: that which can be sized and designed from proven correlations, and that which requires pilot-plant tests to design for the specific application. Most of the latter category fall into the group designated here as special equipment.

Usually, this special equipment does not present problems for simple cost estimation, because the prices can be correlated against some basic parameter, such as surface or diameter. However, it is often difficult to relate those basic parameters to a simplified process design calculation that could be used for project analysis or feasibility studies. Thus it is usually necessary to proceed with empirical relationships or analogies, using the advice of the fabricators where possible.

The special equipment treated in this appendix includes

Dryers (Sec. A8.1)
Crystallizers (Sec. A8.2)
Evaporators (Sec. A8.3)
Filters (Sec. A8.4)
Centrifuges (Sec. A8.5)
Crushers and grinders (Sec. A8.6)
Cyclones (Sec. A8.7)
Conveyors (Sec. A8.8)

Vibrating screens (Sec. A8.9)
Instrumentation (Sec. A8.10)

A8.1 SELECTING, SIZING, AND PRICING DRYERS

A8.1.1 SELECTING A DRYER

Only a specialist can give a reliable decision as to the choice and subsequent size and price for a dryer, which all depend on the characteristics of the product treated. Consequently, it is usually necessary to consult the fabricators for an exact study. Meanwhile, the common types of dryers and their principal applications are summarized in Table A8.1.

A8.1.2 SIZING DRYERS

Despite the difficulties, it is possible to make some preliminary calculations, using the other appendixes where necessary. The discussion will be limited to those types of dryer given in Table A8.1.

A8.1.2.1 Drying by Adsorption

Sizing these dryers is analogous to sizing reactors; the amount of adsorbent determines the required volume of the vessel and its consequent dimensions. Since the drying operation is cyclic, it is necessary to include two or three identical vessels, so that one or two can be on the regenerating cycle while one is on the drying cycle. Also, the associated equipment for regeneration, such as furnaces, exchangers, pumps, etc., must be allowed for.

The optimum cycle duration, which affects the sizes of the regenerating equipment, must be determined by trial-and-error analysis. The primary variables of such an analysis consist of the fluid flow rate and the fluid's water content, plus the adsorbency of the alumina or molecular sieve. The fluid flow rate and water content set the rate at which water is to be removed; the adsorbency and mass of adsorbent set the overall capacity for water, and thus with the removal rate, the cycle time. Larger adsorbent vessels have larger cycle times with smaller regenerating equipment, and vice versa.

The capacity of adsorbents often deteriorates over the course of many regenerations, so that it is necessary to anticipate a reduced adsorbency in sizing calculations. For preliminary analysis, the following capacities and lifetimes may be assumed.

Lifetime:
For alumina = 150 cycles
For molecular sieves = 400 cycles

TABLE A8.1 Applications for Different Types of Dryers

Drying System		Type of Material Dried						
		Gas-Liquid	Suspensions	Slurry	Powders	Granular	Solids	Continuous films
Absorption, adsorption		X X	—	—	—	—	—	—
Refrigeration, compression, coalescence		X X	—	—	—	—	—	—
	Flash	X	X	X X	X X	X X	—	—
Direct	Spray	X X	X X	X X	X	—	—	—
	Fluid bed	X	X	X	X X	X X	—	X
Heat	Tunnel	—	—	X X	X X	X X	X X	—
	Tray	—	X	X	X	X	X	—
	Continuous sheeting	—	—	—	—	—	—	X X
	Cylinder	—	—	—	—	—	—	X X
	Drum	X	X	X	—	—	—	—
Indirect	Steam-tube rotary	—	X	X	X X	X X	—	—
	Screw conveyor	X	X	X X	X X	X X	—	—
Heat	Vibrating tray	—	—	—	X X	X X	—	—
	Vacuum rotary	—	X	X	X X	X X	—	—
	Batch vacuum	—	X	X X	X X	X X	X X	—

X X: Conventional application.
X: Limited application.
—: Rarely or never used.

Average adsorbency over a lifetime, kg/kg:
For alumina = 0.1
For molecular sieves = 0.2

A8.1.2.2 Drying by Heat Exchange

Because of the many different types of dryers, it is impossible not only to to enter into the methods for sizing each type but also to even give the precise information needed by the fabricators for their calculations.

Since dryers have standard dimensions, the volume of a charge and duration of the cycle are treated as primary variables, along with the properties of the material being dried, in dealing with batch dryers. Since drying consists of evaporating water, the basic heat transfer equation is also used:

$$Q = US\,\Delta t$$

where Q = heat transfered, kcal/h
U = overall transfer coefficient, kcal/(h)(m^2)(°C)
S = transfer surface, m^2
Δt = temperature difference, °C

U is assumed to be 100–350 kcal/(h)(m^2)(°C) for solutions of salts and 25–200 kcal/(h)(m^2)(°C) for slurries, powders, and granules.

Since all types of dryers cannot be discussed, Tables A8.2a to A8.2f will give typical performance data for only that equipment used in making bulk chemicals, catalysts, and sometimes mineral products.

Spray dryers The heat consumption of spray dryers is proportional to the thermal efficiency R, where

$$R = \frac{T_e - T_s}{T_s}$$

where T_e is the temperature of the drying gas entering the dryer, and T_s is the temperature of gas plus vapor leaving the dryer.

TABLE A8.2a Heat and Electricity Consumption for Spray Drying a 40% Slurry

Evaporation capacity, kg/h water	750	1,000	1,500	2,500	3,000
Heat duty, 10^6 kcal/h	0.70	0.95	1.40	2.30	2.70
Electricity consumption, kW	85	110	160	270	330

TABLE A8.2b Consumptions of Heat and Electricity for Flash Drying a 40% Slurry

Evaporation Capacity, kg/h	750	1,000	1,500	2,500	3,000
Heat duty, million kcal/h	0.80	1.10	1.70	2.80	3.30
Electricity consumed, kW	45	60	80	130	140

TABLE A8.2c Characteristics of Vacuum Rotary Dryers

Cylindrical						
Diameter, m	0.5	0.7	1	1.3	1.5	1.5
Length, m	1.2	2	3	6.5	8.5	10
Effective capacity, m³	0.1	0.3	1.5	3.5	6.5	8.0
Heating surface, m²	2	4	16.5	30	50	60
Motor hoursepower, cv	1	1.5	5	10	20	25
Double-cone						
Diameter, m	1	1.3	1.5	1.8	2.1	2.4
Effective capacity, m³	0.25	0.60	0.95	1.50	2.50	3.80
Heating surface	2.8	4.8	6.7	9.5	13.1	17.9
Motor horsepower, cv	2.2	3.0	4.0	5.5	11	15

The effective capacity is about 60% of total capacity

TABLE A8.2d Characteristics of Rotary-Drum Dryers

Hot Air Heat							
Diameter, m	1.2	1.4	1.5	1.8	2.1	2.4	3.0
Length, m	7.5	7.5	9	10	12	14	17
Lateral surface, m²	28	33	42	57	79	106	160
Horsepower of motor rotating the drum, cv	3	5	7.5	10	20	25	50
Horsepower of air blower, cv	5	7.5	7.5	10	15	25	30

Heat by Burner Off-Gas

Diameter, m	2.45				2.75		3.05				
Length, m	25	40	60	90	75	90	30	45	75	90	110
Lateral surface, m²	190	310	460	690	650	780	290	140	720	860	1,050
Motor horsepower, cv	20	30	40	60	60	100	40	50	75	100	125

Diameter, m	3.35				3.70				4	4.25	5
Length, m	50	75	110	125	75	100	125	140	150	125	180
Lateral surface, m²	530	790	1,160	1,310	870	1,160	1,450	1,630	1,880	1,670	2,830
Motor horsepower, cv	60	100	125	125	125	150	200	200	250	400	600

TABLE A8.2e Characteristics of Steam-Tube Rotary Dryers

Diameter, m	1		1.4		1.8				2.45				
Length, m	6	9	9	12	9	12	15	18	12	15	18	21	24.5
Heat-exchange surface, m²	31.5	48	95.5	128.5	154	207	260	313	424	530	635	741	846
Free surface, m²	0.6	0.6	1.3	1.3	2.4	2.4	2.4	2.4	4.0	4.0	4.0	4.0	4.0
Motor horsepower, cv	3	5	7.5	7.5	7.5	10	15	20	15	20	20	30	40

TABLE A8.2f Characteristics of Cabinet Dryers

Capacity, m^3	1	1.5	4	7
Surface of one tray, m^2	1	1.5	2.7	4
Maximum energy requirements, kW*	18	18	48	60

*Including heat and agitation.

The electricity consumption, primarily for the blower, is a function of the flowrate of the drying gas; which is in turn dependent on the temperature. Generally, the drying gases are at 150–350 °C, entering the dryer, and at about 80–90°C at the outlet. Table A8.2a gives typical energy consumptions for drying a slurry containing 40% water with a gas entering at 400 °C and leaving at 120 °C.

Flash dryers Table A8.2b gives typical energy consumptions for flash drying a slurry containing 40% water with a gas entering at 400 °C and leaving at 100 °C.

Vacuum rotary dryers Table A8.2c gives typical dimensions and capacities for vacuum rotary dryers. Two types are distinguished: the standard rotary cylinder with a central shaft from which scraping blades are extended to the walls, and the double cones in a V rotating about an axis through the V so as to mix the contents as they are heated against the walls. The basic heat-transfer equation can be used to calculate the steam consumptions and heat duties for these dryers.

Rotary-drum dryers These are in effect furnaces through which a hot gas is circulated among particles tumbled from flights as the drum rotates. The gas can be either heated air or burner off-gas. The drum walls may be lined with refractory for higher temperatures. Their typical dimensions and power consumption are given in Table A8.2d.

Steam-tube rotary dryers As their name implies, these consist of a cylinder rotating about a central shaft from which steam-heated or water-heated tubes project. Typical dimensions and power consumptions are given in Table A8.2e. The basic heat-transfer equation can be used to calculate duties and heat consumption; the overall transfer coefficient U for these dryers is 25–75 kcal/(h)(m^2)(°C).

Cabinet dryers Typical operating characteristics are given in Table A8.2f. Heating is by electricity; the air is circulated at a maximum temperature of 450 °C.

A8.1.3 PRICES FOR DRYERS

Figures A8.1*a* and A8.1*b* give order-of-magnitude prices for dryers for which data are presented in Tables A8.2a through A8.2f. The bases of the curves in the cost charts are the same as those for the tables. These base prices need to be corrected for other operating conditions as well as for materials of construction.

A8.1.3.1 Base-Price Corrections for Dryers

The base price for spray and flash dryers assumes stainless type 304. For other materials of construction but the same operating conditions, this base price can be corrected by the factors given in Table A1.12.

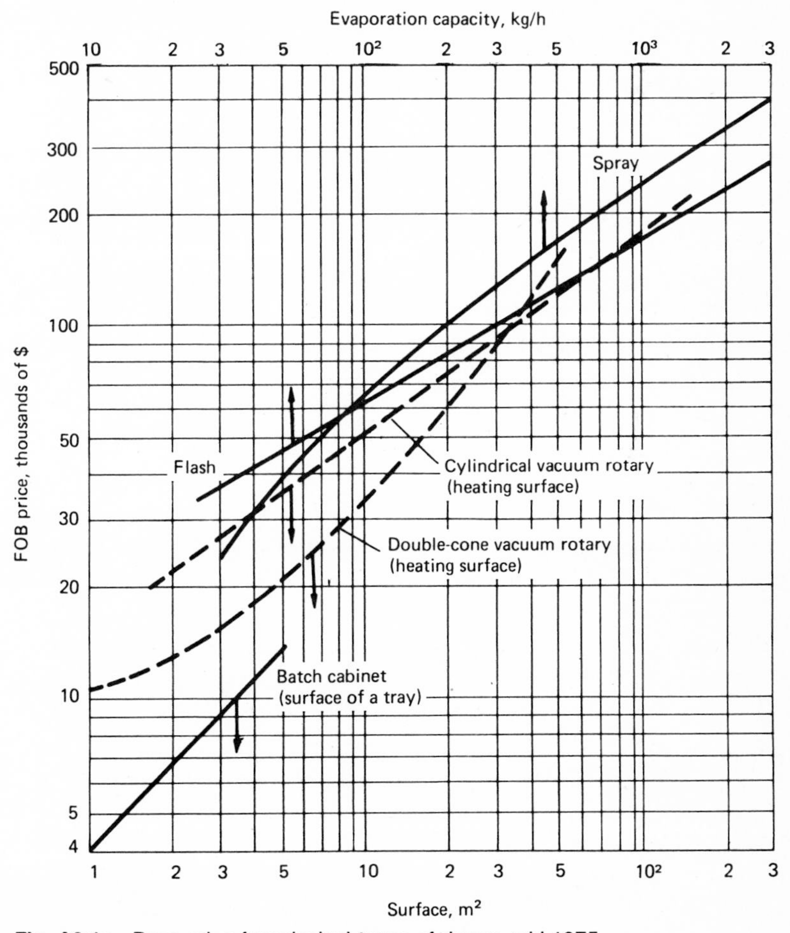

Fig. A8.1*a* Base price for principal types of dryers, mid-1975.

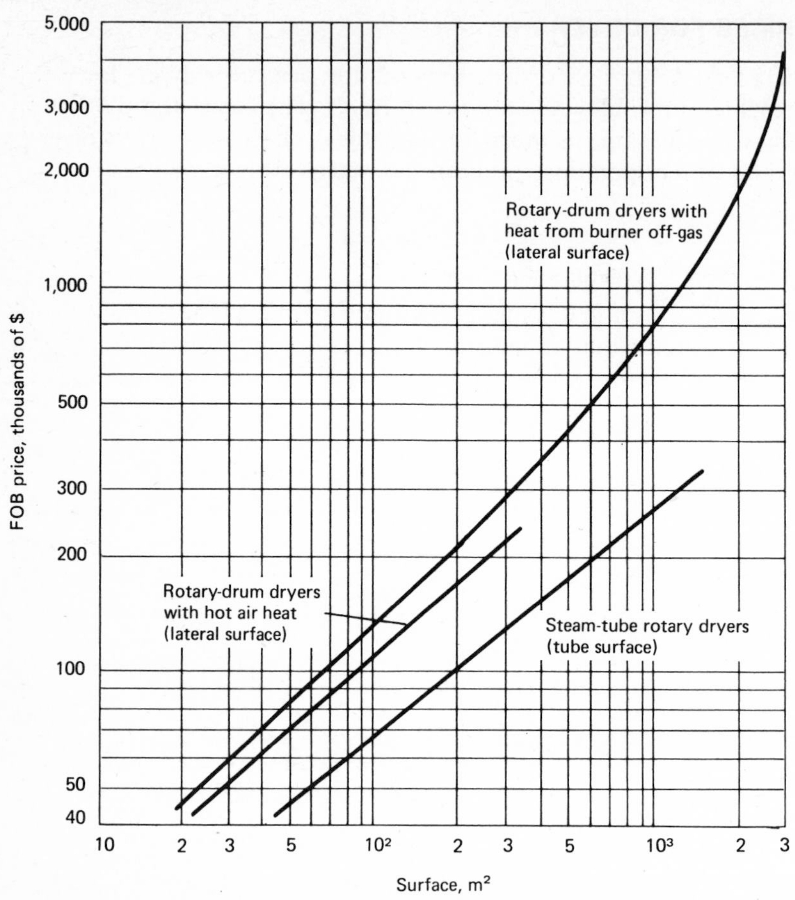

Fig. A8.1*b* Base price for principal types of dryers, mid-1975.

Rotary-vacuum dryers The base price for these dryers assumes construction of stainless type 304, and should be corrected for other materials according to Table A8.3a.

Rotary-drum dryers The base price for these dryers assumes hot air and steel, and should be corrected for other drying gases and materials of construction according to the formula

$$\text{Corrected price} = (\text{base price})\ (1 + f_g + f_m)$$

where f_g is the correction for drying gas and f_m is the correction for materials of construction, as obtained from Table A8.3b.

Steam-tube rotary dryers The base price assumes steel construction and should be corrected for construction in stainless type 304 by multiplying by 1.75.

Cabinet dryers The base price for these dryers assumes atmospheric pressure and steel construction and should be corrected for vacuum operation and construction in stainless type 304 by the formula

$$\text{Corrected price} = (\text{base price})\, f_p f_m$$

where f_p is the pressure-correction factor and f_m the materials-correction factor, as obtained from Table A8.3c.

TABLE A8.3a Base-Price Correction Factors for Construction Materials in Rotary-Vacuum Dryers

Material	Type of Dryer	Effective Capacity, m^3	f_m
Mild steel	Cylindrical	All	0.65
	Double cone	All	0.70
		0.25	1.25
Glass-lined		0.60	1.13
steel	Double cone	0.95	1.05
		1.50	1.00
		2.50	0.95
		3.80	0.90

TABLE A8.3b Base-Price Correction Factors for Drying Gas and Construction Materials in Rotary-Drum Dryers

Drying gas	f_g
Hot air	0.00
Combustion gas (direct contact)	0.12
Combustion gas (indirect contact)	0.35
Materials	f_m
Mild steel	0.00
Lined with stainless 304—20%	0.25
Lined with stainless 316—20%	0.50

TABLE A8.3c Base-Price Correction Factors for Pressure and Construction Material of Cabinet Dryers

Pressure	f_p
Atmospheric pressure	1.0
Vacuum	2.0
Material	f_m
Mild steel	1.0
Stainless type 304	1.4

A8.2 CRYSTALLIZERS

The sizing, choice, and price of a crystallizer are primarily a function of the solution and the crystalline product obtained from that solution, particularly the size of the crystals. Only a fabricator can give reliable decisions for a price application. However, there are some curves that can be used for a first approximation.

A8.2.1 CRYSTALLIZATION EFFICIENCY

The theoretical efficiency of a crystallizer is given by the formula

$$C = R\,\frac{100W_0 - S(H_0 - E)}{100 - S(R - 1)}$$

where C = crystalline product (hydrate or not), kg
R = ratio of the molecular weight of the hydrated product to that of the anhydrous product
S = solubility of anhydrous product, parts per 100 at the final operating temperature
W_0 = weight of anhydrous product in the initial solution, kg
H_0 = weight of solvent in the initial solution, kg
E = amount of solvent evaporated during the operation, kg

When E is not known, it can be determined by the expression

$$E = \frac{(W_0 + H_0)C\,\Delta t[100 - S(R - 1)] + q_cR(100W_0 - SH_0)}{L_w[100 - S(R - 1)] - q_cRS}$$

where C = specific heat of the initial solution, kcal/(kg)(°C)
Δt = temperature difference from inlet to outlet, °C
q_c = latent heat of crystallization of the crystals, kcal/kg
L_w = heat of vaporization of the solution, kcal/kg

When a production capacity is given, these relationships permit calculation of a feed rate to the crystallizer.

A8.2.2 CHOOSING A CRYSTALLIZER

The choice of a crystallizer is essentially a match between one of three general types of operation and one of five principal types of equipment. The types of crystallization operation are

1. Cooling with no or only limited evaporation.
   a. Cooling with ambient air, when the feed is hot.

   b. Cooling with an intermediate fluid, such as can be done with coils or jacketed vessels, either continuously or batch.
2. Evaporation concentration with no cooling, which consists essentially of evaporation.
3. Evaporation and cooling.

The principal types of industrial crystallizers are

1. Forced-external-circulation crystallizers, operating
   a. By either evaporation or cooling.
   b. With or without baffles.
   c. With heat removal by a tubular exchanger or by direct contact.
2. Forced-internal-circulation crystallizers using a draft tube, operating
   a. By either evaporation or cooling.
   b. With or without baffles.
   c. With heat removal by a tubular exchanger or by direct contact.
3. Seeded crystallizers, typically Oslo crystallizers, operating
   a. By either evaporation or cooling.
   b. With heat removal by a tubular exchanger or by direct contact.
4. Scraped surface exchangers.
5. Batch crystallizers, cooled or not, and agitated or not.

A8.2.3 PRICING CRYSTALLIZERS

Figure A8.2 shows the cost of crystallizers as a function of their continuous capacity, t/h, or of their batch capacity, m^3, for seeded crystallizers, draft-tube crystallizers, external forced-circulation crystallizers and vacuum batch crystallizers. This cost is a base price, which assumes

Steel construction

All accessories, such as vacuum equipment, in place

Ninety percent of the product larger than 16 mesh for seeded crystallizers

Ninety percent of the product larger than 20 mesh for forced circulation crystallizers

Price modifications for different performance or different materials of construction can be obtained from Tables A8.4a and A8.4b.

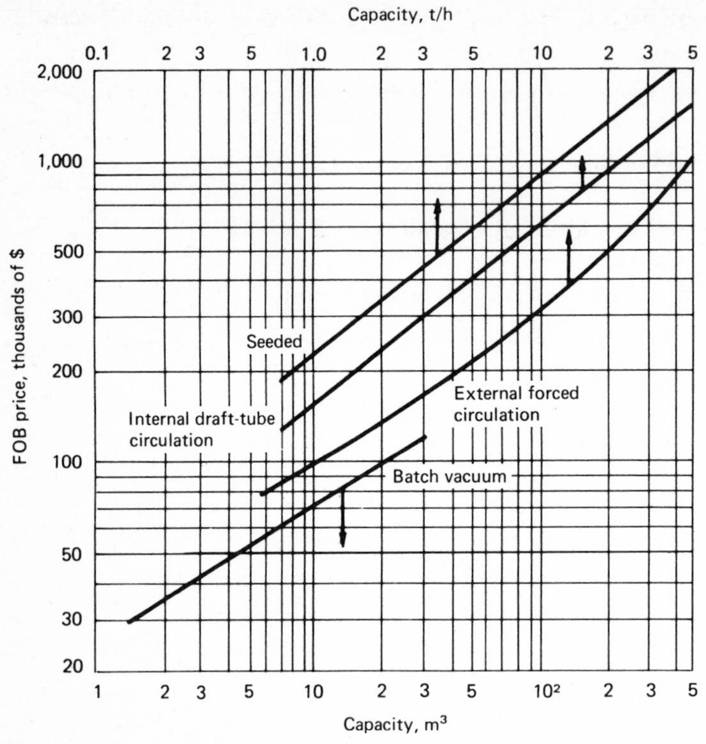

Fig. A8.2 Base price for crystallizers, mid-1975.

TABLE A8.4a Base-Price Correction Factor for Particle Size in Seeded Crystallizers

Dimensions	Correction factor
90% at +16 mesh	1.00
90% at +20 mesh	0.90
90% at +30 mesh	0.75

TABLE A8.4b Base-Price Correction Factor for Construction Materials of Crystallizers

Type	Material	f_m
Forced circulation	Mild steel	1.0
	Stainless type 304	2.5
Vacuum batch	Mild steel	1.0
	Rubber-lined	1.3
	Stainless type 304	2.0

A8.3 EVAPORATORS

All of the evaporators considered here depend on the indirect transfer of heat of evaporation. Consequently, the heat-exchange surface is the principal parameter for establishing a price. First, an evaporator type is selected to establish the type of transfer surface; then a transfer surface is estimated; and finally a price is obtained.

A8.3.1 CHOOSING AN EVAPORATOR TYPE

The principal types of evaporator include climbing film, falling film, shell and tube, jacketed, and thin film.

Climbing-film evaporators are used for clear, relatively dilute solutions far from saturation; falling film serve the same general areas of application and extend them to viscous solutions and heat-sensitive solutions, such as might be found in the food industry.

The shell-and-tube evaporators can be further classified:

1. Short-tube evaporators, with either vertical or horizontal tubes
2. Long-tube vertical evaporators
3. Forced-circulation evaporators

Conventional shell-and-tube evaporators are suited to solutions containing solids and close to saturation, but not viscous or toxic. Some types are equipped with agitators to assure better circulation. The addition of forced circulation, with a large-flow pump, extends the applications of this type to include viscous, foaming, and fouling liquids containing solids or tending to precipitate solids.

The jacketed evaporators are used to treat viscuous or heat-sensible liquids. They have a jacketed cylindrical shell, either horizontal or vertical, in which a shaft turns wall scrapers.

Thin-film evaporators are characterized by very short residence times and small volumes of product. They consist of a vertical tube, jacketed on its lower part, and equipped with a high-speed agitator at the bottom.

A8.3.2 SIZING EVAPORATORS

The basic heat-transfer equation applies here.

$$S = \frac{Q}{U \, \Delta t}$$

where S = surface, m^2
Q = duty, kcal/h

U = overall heat-transfer coefficient, kcal/(h)(m^2)(°C)
Δt = average temperature difference, °C

A8.3.2.1 Determining the Overall Transfer Coefficient

As with dryers, the most difficult term of this equation is the heat-transfer coefficient. Because of the many intervening factors, such as dimensions, agitation, etc., an accurate determination of the heat transfer coefficient for a specific application usually requires the help of an evaporator fabricator. However, different values have been proposed as a function of the temperature, of the viscosity, or of the fluid movement.

Thus the coefficient for thin-film evaporators can be obtained from Table A8.5, and for temperatures between 30 and 120°C, the coefficients for jacketed, forced-circulation, and falling-film evaporators can be obtained from the following formulas:

1. For evaporators with shell and agitator:

$$U = 21t - 355$$

2. For forced-circulation evaporators:

$$U = \begin{cases} 16t + 515 & \text{when the fluid velocity is about 1.5 m/s} \\ 21t + 865 & \text{when the fluid velocity is about 2 m/s} \\ 21t + 1{,}030 & \text{when the fluid velocity is about 3 m/s} \end{cases}$$

3. For falling-film evaporators:

$$U = 33t + 215$$

where t = fluid temperature, °C
U = heat-transfer coefficient, kcal/(h)(m^2)(°C)

TABLE A8.5 Heat-Transfer Coefficients for Thin-Film Evaporators

Applications	U, kcal/(h)(m^2)(°C)
Steam heated	
Concentration of aqueous solutions	1,100–1,600
Dehydration of organic solutions	700–950
Distillation of organic liquids	700–1,000
Stripping of organic liquids	400–600
Reboiler service	850–1,000
Solvent recovery	750–800
Heated by heat-transfer fluid	
Distillation of high-boiling-point liquids	350–500
Stripping of organic liquids	200–350
Reboiler service	500–550

TABLE A8.6a Horsepower Requirements for Scraped-Film Jacketed Evaporators

Horizontal		Vertical	
Heated surface, m^2	Agitation power, cv	Heated surface, m^2	Agitation power, cv
0.2	2.5–4	2	7.5–15
0.5	3.5–7	3	10–20
1.0	5–10	4	10–20
2.0	7.5–15	6	15–25
3.0	10–20	8	15–30
4.0	10–20	12	15–30
5.0	15–25	18	25–40

TABLE A8.6b Characteristics of Forced-Circulation Evaporators

Fluid velocity, m/s	1.5	2	2.5	3
Heat-transfer coefficient, kcal/(h)(m^2)(°C)	2,800	3,400	3,900	4,200
Heating surface, m^2	170	155	145	140
Pump power, cv*	26	32	40	50

*For a flow of 1,500 m^3/h.

A8.3.2.2 Operating Characteristics

The operating characteristics typical of scraped-film jacketed evaporators and forced-circulation evaporators are shown in Tables A8.6a and A8.6b, respectively.

A8.3.3 EVAPORATOR COSTS

A8.3.3.1 Base Price

Figure A8.3 gives the base prices for different evaporators as a function of the heat-transfer surface. These prices assume the following materials of construction:

Jacketed evaporator: stainless type 316L

Thin-film evaporator: stainless type 316L

Shell-and-tube evaporator: shell in steel and tubes in copper

Forced-circulation evaporator: shell in steel and tubes in copper

Falling-film evaporator: internals in stainless type 316L, externals in steel

When a specific evaporator differs from these materials, its base price should be multiplied by a correction factor obtained from Table A8.7a through A8.7c.

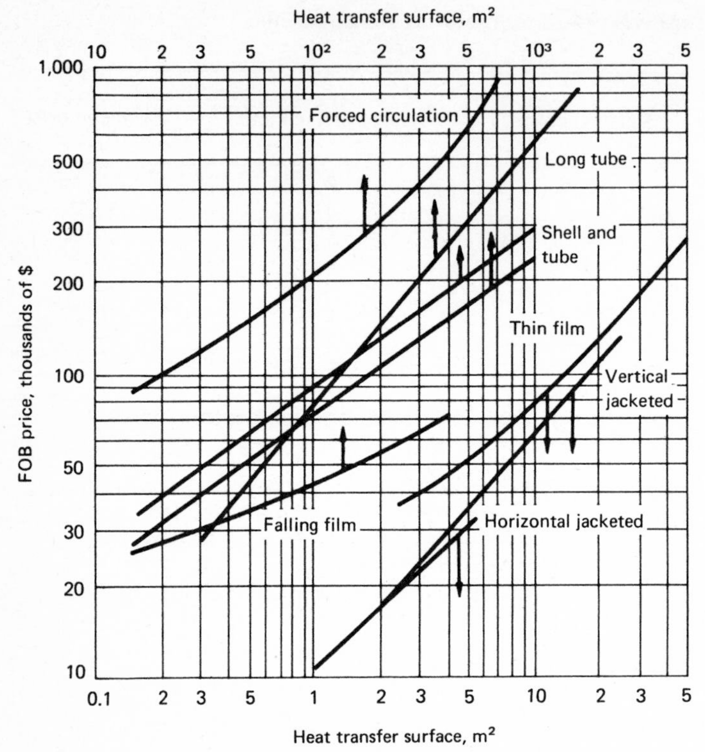

Fig. A8.3 Base prices for various types of evaporator, mid-1975.

A8.4 FILTERS

The choice of a filter depends on the concentration of solids, the fineness of the solids particles, and the type of operation imposed by the process. Its size, which is essentially its surface, depends on the required filtration rate, and its cost depends on the size and the type of filter required.

A8.4.1 CHOOSING A FILTER

The many kinds of filter can be classified as either batch or continuous. Some varieties of batch filters are

Cartridge
Basket
Sand
Bag
Sheet

Candle
Vertical-leaf
Level-leaf
Horizontal-leaf
Plate and frame

and some varieties of continuous filter are

Rotary pressure
Rotary vacuum
Belt
Horizontal-disk vacuum

A choice among these will not only take into account whether the operation should be continuous or batch but also the method for recovering the cake and the manner of its washing, etc.

A8.4.2 SIZING FILTERS

The data for calculating a required filter surface require pilot-plant experiments; and it is practically impossible to give any general rule for estimat-

TABLE A8.7 Base-Price Correction Factors

a. Jacketed Evaporators	
Construction material	f_m
Mild steel	0.70
Stainless 304 L	0.95
Stainless 316 L	1.00
Hastelloy G	2.50

b. Long-Tube Evaporators	
Construction Material: Shell/Tube	f_m
Steel/ copper	1.0
Steel/ steel	0.6
Steel/ aluminum	0.7
Nickel/ nickel	3.3

c. Forced-Circulation Evaporators	
Construction Material: Shell/Tube	f_m
Steel/ copper	1.00
Monel/ cupronickel	1.35
Nickel/ nickel	1.80

TABLE A8.8 Operating Characteristics of Common Types of Filters

Type of filter	Solids Concentration, %	Minimum Particle Size, microns	Cake Handling*		
			Recovered	Washing	Drying
Vertical leaf	<1–5	1–5 to 25–50	X X	X	X
Level leaf	<5–10	1–5 to 25–50	X X	X X	X X
Automatic frame	1 to 50	1–5 to 25–50	X X	X X	X X
Rotary vacuum	5 to 25	5–50 to 150–250	X X	X	X
Belt	5–10 to 50	15–50 to 150–250	X X	X X	X
Horizontal vacuum disk	5–10 to 50	15–50 to 150–250	X X	X X	X

*X X: conventional; X: possible.

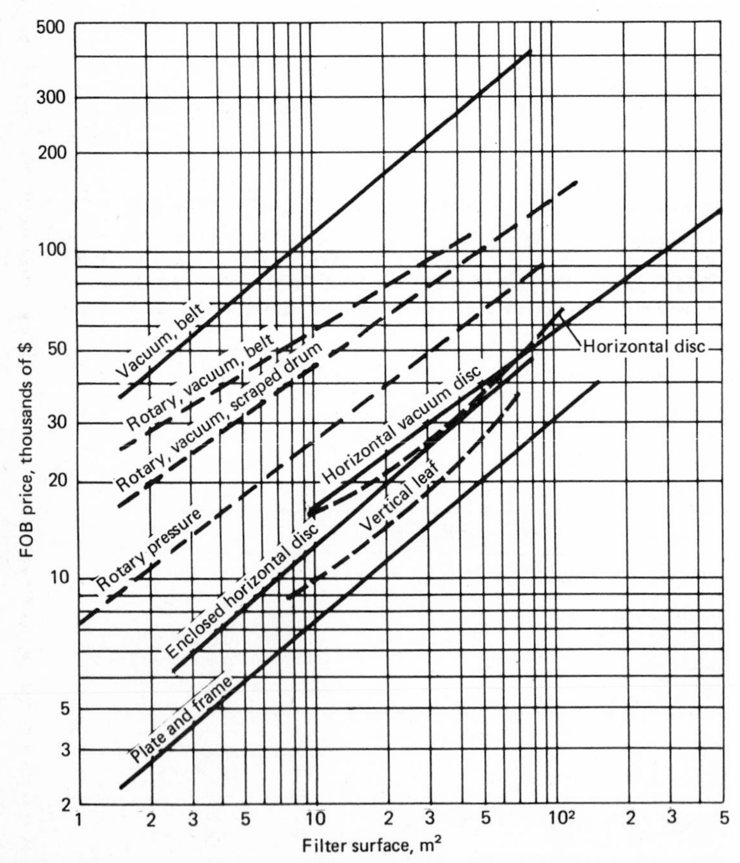

Fig. A8.4 Base prices in mid-1975 for various types of filter.

TABLE A8.9 Base-Price Correction for Construction Materials

Material	Correction Factor
a. Leaf Filters	
Mild steel (shell and leaves)	0.4
Stainless 304 (leaves only)	0.6
Stainless 304 (shell and leaves)	1.0
Stainless 316 (leaves only)	0.9
Stainless 316 (shell and leaves)	1.3
b. Automatic Plate and Frame Filters	
Mild steel	1.00
Aluminum	1.75
PVC lined	2.00
Rubber lined	2.50
Stainless 304	2.80
Stainless 316	5.00

ing filter surface requirements. Table A8.8 gives operating characteristics for the types of filter shown in pricing graph Fig. A8.4, but no indication as to sizes. The size of the motors used to drive belt filters ranges from 4 to 10 cv.

A8.4.3 PRICING FILTERS

Base prices are given as a function of the filter surface for nine types of filter in Fig. A8.4. This base price includes the cloths and pump for the automatic filter presses, and it assumes the following materials of construction:

Stainless type 304 for the rotary vacuum

Steel for the rotary pressure

Stainless type 316 with a rubber belt for belt filters

Stainless type 304 for the horizontal disk vacuum

Stainless type 316 for the enclosed horizontal disk

For other materials of construction, the base price should be multiplied by a correction factor obtained from Table A8.9a and b.

A8.5 CENTRIFUGES

A8.5.1 OPERATING CHARACTERISTICS

Refer to Table A8.10 for the operating characteristics of the main types.

TABLE A8.10 Typical Operating Characteristics of Conventional Centrifuges

Type	Use*	Solids, %	Particle Diameter, microns
Basket centrifuges:			
Batch			
With pusher	c	20–75	10–5,000
With knives	b (?) and c	5–60	0.7–1,000
Free-casing	c	5–60	0.7–1,000
Continuous helical conveyor	b (?) and c	2–40	0.3–3,000
High-speed disk separators:			
Nozzle discharge	b and c	5–25	0.07–70
Self-opening	b and c	1–10	0.07–70
Hermetic bowl	a, b, and c	0–1	0.07–70

*a: liquid-liquid separations; b: liquid-liquid-solid separations; c: liquid-solid separations.

A8.5.2 SIZING CENTRIFUGES

Refer to Fig. A8.5*a* to obtain bowl or basket diameters for the main types of centrifuge as a function of capacity or power. It should be borne in mind that these sizes are only approximate and that a fabricator should be consulted for reliable information.

A8.5.3 PRICING CENTRIFUGES

Prices for basket centrifuges are shown in Fig. A8.5*b*, as a function of the basket diameter. These curves assume stainless-type-316 construction, which is most common for this machinery. If steel is to be used, the stainless steel price should be divided by 2. The prices for high-speed disk separators, also in stainless type 316, are given in Fig. A8.5*c*.

A8.6 CRUSHERS AND GRINDERS

A8.6.1 CHOICE

Table A8.11, p. 367, permits a quick choice of type of crusher or grinder according to the properties of the material to be processed.

A8.6.2 PRICING CRUSHERS AND GRINDERS

Base prices are given in Fig. A8.6, p. 367. This base price assumes construction in steel; it should be modified according to Tables A8.12a through A8.12g, p. 368–369. The cost of bars for a bar mill was $420 per metric ton in mid-1975. Motors can be priced from Figs. A5.1 and A5.2.

A8.7 CYCLONE DUST SEPARATORS

Cyclone dust separators are mainly used to separate particles ranging between 10 μ and 1 m in diameter. Finer particles are removed by means of small bore multicyclones operating in series with higher pressure drops. When the particles or the gas to purify are corrosive, appropriate materials of construction should be used. Otherwise cyclones are usually of steel. Figure A8.7*a*, p. 370,

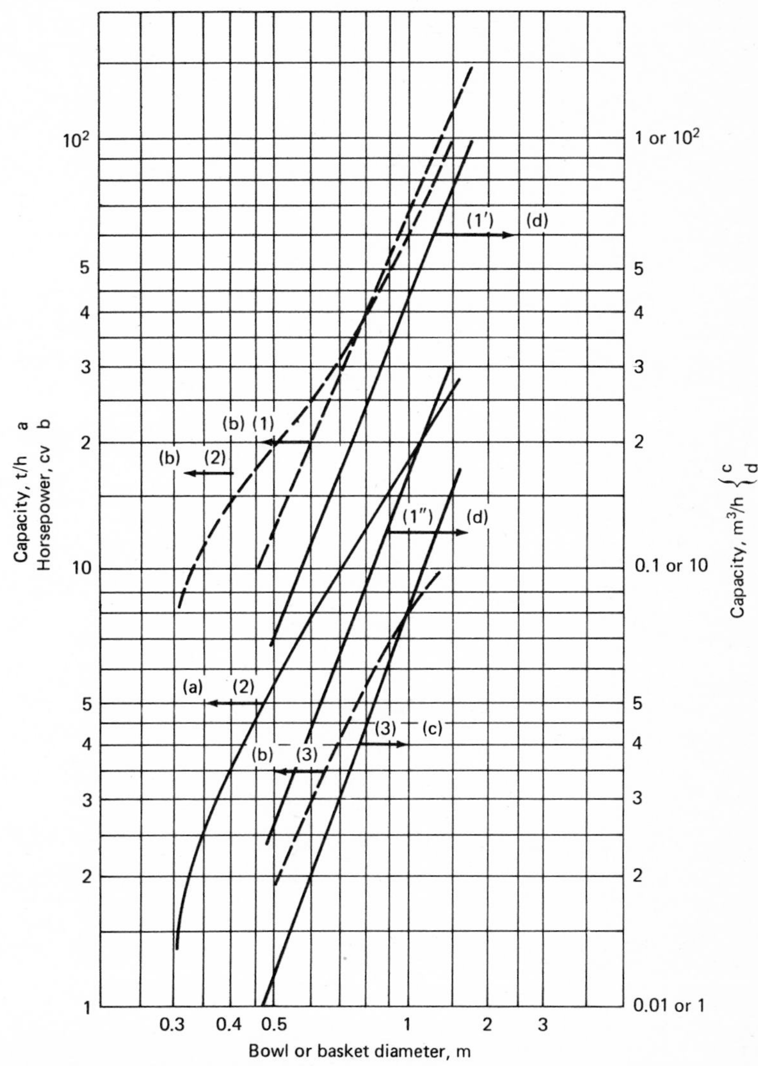

Fig. A8.5*a* Bowl or basket sizes for centrifuges. (1) Continuous helical conveyor; (1′) capacity in solids; (1″) maximum liquid flow rate; (2) centrifuges with pusher; (3) bottom-drive centrifuges, batch.

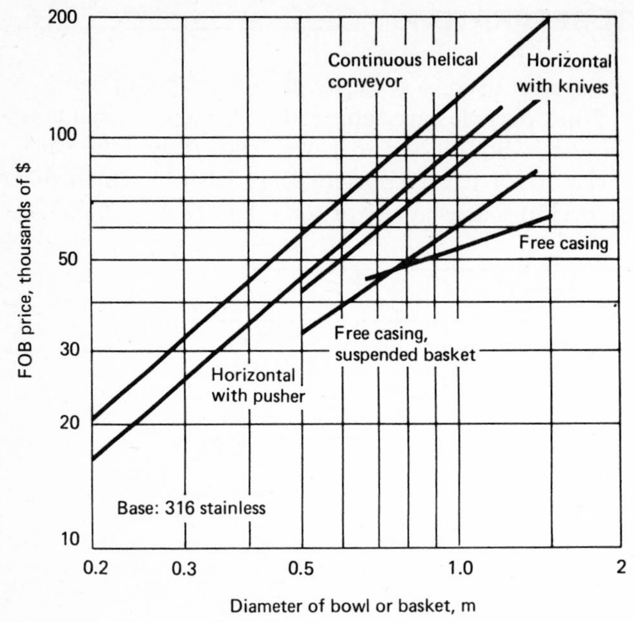

Fig. A8.5*b* Price in mid-1975 for centrifuges.

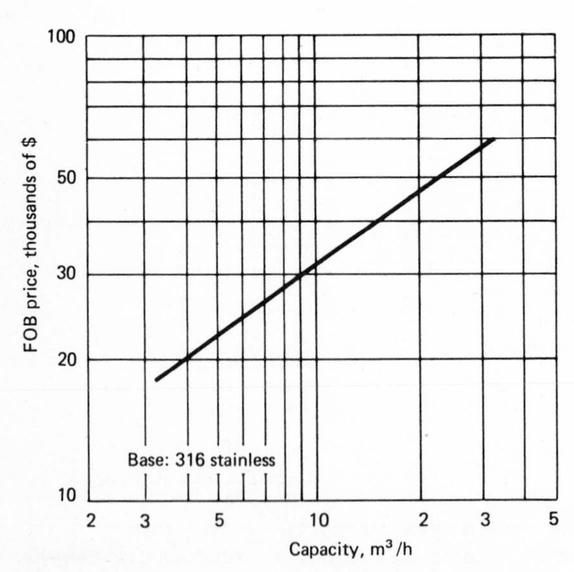

Fig. A8.5*c* Price in mid-1975 for self-opening, high-speed disk separators.

TABLE A8.11 Classification of Crushers and Grinders

Operation	Hardness of Product	Dimensions, cm Feed	Product	Type of Equipment
Crushing				
Primary	Hard	150–30	50–10	Jaw and gyratory crushers
		50–10	12–2.5	Roller and bar mills
Secondary	Hard	12–2.5	2.5–0.5	Jaw and gyratory crushers
		4–0.6	0.5–0.08	Roller, bar, disk, and hammer mills
	Weak	50–10	5–1	Roller, disk, hammer, and rotating knife mills
Grinding				
Coarse	Hard	0.5–0.08	0.06–0.008	Roller, disk, hammer, and rotating knife mills Mills with low and medium tangential speeds
Fine	Hard	0.12–0.015	0.008–0.001	Mills with low, medium, and high tangential speeds Fluid mills
Particle separation				
Coarse	Weak	1.2–0.17	0.06–0.008	High-speed mills or fluid mills
Fine	Weak	0.4–0.05	0.008–0.001	High-speed mills or fluid mills

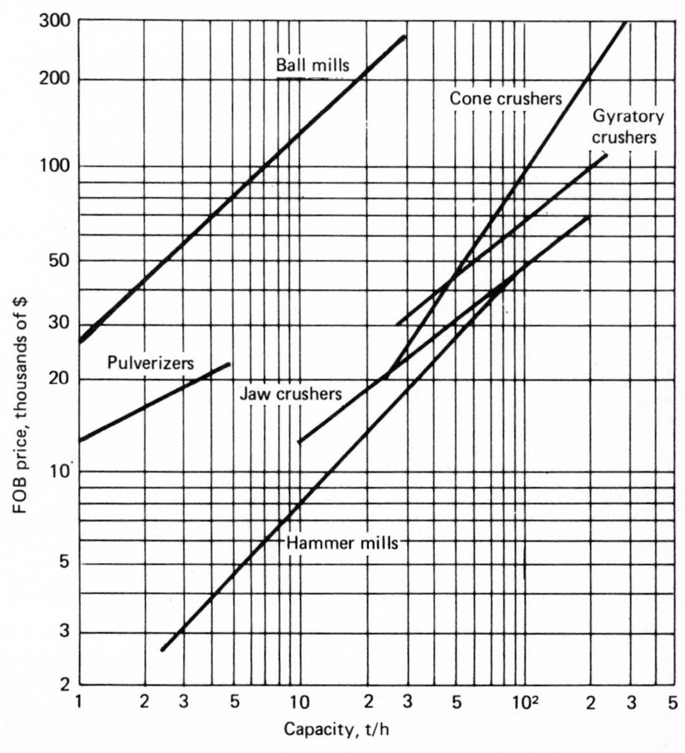

Fig. A8.6 Base price in mid-1975 for crushers and grinders

TABLE A8.12a Bar-Mill Base-Price Correction for Particle Size

Particle Size		
Mesh	Opening, mm	Correction Factor
10	1.651	0.6
48	0.294	1.0
100	0.147	1.7
98% minimum at 325 mesh	—	2.5

TABLE A8.12b Effect of Capacity on Bar Loading and Horsepower Requirements in Bar Mills

Capacity, t/h	3	10	25
Bar loading, tons	12	30	70
Brake horsepower, cv	125	400	960

TABLE A8.12c Effect of Capacity on Horsepower Requirements of Hammer Mills

Average capacity, t/h	25	40	100	150
Brake horsepower, cv	90	150	300	400

TABLE A8.12d Effect of Product Size and Capacity on Power Requirements for Jaw Crushers

Maximum and minimum product size, cm	2–6	2.5–7	3–8	5–11	6–12
Capacity range, t/h	4–9	7–17	15–30	30–70	50–90
Motor horsepower, cv	13	25	35	50	60

TABLE A8.12e Effect of Product Size and Capacity on Power Requirements for Cone Crushers

Product size, cm	0.6	0.9	1.0	1.2
Average capacity, t/h	50	75	120	165
Motor horsepower, cv	70	110	180	270

TABLE A8.12f Effect of Product Size and Capacity on Power Requirements for Gyratory Crushers

Product size, cm	0.75	1.00	1.25	1.50
Average capacity, t/h	30	90	160	260
Motor horsepower, cv	60	125	220	270

TABLE A8.12g Effect of Capacity on Power Requirements for Pulverizers

Average capacity, t/h	1.0	1.6	2.2	2.7	3.2	3.9
Motor horsepower, cv	25	30	40	50	60	75

gives the efficiency of cyclones in terms of particle size. Figure A8.7*b*, p. 370, gives prices as a function of the gas capacity.

A8.8 VIBRATING SCREENS

Vibrating screens are the most easily priced of the equipment for separating solids according to particle size. The essential variable is screen surface, which in turn depends on the characteristics of the solid particles and the type of screen used. Despite the standardized prices, however, it is difficult to calculate a surface for a vibrating screen with a simplified method; and a fabricator should generally be consulted.

Figure A8.8 gives typical prices for single-stage vibrating screens as a function of surface. The effect of two or three stages can be obtained from Table A8.13, and motor horsepowers can be obtained from Table A8.14.

A8.9 CONVEYORS

Conveyors can be belt, screw, or bucket elevators. Belt conveyors are characterized by the large carrying belt, which can be flat or formed into a trough by idling rollers, and which can be level or inclined. The capacity and price of belt conveyors depend on the belt width, the distance between the roller drums, the linear speed, and the density of the transported product. Screw conveyors can also be level or inclined, and their capacity and price depend on the screw diameter, speed of rotation, motor horsepower, and properties of the solids transported. The capacity and price of bucket elevators depend on the bucket capacity, the elevation, the linear speed of the buckets, the driver power, and the properties of the solid lifted.

Figure A8.9, p. 372, gives base prices for conveyors as a function of the height or distance conveyed. This base price (Table A8.15, p. 372) assumes

1. For belt conveyors:
   a. A belt width of 0.5 m
   b. A linear speed of 1 m/s
2. For screw conveyors:
   a. A screw diameter of 0.4 m
   b. A rotor speed of 60 rpm

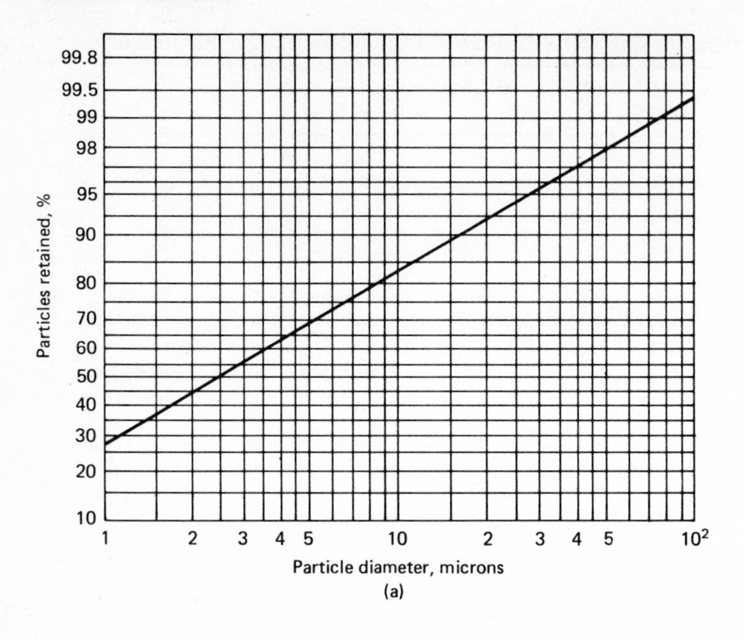

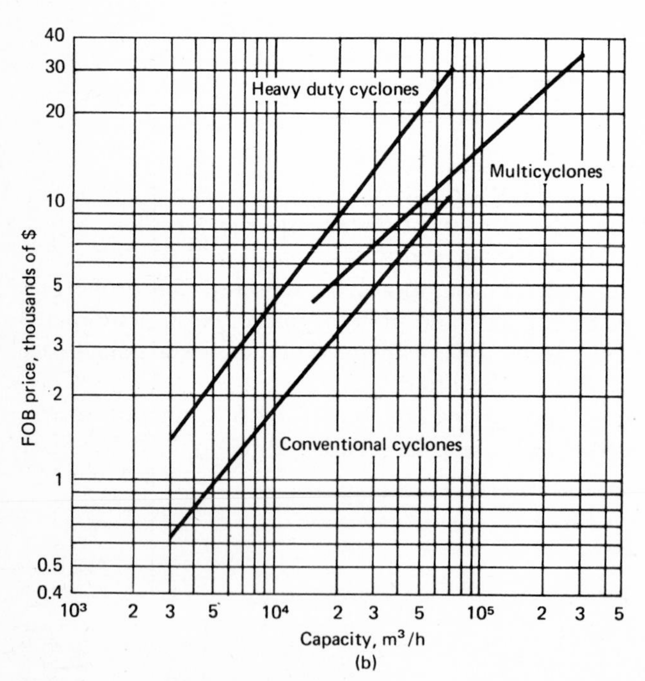

Fig. A8.7 (*a*) Efficiency of cyclone dust separators. (*b*) Prices in mid-1975 of cyclone dust separators.

TABLE A8.13 Effect of Multiple Stages on the Price for Vibrating Screens

Number of Stages	Multiplying Factor
1	1.00
2	1.20
3	1.35

TABLE A8.14 Sizes and Power Consumption for Vibrating Screens

Dimensions, m	Number of Stages	Horsepower, cv
0.30–0.90	1	0.50
	2	0.75
0.45–0.90	1	0.50
	2	0.75
0.60–1.20	1	0.75
	2	1.0
0.90–1.80	1	2.0
	2	3.0
1.20–2.40	1	3.0
	2	5.0
1.50–3.00	1	5.0
	2	7.5

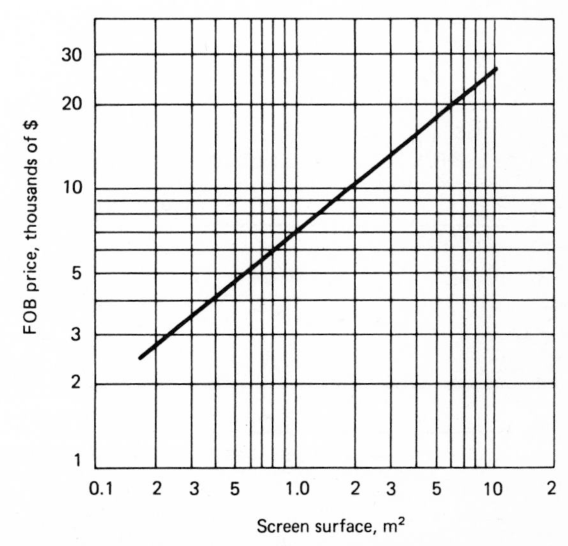

Fig. A8.8 Prices in mid-1975 for vibrating screens.

3. For bucket elevators:
   a. A capacity of 80 t/h
   b. Buckets in stainless type 304L

Prices for other widths, diameters, capacities, and materials may be obtained by multiplying by the correction factors found in Tables A8.16, A8.18, and

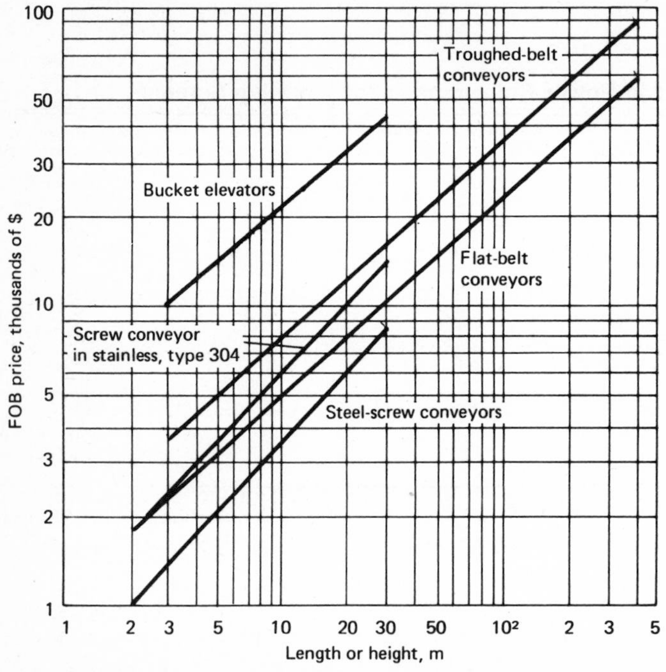

Fig. A8.9 Base price in mid-1975 for different conveyors.

TABLE A8.15 Characteristics of Belt Conveyors (Specific Gravity ≅ 1)

a. Capacity, t/h

	Speed, m/s			
Width, m	0.8	1.0	1.5	2.0
0.500	80	100	150	200
0.650	150	180	270	360
0.800	200	250	380	500

b. Power requirements, cv, for 1 m/s

Width, m	Ten-Meter Length	Hundred-Meter Length
0.500	3.0	7.5
0.650	4.5	10.0
0.800	5.5	12.5

TABLE A8.16 Base-Price Correction Factor for Width of Belt Conveyors

Width, m	Correction Factor
0.500	1.00
0.650	1.20
0.800	1.30
1.000	1.50
1.200	1.70

TABLE A8.17 Typical Operations of Screw Conveyors

Screw Diameter, m	Length, m	Solid Conveyed, t/h		Power, cv
		Grain	Sand	
0.300	5	45	100	1.5
0.400	5	45	100	3.0
0.400	15	45	100	7.5
0.400	5	110	250	5.0
0.400	15	110	250	12.0

TABLE A8.18 Base-Price Correction Factor for the Diameter of Screw Conveyors

Diameter, m	Correction Factor	
	Conveyor in Mild Steel	Conveyor in Stainless 304L
0.150	0.65	0.60
0.300	0.85	0.80
0.400	1.00	1.00
0.500	1.15	1.20

A8.19. Relative sizes and other operating characteristics may be obtained from Table A8.15a and b and Table A8.17.

A8.10 INSTRUMENTATION

The factor f_g given in Tables 4.16 and 4.17 for the installed cost of equipment took into account locally mounted instrumentation. On the other hand, no provision has been made for automatic control systems. As a first approximation, each such system may be priced at $6,250 in mid-1975.

TABLE A8.19 Base-Price Correction Factor for Capacity of Bucket Elevator

Capacity, t/h*	Correction Factor
0.5	0.30
25	0.75
80	1.00
120	1.20

*Specific gravity ≅ 1.

Appendix 9 Process Design Estimation: Utilities

Two types of investment cost are incurred by a process plant's demand for utilities: production and distribution.

The production units correspond to complete plants, ready to operate, and designed and equipped to meet the needs of the process plants they serve through the interconnected distribution system, which comprises the second investment cost.

A9.1 UTILITY-PRODUCTION UNITS

These units perform the functions of

Producing steam
Generating electricity
Cooling circulating cooling water
Cooling circulating refrigeration fluids

A9.1.1 STEAM PRODUCTION

Steam for a process plant can be supplied by a packaged boiler, which is generally limited to 150 t/h maximum capacity or by field-erected boilers.

Since a steam boiler consists of a complete plant with its own roster of equipment, no short-cut method of sizing boilers will be included. Rather, the investment costs are given as a function of the process plant requirements. In the event the requirements of a study differ from the conditions of the base price, correction factors are used to account for degrees of superheat and pressure ratings.

When obtaining the base price, the process demand should be increased by a margin of 10% for safety. Then the estimated investment cost would be

$$\text{Cost of complete installation} = (\text{base price})\ (1 + f_t + f_p)$$

where f_t = correction factor for degrees of superheat
f_p = correction factor for pressure rating

A9.1.1.1 Packaged Boilers

Base price Figure A9.1*a* gives the price of a packaged boiler rated at 30 bars for producing saturated steam, installed, with all accessories, including economiser, feed pumps, regulation and control instrumentation, extraction for analysis, condensate recovery, return condensate degaser, storage and injection system for treating chemicals, stack, etc.

Correction factors This base price is modified for the required degrees of superheat and pressure ratings other than 30 bars by the correction factors given in Tables A9.1a and b, respectively.

A9.1.1.2 Field-Erected Boilers

Base price Figure A9.1*b* gives the price of a field-erected boiler rated at 20 bars for producing saturated steam, with all accessories, and including all field installation costs such as civil engineering, mechanical construction, erection, and placing the equipment.

Correction factors This base price is modified for the required degrees of superheat and pressure ratings other than 20 bars by the correction factors given in Table A9.2a and b, respectively.

A9.1.2 ELECTRICITY GENERATION

It is possible to consider electricity production alone, or in combination with steam. All in all there are three possibilities:

1. A boiler and condensing steam committed solely to the production of electricity

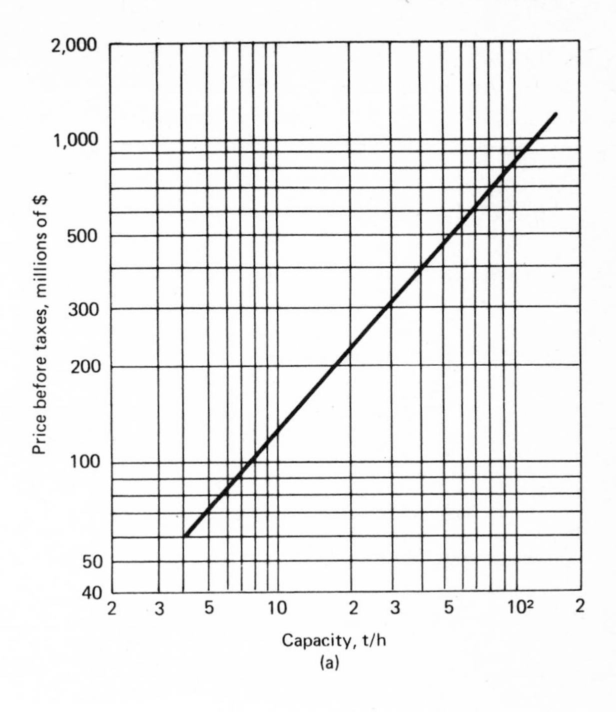

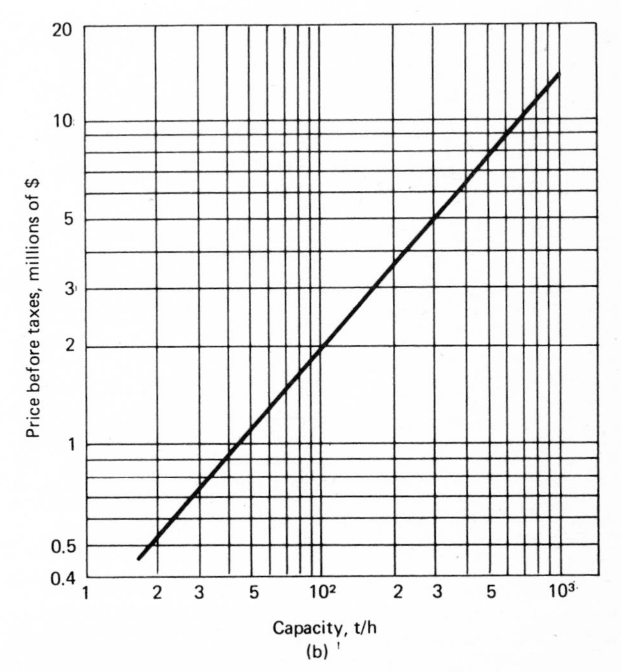

Fig. A9.1 Investment costs in mid-1975 for (*a*) packaged steam generators and (*b*) field-erected steam generators.

TABLE A9.1 Investment-Correction Factors for Packaged Steam Generators

a. Superheat

Degrees Superheat, °C	f_t
Saturation	0.00
50	0.10
100	0.17
150	0.22
200	0.25

b. Pressure

Pressure Rating, bar	f_p*
15	−0.15
20	−0.10
25	−0.05
30	0.00
40	0.10
50	0.20
75	0.40
100	0.60

*For mixed firing, use $f_p = 0.07$.

TABLE A9.2 Investment-Correction Factors for Field-Erected Steam Generators

a. Superheat

Degrees Superheat, °C	f_t
Saturation	0.00
50	0.10
100	0.16
150	0.20
200	0.24

b. Pressure

Pressure Rating, bar	f_p
20	0.00
30	0.05
40	0.10
50	0.20
100	0.40
200	0.60

2. A boiler and a back pressure turbine for producing both electricity and steam

3. A gas turbine with or without a boiler for the production of steam

In addition to the construction materials, the manufacturing cost is also affected by the relative demands of both steam and electricity in the process plant. Generally, the most economical system is coproduction of both electricity and steam. In project evaluation, however, it is necessary to delay the study of such improvements until after the essential decisions have been made.

Therefore, in this section we include only the production of electricity by the conventional method, using a condensing steam turbine.

Figure A9.2 gives, as a function of the demand, the total investment cost for a plant to generate electricity, from steam boiler through the condensing turbines to the generators, and including the field costs. A 10% margin should be added to the estimated electrical demand when using this figure.

A9.1.3 COOLING CIRCULATING COOLING WATER

The large quantities of cooling water required by process plants generally require some means of recooling and recycling the hot cooling-water effluent.

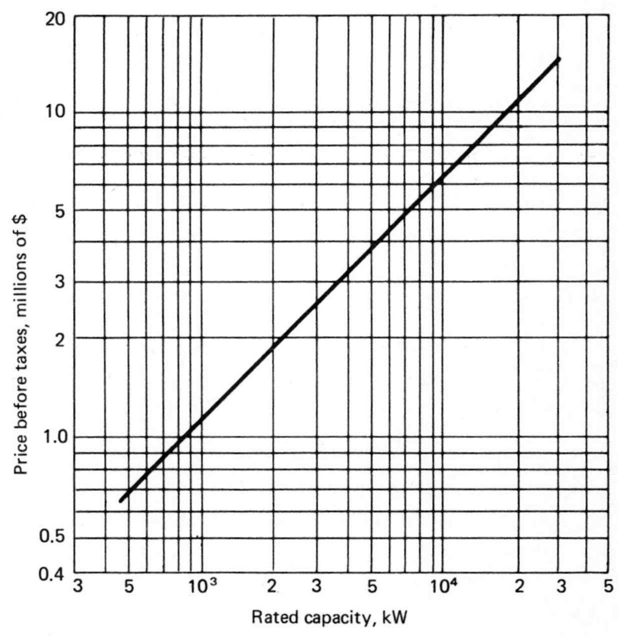

Fig. A9.2 Investment costs in mid-1975 for a complete steam-electric power generating station.

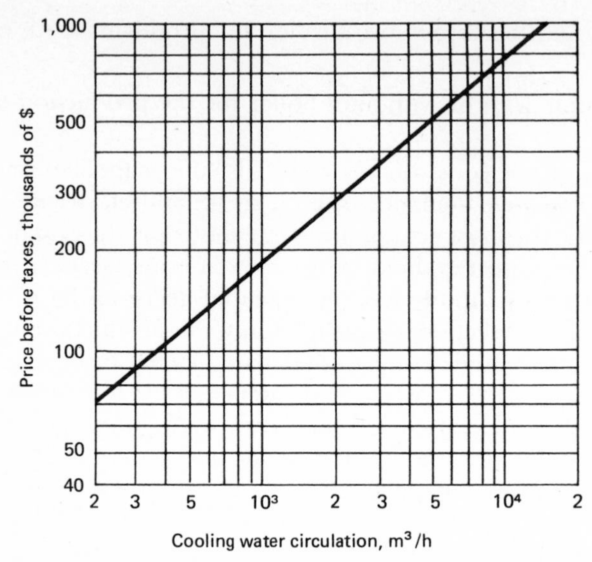

Fig. A9.3 Base price in mid-1975 for concrete water-cooling towers.

TABLE A9.3 Base-Price Correction for Temperature Drop across Cooling Tower

Δt, °C	Correction Factor
10	1.0
12	1.5
15	2.0

Usually the cooling-water supply enters the process plant at 20–25°C, and the hot cooling-water effluent leaves at about 35°C. Three types of cooling are used:

1. Spray ponds
2. Natural-draft cooling towers, in wood or concrete
3. Forced-draft cooling towers

Figure A9.3 gives the base price for a concrete natural draft cooling tower for reducing water temperatures from 35 to 25 °C, including the basin, circulation pumps, and drivers, and the installation costs. The circulating-water requirements should be increased by a safety factor of 15% before using this curve. Cooling for other than the base 10°C temperature drop can be allowed for by the correction factors in Table A9.3.

A9.1.4 REFRIGERATION

Process plants frequently have need of the cooling afforded by compression-condensing-flashing cycles of refrigeration. The equipment and consequently the investment for such refrigeration varies considerably with the temperature level of the refrigerant.

Figure A9.4 gives costs for refrigeration at 0°C as a function of the capacity using steel equipment, including centrifugal compressors, evaporators, condensers, etc., plus the installation costs. Corrections for other temperature levels can be made to this base price with the factors obtained from Table A9.4.

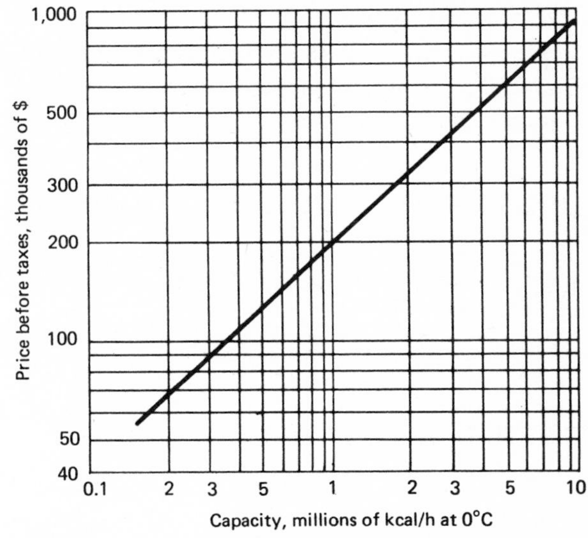

Fig. A9.4 Base price in mid-1975 for complete refrigeration units, installed.

TABLE A9.4 Base-Price Correction for Temperature Level of Refrigeration Units

Temperature Level	Correction Factor
0	1.00
−10	1.55
−20	2.10
−30	2.65
−40	3.20
−50	4.00

A9.2 UTILITY DISTRIBUTION

Although the utility-distribution system may require a neglegible investment compared to the generating plant, when that is close to the process unit, there are frequent instances when costs for insulation and so forth for a remotely located process plant causes the distribution costs to rival the generation costs. Consequently, the project evaluator needs some method for estimating utility distribution costs, which are given below for steam, cooling water, and electricity.

A9.2.1 STEAM DISTRIBUTION

In mid-1975, the investment for a steam distribution system varied between \$3,100 per t/h and \$8,300 per t/h according to the distance, capacity, and complexity of the distribution system.

A9.2.2 COOLING-WATER DISTRIBUTION

If the cooling water is circulated through a cooling tower, the mid-1975 cost of the distribution system varies between \$90 and \$380 per m^3/h; if the water comes from a river in a once-through system, these costs become \$63–2,100 per m^3/h.

A9.2.3 DISTRIBUTION OF ELECTRICITY

Three cases are allowed for

1. General current for lighting, etc., for which the mid-1975 cost is \$150–210 per kW

2. The primary transformer, for which the mid-1975 cost is \$62–104/per kVA, with corrections as obtained from Table A9.5

TABLE A9.5 Base-Price Correction for Size of Primary Transformers

Capacity, KVA	Correction Factor
3,000	1.0
5,000	0.6
10,000	0.4
20,000	0.3

3. The secondary transformer, for which the mid-1975 cost is $52–83 per kVA, with corrections as obtained from Table A9.6

TABLE A9.6 Base-Price Correction for Size of Secondary Transformers

Capacity, KVA	Correction Factor
600	1.0
1,000	0.7
1,500	0.5
2,000	0.5

A9.3 MISCELLANEOUS UTILITIES

Process plants also require compressed air, treated (or process) water, and fuel oil. The equipment to supply these utilities can be sized and priced according to the process plant's demand by the methods given in Apps. 1 to 8. Thus plant air would require a filter, compressor and dryer; process water might require a filter, distillation tower and associated equipment, while fuel oil would require a storage tank and pump. The investment cost for distribution networks for these utilities can be estimated as follows:

1. For compressed air, 0.03–0.12 times the cost of the primary equipment.
2. For treated water

   $83–160 (in mid-1975) per m^3 of water that is filtered and softened.

   $310–630 per m^3 (mid-1975) for distilled water.
3. For fuel,

   0.015–0.100 times the cost of primary equipment, for fuel oil.

   0.025–0.120 times the cost of the primary equipment for fuel gas.

Appendix 10 Process Design Estimation: Storage Tanks

Depending on their vapor pressures, liquids are stored in atmospheric or pressurized tanks; and depending on the amount of pressure, the capacity and economics, liquids under pressure are stored in cylinders or spheres. Base prices for atmospheric, pressurized cylindrical, and pressurized spherical storage tanks are given in Figs. A10.1*a* and *b*, A10.2, and A10.3, respectively. These base prices are corrected for design features and materials of construction, as follows:

A10.1 ATMOSPHERIC PRESSURE TANKS

A10.1.1 TANKS WITH LESS THAN 100-m^3 CAPACITY

The base price for these tanks is corrected for materials of construction by multiplying with a factor obtained from Table A10.1.

A10.1.2 TANKS WITH MORE THAN 100-m^3 CAPACITY

The base price for these tanks (Fig. A10.1*b*) assumes a conical roof and construction in mild steel. This price is corrected for other roof types and other

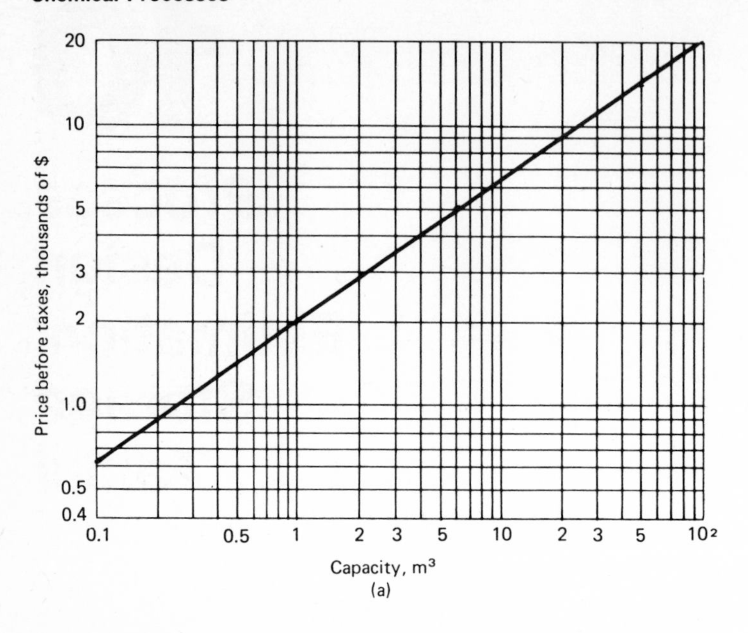

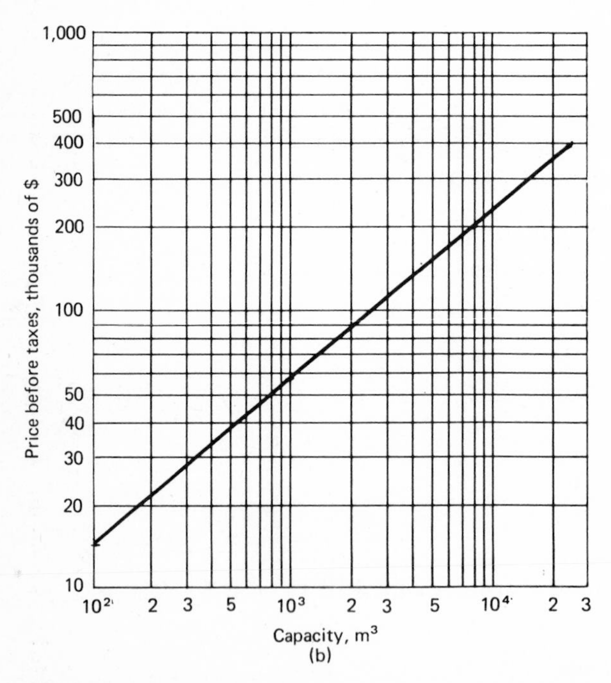

Fig. A10.1 Base price in mid-1975 for cylindrical atmospheric storage tanks (*a*) of less than 100-m^3 capacity and (b) of more than 100-m^3 capacity.

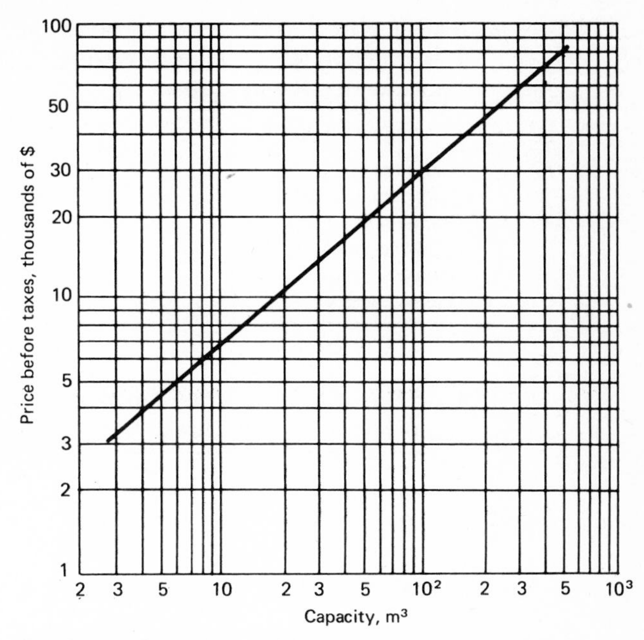

Fig. A10.2 Base price in mid-1975 for horizontal pressurized storage tanks.

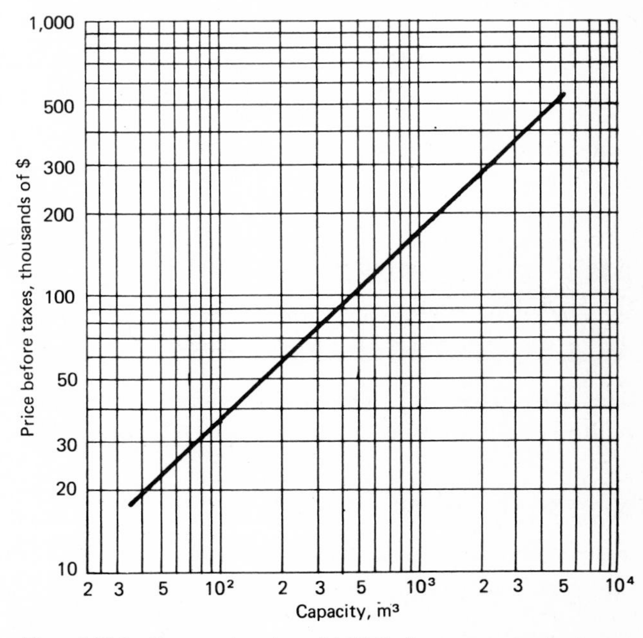

Fig. A10.3 Base price in mid-1975 for pressurized storage spheres.

TABLE A10.1 Base-Price Correction for Construction Materials of Atmospheric Storage Tanks under 100-m³ Capacity

Construction Material	Correction Factor
Mild steel	1.00
Aluminum	1.40
Rubber-lined steel	1.50
Lead-lined steel	1.55
Stainless steel	3.20
Glass-lined steel	3.25

TABLE A10.2 Base-Price Correction Factors for Atmospheric Storage Tanks Larger Than 100 m³

a. Type of Roof

Type	f_d
Cone roof	1.0
Floating roof	1.1
Lifter roof	1.3

b. Materials of Construction

Materials of Construction	f_m
Mild steel	1.0
Rubber lined	1.5
Lead lined	1.7
Stainless type 304	2.8

materials of construction by multiplying with the factors found in Table A10.2a and b.

A10.2 PRESSURIZED STORAGE TANKS

A10.2.1 HORIZONTAL CYLINDERS

The base price (Fig. A10.2) is modified for other than 10 bars pressure by multiplying with a factor from Table A10.3

A10.2.2 STORAGE SPHERES

The base price (Fig. A10.3) is modified for other than 3 bars pressure by multiplying with a factor from Table A10.4.

TABLE A10.3 Base-Price Correction for Pressure of Horizontal Storage Tanks

Pressure, bar	f_p
<10	1.00
10–15	1.20
15–20	1.50

TABLE A10.4 Base-Price Correction for Pressure of Storage Spheres

Pressure, bar	f_p
3	1.0
5	1.3
7	1.5
10	1.8
15	2.3

Appendix 11

Process Design Estimation: Heats of Reaction

Heats of reaction are calculated as the difference between the sums of the enthalpies of formation of the reactants and the products, following the convention that the enthalpies of formation of the elements are zero. Experimental values of enthalpy of formation are always to be preferred for these calculations, when such values are available in the various tables that have been prepared.

A11.1 ENTHALPIES OF FORMATION OF ORGANIC COMPOUNDS

A sampling is shown in Table A11.1 from the enthalpies of formation that have been published in the references cited in the bibliography. When those sources are not available for compounds not in Table A11.1, the enthalpies of formation may be estimated by the method of K. K. Verma and L. K. Doraiswamy.

This method consists of calculating the enthalpy of formation at a given temperature, for a compound, as the sum of enthalpies of formation of structural groups forming that compound, plus corrections when corrections are necessary. The enthalpies of formation of the structural groups are calculated as

$$(\Delta H^0_f)_T = A + BT$$

TABLE A11.1 Experimental Enthalpies of Formation $(AH^{\circ}_{f})_{T}$ of Organic Compounds, kcal/(g)(mole)

Compound	Temperature, °K								
	298	300	400	500	600	700	800	900	1000
Methane	−17.89	−17.90	−18.63	−19.30	−19.90	−20.40	−20.82	−21.15	−21.40
Ethane	−20.24	−20.26	−21.42	−22.44	−23.29	−23.99	−24.54	−24.97	−25.28
Propane	−24.82	−24.85	−26.36	−27.62	−28.66	−29.48	−30.11	−30.58	−30.89
Butane	−30.15	−30.19	−31.99	−33.51	−34.73	−35.68	−36.41	−36.93	−37.25
Ethylene	12.50	12.49	11.77	11.14	10.60	10.15	9.77	9.45	9.21
Propylene	4.88	4.86	3.76	2.80	1.98	1.31	0.77	0.35	0.04
Butene-1	−0.03	−0.06	−1.49	−2.70	−3.71	−4.51	−5.14	−5.62	−5.94
Butene-2-*cis*	−1.67	−1.70	−3.30	−4.68	−5.82	−6.75	−7.49	−8.05	−8.44
Butene-2-*trans*	−2.67	−2.70	−4.11	−5.33	−6.37	−7.20	−7.87	−8.38	−8.73
Acetylene	54.19	54.19	54.13	54.04	53.92	53.78	53.62	53.45	53.29
Propadiene	45.92	45.91	45.35	44.85	44.41	44.04	43.71	43.44	42.23
Butadiene-1,2	38.77	38.75	37.82	37.00	36.30	35.72	35.24	34.87	34.61
Butadiene-1,3	26.33	26.31	25.42	24.70	24.11	23.63	23.25	22.95	22.74
Piperylene:									
Pentadiene-1,3-*cis*	8.70	18.67	17.28	16.14	15.22	14.48	13.90	13.47	13.17
Pentadiene-1,3-*trans*	18.60	18.57	17.38	16.39	15.58	14.94	14.43	14.05	13.80
Isoprene:									
Methyl-2 butadiene-1,3	18.10	18.07	16.92	15.99	15.23	14.63	14.16	13.82	13.60
Benzene	19.82	19.79	18.56	17.54	16.71	16.04	15.51	15.10	14.82
Toluene	11.95	11.92	10.34	9.05	8.02	7.24	6.65	6.24	6.01
o-xylene	4.54	4.50	2.72	1.19	−0.07	−1.07	−1.85	−2.43	−2.79
m-xylene	4.12	4.08	2.18	0.57	−0.75	−1.79	−2.60	−3.19	−3.58
p-xylene	4.29	4.25	2.32	0.68	−0.67	−1.75	−2.59	−3.21	−3.61
Ethylbenzene	7.12	7.08	5.23	3.71	2.48	1.53	0.80	0.27	−0.05
Methanol	−48.08	−48.10	−48.95	−49.70	−50.34	−50.88	−51.31	−51.66	−51.93
Ethanol	−56.12	−56.14	−57.32	−58.31	−59.11	−59.76	−60.27	−60.65	−60.93
Acetaldehyde	−39.76	−39.78	−40.57	−41.27	−41.86	−42.35	−42.74	−43.05	−43.27
Acetone	−52.00	−52.02	−53.20	−54.22	−55.07	−55.74	−56.28	−56.67	−56.93
Formic acid	−90.49	−90.50	−90.96	−91.34	−91.64	−91.87	−92.05	−92.18	−92.27
Acetic acid	−103.93	−103.95	−104.76	−105.42	−105.94	−106.34	−106.64	−106.87	−107.01
Carbon tetrachloride	−24.00	−24.00	−23.79	−23.55	−23.29	−23.04	−22.79	−22.54	−22.29
Chloroform	−24.20	−24.20	−24.36	−24.44	−24.48	−24.49	−24.47	−24.43	−24.37

where $(\Delta H^0_f)_T$ = enthalpy of formation for the group, kcal/(g) (mole)
A, B = constants determined from Table A11.2
T = temperature,

The problems of resonance do not enter into this method, since the different groups are identified and assigned values according to the type responsible for resonance. Other corrections, also made in the form $A + BT$ have to do with the branching on linear and cyclic carbon chains. In the event of joined carbon rings, the branching correction should be made for the ortho position.

The use of the method and the values in Table A11.2 does have some limitations, particularly with respect to the heats of formation of compounds such as methane, butadiene, isoprene, and cyclopentadiene.

In summary, the procedure should procede in five steps:

1. Determine the temperature for the calculation.
2. Consult Table A11.1.
3. For those compounds not found in Table A11.1, consult Table A11.2, and determine the values of A for all the groups and corrections pertinent to that compound, adding these values of A.
4. Similarly to step 3, determine the values of B and add those.
5. Calculate the enthalpy of formation of the compound as

$$(\Delta H^0_f)_T = \Sigma A + T\Sigma B$$

A11.2 ENTHALPIES OF FORMATION OF INORGANIC COMPOUNDS

Use Table A11.3, p. 400, or other experimental values published in the references cited.

TABLE A11.2 Group Contribution Constants, *A* and *B*,* to the Enthalpy of Formation of Organic Compounds

a. Aliphatic Hydrocarbons

GROUP	300–850 K		850–1,500 K		Maximum Temperature, K
	A	$B \times 10^2$	A	$B \times 10^2$	
$-CH_3$	−8.948	−0.436	−12.800	0.000	1,500
$-CH_2-$	−4.240	−0.235	−6.720	0.090	1,500
$-CH<$	−1.570	0.095	−2.200	0.172	1,500
$>C<$ (*a*)	−0.650	0.425	0.211	0.347	1,500
$=CH_2$	7.070	−0.295	4.599	−0.0114	1,500
$-C\equiv$	27.276	0.036	27.600	−0.010	1,500
$\equiv CH$ (*b*)	27.242	−0.046	27.426	−0.077	1,500
$=C=$	33.920	−0.563	33.920	−0.563	1,500
$H(-)C=CH_2$	16.323	−0.437	12.369	0.128	1,500
$>C=CH_2$	16.725	−0.150	15.837	0.038	1,500
$>C=C<$	29.225	0.415	30.129	0.299	1,500
$>C=CH(-)$	20.800	−0.100	19.360	0.080	1,500
$-CH=CH-$	20.100	0.000	19.212	0.102	1,500
$HC=CH$ (*cis*)	19.088	−0.378	17.100	0.000	1,500

TABLE A11.2 Group Contribution Constants, *A* and *B*,* to the Enthalpy of Formation of Organic Compounds *(Continued)*

a. Aliphatic Hydrocarbons					
GROUP	300–850 K		850–1,500 K		Maximum Temperature, K
	A	$B \times 10^2$	A	$B \times 10^2$	
(trans) H–C=C–H, $>C=C<$	18.463	−0.211	16.850	0.000	1,500
$>C=C=CH_2$	51.450	−0.050	50.200	0.100	1,500
$(H)C=C=CH_2$	50.163	−0.233	48.000	0.000	1,500
$(H)C=C=C(H)$	54.964	0.027	53.967	0.133	1,500

TABLE A11.2 Group Contribution Constants, *A* and *B*,* to the Enthalpy of Formation of Organic Compounds *(Continued)*

b. Aromatic Hydrocarbons

GROUP	300–750 K A	300–750 K B×10²	850–1,500 K A	850–1,500 K B×10²	Maximum Temperature, K
HC⟨	3.768	−0.167	2.616	−0.016	1,500
−C⟨	5.437	0.037	5.279	0.058	1,500
⟷C⟨	4.208	0.092	4.050	0.100	1,500

c. Linkages on a Paraffinic Chain

GROUP	300–750 K A	300–750 K B×10²	750–1,500 K A	750–1,500 K B×10²	Maximum Temperature, K
Side chain with 2 or more carbon atoms	0.800	0.000	0.800	0.000	1,500
3 groups −CH connected (with vertical bonds above and below C)	−1.200	0.000	—	—	1,500
−CH and −C− connected (with vertical bonds above and below each C)	0.600	0.000	0.600	0.000	1,500

d. Linkages on Cycloparaffins

GROUP	300–850 K A	300–850 K B×10²	850–1,000 K A	850–1,000 K B×10²	Maximum Temperature, K
On 6-Carbon Rings					
Single	0.00	0.00	2.85	−0.40	1,500
Double, as follows:					
1,1	1.10	0.45	−0.40	0.00	1,500
1,2-*cis*	3.05	−1.09	1.46	−0.13	1,500
1,2-*trans*	−0.90	−0.60	−1.50	0.00	1,500
1,3−*d*-*cis*	0.00	−1.00	−2.60	0.00	1,500
1,3−*l*-*trans*	0.00	−0.16	2.80	−0.32	1,500
1,4-*cis*	0.00	−0.16	2.80	−0.32	1,500
1,4-*trans*	0.00	−1.00	−2.60	0.00	1,500
On 5-Carbon Rings					
Single	0.00	0.00	1.40	−0.20	1,500
Double, as follows:					
1,1	0.30	0.00	1.90	−0.25	1,500
1,2-*cis*	0.70	0.00	0.00	0.00	1,500
1,2-*trans*	−1.10	0.00	−1.60	0.0	1,500
1,3-*cis*	−0.30	0.00	0.15	0.00	1,500
1,3-*trans*	−0.90	0.00	−1.40	0.00	1,500

e. Linkages on Aromatic Rings					
GROUP	300–850 K		850–500 K		Maximum Temperature, K
	A	$B \times 10^2$	A	$B \times 10^2$	
Two side chains:					
1,2 positions	0.85	0.03	0.85	0.03	1,500
1,3 positions	0.56	−0.06	0.56	−0.06	1,500
1,4 positions	1.00	−0.14	1.40	−0.12	1,500
Three side chains					
1,2,3 positions	2.01	−0.07	1.50	0.00	1,500
1,2,4 positions	1.18	−0.25	1.50	−0.10	1,500
1,3,5 positions	1.18	−0.25	1.80	−0.08	1,500

f. Ring Formations					
GROUP	300–750 K		750–1,500 K		Maximum Temperature, K
	A	$B \times 10^2$	A	$B \times 10^2$	
C_3 cycloparaffin	24.850	−0.240	24.255	−0.174	1,500
C_4 cycloparaffin	19.760	−0.440	17.950	−0.231	1,500
C_5 cycloparaffin	7.084	−0.552	4.020	−0.140	1,500
C_6 cycloparaffin	0.378	−0.382	4.120	0.240	1,500

g. Groups Containing Oxygen					
GROUP	300–850 K		850–1,500 K		Maximum Temperature, K
	A	$B \times 10^2$	A	$B \times 10^2$	
>C=O	−31.505	0.007	−32.113	0.073	1,500
−O− (−C−O−C)	−24.200	0.000	−24.200	0.000	1,000
−O− (H_3C−C−O−C−CH_3)	−30.500	0.000	−30.500	0.000	1,000
O (aromatic ring)	−21.705	0.030	−21.600	0.020	1,500
−CHO (*c*)	−29.167	−0.183	−30.500	0.000	1,000
−C(=O)OH (*c*)	−94.488	−0.063	−94.880	0.000	1,500

TABLE A11.2 Group Contribution Constants, *A* and *B*,* to the Enthalpy of Formation of Organic Compounds *(Continued)*

g. Groups Containing Oxygen *(Continued)*

GROUP	300–850 K		850–1,500 K		Maximum Temperature, K
	A	$B \times 10^2$	*A*	$B \times 10^2$	
Contributions of $-OH$ Groups					
HO--CH_3	−37.207	−0.259	−37.993	−0.136	1,000
HO--CH_2—	−40.415	−0.267	−41.265	−0.116	1,000
HO--CH< (one H, two bonds)	−43.200	−0.200	−43.330	−0.143	1,000
HO--C— (three bonds)	−46.850	−0.250	−47.440	−0.146	1,000
HO--C< (aromatic)	−44.725	−0.125	−45.220	−0.021	1,000

h. Groups Containing Nitrogen and Sulfur†

GROUP	300–750 K		750–1,500 K		Maximum Temperature, K
	A	$B \times 10^2$	*A*	$B \times 10^2$	
$-NO_2$ (aliphatic)	−7.813	−0.043	−9.250	0.143	1,500
−C≡N	36.580	0.080	37.170	0.000	1,000
$-NH_2$ (aliphatic)	3.832	−0.208	2.125	0.002	1,500
=NH (aliphatic)	13.666	−0.067	12.267	0.133	1,500
≡N (aliphatic)	18.050	0.300	18.050	0.300	1,500
$-NH_2$ (aromatic)	−0.713	−0.188	−1.725	0.000	1,000
=NH (aromatic)	9.240	−0.250	8.460	−0.140	1,000
≡N (aromatic)	18.890	0.110	16.200	0.250	1,000
−SH (*c*)	−10.580	−0.080	—	—	1,000
−S− (*d*)	−4.725	0.160	—	—	1,000

i. Groups Containing Halogens

GROUP	300–750 K		750–1,500 K		Maximum Temperature, K
	A	$B \times 10^2$	*A*	$B \times 10^2$	
H_2C(H)---Cl (H–C(H)(H)---Cl)	−9.322	−0.045	−9.475	−0.025	1,000
H_3C–C(H)(H)---Cl	−10.007	−0.033	−10.438	0.029	1,000

i. Groups Containing Halogens *(Continued)*					
	300–750 K		750–1,500 K		Maximum
GROUP	*A*	*B* × 10²	*A*	*B* × 10²	Temperature, K
$H_3C-C(Cl)_2-H$	−14.780	−0.040	−14.780	−0.040	1,500
$H-C(H)(Cl)-Cl$	−13.222	−0.029	−13.222	−0.029	1,500
$H-C(Cl)(Cl)-Cl$	−6.684	−0.033	−6.684	−0.033	1,500
$Cl-C(Cl)(Cl)-Cl$	−6.400	−0.050	−6.400	−0.050	1,500
$H_2C=C(Cl)H$	−7.622	0.029	−7.390	0.000	1,500
$H_2C=CCl_2$	−6.171	−0.029	−6.171	−0.029	1,500
$HClC=CClH$ *(1,2-cis)*	−5.916	0.071	−5.386	−0.007	1,500
$ClHC=CHCl$ *(1,2-trans)*	−6.532	0.233	−5.480	0.106	1,500
$Cl_2C=CCl_2$	−6.047	0.236	−6.047	0.236	1,500

*A = kcal/gmole
B = kcal/(gmole)(K)

†The standard state of sulfur is S_2 gas at 298 K; the factor is 15.42 kcal/gmole to convert to rhombic.

(*a*) For 300-1100 K and 1100-1500 K
(*b*) For 300-600 K and 600-1500 K
(*c*) For 300-600 K and 600-1000 K
(*d*) For 300-1000 K

TABLE A11.3 Experimental Enthalpies of Formation $(AH^\circ_f)_T$ of Inorganic Compounds

kcal/gmole

Compound	Temperature, K								
	298	300	400	500	600	700	800	900	1,000
Hydrogen bromide	8.66	−8.67	−12.44	−12.53	−12.62	−12.70	−12.77	−12.83	−12.88
Hydrogen chloride	−22.06	−22.06	−22.13	−22.20	−22.29	−22.36	−22.44	−22.50	−22.56
Hydrogen fluoride	−64.80	−64.80	−64.84	−64.89	−64.96	−65.04	−65.12	−65.20	−65.29
Hydrogen iodide	6.30	6.29	4.06	−1.35	−1.43	−1.49	−1.55	−1.58	−1.61
Hydrogen sulfide	−4.82	−4.83	−5.79	−6.55	−7.20	−7.75	−21.28	−21.39	−21.47
Hydrogen cyanide	31.20	31.20	31.16	31.11	31.06	31.00	30.94	30.88	30.83
Nitric acid	−32.02	−32.03	−32.38	−32.56	−32.61	−32.58	−32.48	−32.34	−32.16
Sulfuric acid	−194.45	−194.44	−194.14	−193.46	−192.59	−191.60	—	—	—
Carbon monoxide	−26.42	−26.42	−26.32	−26.30	−26.33	−26.41	−26.52	−26.64	−26.77
Carbon dioxide	−94.05	−94.05	−94.07	−94.09	−94.12	−94.17	−94.22	−94.27	−94.32
Nitric oxide	21.60	21.60	21.61	21.62	21.62	21.62	21.63	21.63	21.64
Nitrogen dioxide	8.09	8.09	7.95	7.87	7.83	7.82	7.83	7.85	7.89
Nitrous oxide	19.49	19.49	19.41	19.40	19.44	19.50	19.59	19.69	19.79
Sulfur dioxide	−70.95	−70.96	−71.77	−72.36	−72.83	−73.21	−86.60	−86.58	−86.56
Carbon disulfide	27.98	27.98	26.67	25.69	24.87	24.17	−2.59	−2.58	−2.57
Cyanogen	73.84	73.84	74.08	74.26	74.39	74.48	74.54	74.57	74.59
Phosgene	−52.80	−52.80	−52.77	−52.73	−52.69	−52.65	−52.63	−52.60	−52.58
Carbonyl sulfide	−33.08	−33.08	−33.74	−34.22	−34.64	−25.00	−48.40	−48.41	−48.42
Nitrosyl chloride	12.57	12.57	12.55	12.56	12.60	12.64	12.70	12.76	12.83
Thionyl chloride	−85.40	−85.40	−86.12	−86.58	−86.91	−87.17	−100.43	−100.30	−100.17
Ammonia	−10.92	−10.93	−11.43	−11.87	−12.83	−12.53	−12.77	−12.96	−13.11
Hydrazine	22.75	22.73	22.04	21.54	21.19	20.95	20.80	20.73	20.71
Water	−57.80	−57.80	−58.04	−58.28	−58.50	−58.71	−58.91	−59.08	−59.24
Oxygenated water	−32.53	−32.54	−32.84	−33.06	−33.22	−33.35	−33.44	−33.52	−33.57
Ozone	34.00	34.00	33.93	33.92	33.94	33.98	33.03	34.09	34.14

Appendix 12

General Tables

This appendix includes

- Tables for converting °F to °C: Table A12.1a to c
- Tables for converting English and metric units: Table A12.2a to r
- A table relating international standards for steel: Table A12.3
- A table indicating the resistance of contruction materials to various compounds and solutions, etc.: Table A12.4

TABLE A12.1 Temperature Conversions

a. Formulas

	°C	°F	°R	K
°C	1	1.8 (°C)+32	1.8(°C)+491.7	°C+273.2
°F	(°F − 32)/1.8	1	°F+459.7	(°F+459.7)/1.8
°R	(°R−491.7)/1.8	°R − 459.7	1	°R/1.8
K	K − 273.2	1.8(K) − 459.7	1.8 K	1

b. Equivalent Temperatures

(Use the center column for the known temperature and read left or right to obtain the equivalent degrees centigrade or Fahrenheit, respectively.)

°C		°F	°C		°F	°C		°F	°C		°F
−273	**−459.4**		−168	**−270**	−454	−57	**−70**	−94	−35.6	**−32**	−25.6
−268	**−450**		−162	**−260**	−436	−51	**−60**	−76	−35.0	**−31**	−23.8
−262	**−440**		−157	**−250**	−418	−45.6	**−50**	−58.0	−34.4	**−30**	−22.0
−257	**−430**		−151	**−240**	−400	−45.0	**−49**	−56.2	−33.9	**−29**	−20.2
−251	**−420**		−146	**−230**	−382	−44.4	**−48**	−54.4	−33.3	**−28**	−18.4
−246	**−410**		−140	**−220**	−364	−43.9	**−47**	−52.6	−32.8	**−27**	−16.6
−240	**−400**		−134	**−210**	−346	−43.3	**−46**	−50.8	−32.2	**−26**	−14.8
−234	**−390**		−129	**−200**	−328	−42.8	**−45**	−49.0	−31.7	**−25**	−13.0
−229	**−380**		−123	**−190**	−310	−42.2	**−44**	−47.2	−31.1	**−24**	−11.2
−223	**−370**		−118	**−180**	−292	−41.7	**−43**	−45.4	−30.6	**−23**	−9.4
−218	**−360**		−112	**−170**	−274	−41.1	**−42**	−43.6	−30.0	**−22**	−7.6
−212	**−350**		−107	**−160**	−256	−40.6	**−41**	−41.8	−29.4	**−21**	−5.8
−207	**−340**		−101	**−150**	−238	−40.0	**−40**	−40.0	−28.9	**−20**	−4.0
−201	**−330**		−96	**−140**	−220	−39.4	**−39**	−38.2	−28.3	**−19**	−2.2
−196	**−320**		−90	**−130**	−202	−38.9	**−38**	−36.4	−27.8	**−16**	−0.4
−190	**−310**		−84	**−120**	−184	−38.3	**−37**	−34.6	−27.2	**−17**	1.4
−184	**−300**		−79	**−110**	−166	−37.8	**−36**	−32.8	−26.7	**−16**	3.2
−179	**−290**		−73	**−100**	−148	−37.2	**−35**	−31.0	−26.1	**−15**	5.0
−173	**−280**		−68	**−90**	−130	−36.7	**−34**	−29.2	−25.6	**−14**	6.8
−169	**−273**	−459.4	−62	**−80**	−112	−36.1	**−33**	−27.4	−25.0	**−13**	8.6

TABLE A12.1 Temperature Conversions *(Continued)*

b. Equivalent Temperatures *(Continued)*
(Use the center column for the known temperature and read left or right to obtain the equivalent degrees centigrade or Fahrenheit, respectively.)

°C		°F	°C		°F	°C		°F	°C		°F
−24.4	**−12**	10.4	−8.9	**16**	60.8	6.7	**44**	111.2	22.2	**72**	161.6
−23.9	**−11**	12.2	−8.3	**17**	62.6	7.2	**45**	113.0	22.8	**73**	163.4
−23.3	**−10**	14.0	−7.8	**18**	64.4	7.8	**46**	114.8	23.3	**74**	165.2
−22.8	**−9**	15.8	−7.2	**19**	66.2	8.3	**47**	116.6	23.9	**75**	167.0
−22.2	**−8**	17.6	−6.7	**20**	68.0	8.9	**48**	118.4	24.4	**76**	168.8
−21.7	**−7**	19.4	−6.1	**21**	69.8	9.4	**49**	120.2	25.0	**77**	170.6
−21.1	**−6**	21.2	−5.6	**22**	71.6	10.0	**50**	122.0	25.6	**78**	172.4
−20.6	**−5**	23.0	−5.0	**23**	73.4	10.6	**51**	123.8	26.1	**79**	174.2
−20.0	**−4**	24.8	−4.4	**24**	75.2	11.1	**52**	125.6	26.7	**80**	176.0
−19.4	**−3**	26.6	−3.9	**25**	77.0	11.7	**53**	127.4	27.2	**81**	177.8
−18.9	**−2**	28.4	−3.3	**26**	78.8	12.2	**54**	129.2	27.8	**82**	179.6
−18.3	**−1**	30.2	−2.8	**27**	80.6	12.8	**55**	131.0	28.3	**83**	181.4
−17.8	**0**	32.0	−2.2	**28**	82.4	13.3	**56**	132.8	28.9	**84**	183.2
−17.2	**1**	33.8	−1.7	**29**	84.2	13.9	**57**	134.6	29.4	**85**	185.0
−16.7	**2**	35.6	−1.1	**30**	86.0	14.4	**58**	136.4	30.0	**86**	196.8
−16.1	**3**	37.4	−0.6	**31**	87.8	15.0	**59**	138.2	30.6	**87**	188.6
−15.6	**4**	39.2	0.0	**32**	89.6	15.6	**60**	140.0	31.1	**88**	190.4
−15.0	**5**	41.0	0.6	**33**	91.4	16.1	**61**	141.8	31.7	**89**	192.2
−14.4	**6**	42.8	1.1	**34**	93.2	16.7	**62**	143.6	32.2	**90**	194.0
−13.9	**7**	44.6	1.7	**35**	95.0	17.2	**63**	145.4	32.8	**91**	195.8
−13.3	**8**	46.4	2.2	**36**	96.8	17.8	**64**	147.2	33.3	**92**	197.6
−12.8	**9**	48.2	2.8	**37**	98.6	18.3	**65**	149.0	33.9	**93**	199.4
−12.2	**10**	50.0	3.3	**38**	100.4	18.9	**66**	150.8	34.4	**94**	201.2
−11.7	**11**	51.8	3.9	**39**	102.2	19.4	**67**	152.6	35.0	**95**	203.0
−11.1	**12**	53.6	4.4	**40**	104.0	20.0	**68**	154.4	35.6	**96**	204.8
−10.6	**13**	55.4	5.0	**41**	105.8	20.6	**69**	156.2	36.1	**97**	206.6
−10.0	**14**	57.2	5.6	**42**	107.6	21.1	**70**	158.0	36.7	**98**	208.4
−9.4	**15**	59.0	6.1	**43**	109.4	21.7	**71**	159.8	37.2	**99**	210.2

TABLE A12.1 Temperature Conversions *(Continued)*

b. Equivalent Temperatures *(Continued)*
(Use the center column for the known temperature and read left or right to obtain the equivalent degrees centigrade or Fahrenheit, respectively.)

°C		°F	°C		°F	°C		°F	°C		°F
37.8	**100**	212.0	53.3	**128**	262.4	68.9	**156**	312.8	84.4	**184**	363.2
38.3	**101**	213.8	53.9	**129**	264.2	69.4	**157**	314.6	85.0	**185**	365.0
38.9	**102**	215.6	54.4	**130**	266.0	70.0	**158**	316.4	85.6	**186**	366.8
39.4	**103**	217.4	55.0	**131**	267.8	70.6	**159**	318.2	86.1	**187**	368.6
40.0	**104**	219.2	55.6	**132**	269.6	71.1	**160**	320.0	86.7	**188**	370.4
40.6	**105**	221.0	56.1	**133**	271.4	71.7	**161**	321.8	87.2	**189**	372.2
41.1	**106**	222.8	56.7	**134**	273.2	72.2	**162**	323.6	87.8	**190**	374.0
41.7	**107**	224.6	57.2	**135**	275.0	72.8	**163**	325.4	88.3	**191**	375.8
42.2	**108**	226.4	57.8	**136**	276.8	73.3	**164**	327.2	88.9	**192**	377.6
42.8	**109**	228.2	58.3	**137**	278.6	73.9	**165**	329.0	89.4	**193**	379.4
43.3	**110**	230.0	58.9	**138**	280.4	74.4	**166**	330.8	90.0	**194**	381.2
43.9	**111**	231.8	59.4	**139**	282.2	75.0	**167**	332.6	90.6	**195**	383.0
44.4	**112**	233.6	60.0	**140**	284.0	75.6	**168**	334.4	91.1	**196**	384.8
45.0	**113**	235.4	60.6	**141**	285.8	76.1	**169**	336.2	91.7	**197**	386.6
45.6	**114**	237.2	61.1	**142**	287.6	76.7	**170**	338.0	92.2	**198**	388.4
46.1	**115**	239.0	61.7	**143**	289.4	77.2	**171**	339.8	92.8	**199**	390.2
46.7	**116**	240.8	62.2	**144**	291.2	77.8	**172**	341.6	93.3	**200**	392.0
47.2	**117**	242.6	62.8	**145**	293.0	78.3	**173**	343.4	93.9	**201**	393.8
47.8	**118**	244.4	63.3	**146**	294.8	78.9	**174**	345.2	94.4	**202**	395.6
48.3	**119**	246.2	63.9	**147**	296.6	79.4	**175**	347.0	95.0	**203**	397.4
48.9	**120**	248.0	64.4	**148**	298.4	80.0	**176**	348.8	95.6	**204**	399.2
49.4	**121**	249.8	65.0	**149**	300.2	80.6	**177**	350.6	96.1	**205**	401.0
50.0	**122**	251.6	65.6	**150**	302.0	81.1	**178**	352.4	96.7	**206**	402.8
50.6	**123**	253.4	66.1	**151**	303.8	81.7	**179**	354.2	97.2	**207**	404.6
51.1	**124**	255.2	66.7	**152**	305.6	82.2	**180**	356.0	97.8	**208**	406.4
51.7	**125**	257.0	67.2	**153**	307.4	82.8	**181**	357.8	98.3	**209**	408.2
52.2	**126**	258.8	67.8	**154**	309.2	83.3	**182**	359.6	98.9	**210**	410.0
52.8	**127**	260.6	68.3	**155**	311.0	83.9	**183**	361.4	99.4	**211**	411.8

TABLE A12.1 Temperature Conversions *(Continued)*

b. Equivalent Temperatures *(Continued)*

(Use the center column for the known temperature and read left or right to obtain the equivalent degrees centigrade or Fahrenheit, respectively.)

°C		°F	°C		°F	°C		°F	°C		°F
100.0	**212**	413.6	115.6	**240**	464.0	221	**430**	806	377	**710**	1310
100.6	**213**	415.4	116.1	**241**	465.8	227	**440**	824	382	**720**	1328
101.1	**214**	417.2	116.7	**242**	467.6	232	**450**	842	388	**730**	1346
101.7	**215**	419.0	117.2	**243**	469.4	238	**460**	860	393	**740**	1364
102.2	**216**	420.8	117.8	**244**	471.2	243	**470**	878	399	**750**	1382
102.8	**217**	422.6	118.3	**245**	473.0	249	**480**	896	404	**760**	1400
103.3	**218**	424.4	118.9	**246**	474.8	254	**490**	914	410	**770**	1418
103.9	**219**	426.2	119.4	**247**	476.6	260	**500**	932	416	**780**	1436
104.4	**220**	428.0	120.0	**248**	478.4	266	**510**	950	421	**790**	1454
105.0	**221**	429.8	120.6	**249**	480.2	271	**520**	968	427	**800**	1472
105.6	**222**	431.6	121	**250**	482	277	**530**	986	432	**810**	1490
106.1	**223**	433.4	127	**260**	500	282	**540**	1004	438	**820**	1508
106.7	**224**	435.2	132	**270**	518	288	**550**	1022	443	**830**	1526
107.2	**225**	437.0	138	**280**	536	293	**560**	1040	449	**840**	1544
107.8	**226**	438.8	143	**290**	554	299	**570**	1058	454	**850**	1562
108.3	**227**	440.6	149	**300**	572	304	**580**	1076	460	**860**	1580
108.9	**228**	442.4	154	**310**	590	310	**590**	1094	466	**870**	1598
109.4	**229**	444.2	160	**320**	608	316	**600**	1112	471	**880**	1616
110.0	**230**	446.0	166	**330**	626	321	**610**	1130	477	**890**	1634
110.6	**231**	447.8	171	**340**	644	327	**620**	1148	482	**900**	1652
111.1	**232**	449.6	177	**350**	662	332	**630**	1166	488	**910**	1670
111.7	**233**	451.4	182	**360**	680	338	**640**	1184	493	**920**	1688
112.2	**234**	453.2	188	**370**	698	343	**650**	1202	499	**930**	1706
112.8	**235**	455.0	193	**380**	716	349	**660**	1220	504	**940**	1724
113.3	**236**	456.8	199	**390**	734	354	**670**	1238	510	**950**	1742
113.9	**237**	458.6	204	**400**	752	360	**680**	1256	516	**960**	1760
114.4	**238**	460.4	210	**410**	770	366	**690**	1274	521	**970**	1778
115.0	**239**	462.2	216	**420**	788	371	**700**	1292	527	**980**	1796

TABLE A12.1 Temperature Conversions *(Continued)*

b. Equivalent Temperatures *(Continued)*

(Use the center column for the known temperature and read left or right to obtain the equivalent degrees centigrade or Fahrenheit, respectively.)

°C		°F	°C		°F	°C		°F	°C		°F
532	**990**	1814	604	**1120**	2048	677	**1250**	2282	749	**1380**	2516
538	**1009**	1832	610	**1130**	2066	682	**1260**	2300	754	**1390**	2534
543	**1010**	1850	616	**1140**	2084	683	**1270**	2318	760	**1400**	2552
549	**1020**	1863	621	**1150**	2102	693	**1280**	2336	766	**1410**	2570
554	**1030**	1866	627	**1160**	2120	699	**1290**	2354	771	**1420**	2588
560	**1040**	1904	632	**1170**	2138	704	**1300**	2372	777	**1430**	2606
566	**1050**	1922	638	**1180**	2156	710	**1310**	2390	782	**1440**	2624
571	**1060**	1940	643	**1190**	2174	716	**1320**	2408	788	**1450**	2642
577	**1070**	1958	649	**1200**	2192	721	**1330**	2426	793	**1460**	2660
582	**1080**	1976	654	**1210**	2210	727	**1340**	2444	799	**1470**	2678
588	**1090**	1994	660	**1220**	2228	732	**1350**	2462	804	**1480**	2696
593	**1100**	2012	666	**1230**	2246	738	**1360**	2480	810	**1490**	2714
599	**1110**	2030	671	**1240**	2264	743	**1370**	2498	816	**1500**	2732

c. Interpolations for Equivalent-Temperature Table

°C	0.56	1.11	1.67	2.22	2.78	3.33	3.89	4.44	5	5.56	6.11	6.67	7.22	7.78	8.33	8.89	9.44	10	10.6	11.1
	1	2	3	4	5	6	7	8	9	10	11	12	13	14	15	16	17	18	19	20
°F	1.8	3.6	5.4	7.2	9	10.8	12.6	14.4	16.2	18	19.8	21.6	23.4	25.2	27	28.8	30.6	32.4	34.2	36

TABLE A12.2 Conversion Factors

a. Lengths

Length	cm	m	in	ft	yd	mile
1 centimeter	1	0.01	0.3937	0.03281	0.01094	621.36×10^{-8}
1 meter	100	1	39.37	3.2808	1.0936	621.36×10^{-6}
1 inch (pouce)	2.54	0.0254	1	0.08333	0.02778	157.83×10^{-7}
1 foot (pied)	30.48	0.3048	12	1	0.33333	189.39×10^{-6}
1 yard	91.44	0.9144	36	3	1	568.186×10^{-6}
1 mile	161×10^3	1,609.35	63,360	5,279.95	1,759.985	1

b. Surfaces

Area	cm^2	m^2	in^2	ft^2
1 square centimeter	1	0.0001	0.155	10.76×10^{-4}
1 square meter	10,000	1	1,550	10.76
1 square inch	6.451	6.45×10^{-4}	1	69.44×10^{-4}
1 square foot	929	0.0929	144	1

c. Volumes

Volume	cm^3	L	in^3	ft^3	gal (US)	gal (Imp.)	bbl
1 cubic centimeter	1	0.001	0.06102	35.31×10^{-6}	264.2×10^{-6}	220.1×10^{-6}	6.29×10^{-6}
1 liter	10,000	1	61.02	0.03531	0.2642	0.2201	6.29×10^{-3}
1 cubic inch	16.39	0.01639	1	578.7×10^{-6}	432.9×10^{-5}	360.7×10^{-5}	10.31×10^{-5}
1 cubic foot	28320	28.32	1728	1	7.481	6.232	0.1781
1 gallon (US)	3785	3.785	231	0.1337	1	0.8326	0.02381
1 Imperial gallon	4543	4.543	277.3	0.1605	1.2	1	0.02857
1 barrel	159×10^3	159	9700	5.614	42	34.97	1

TABLE A12.2 Conversion Factors *(Continued)*

d. Weights

Weight	g	lb	t (US)	long t	t
1 gram	1	22.05×10^{-4}	110.2×10^{-8}	98.4×10^{-8}	10^{-6}
1 pound	453.59	1	0.0005	446.4×10^{-6}	453.6×10^{-6}
1 short ton (US)	$9{,}071.84\times10^{2}$	2,000	1	0.8929	0.9072
1 long ton	$10{,}160.47\times10^{2}$	2,240	1.12	1	1.016
1 metric ton	10^{6}	2,205	1.102	0.9842	1

e. Densities

Density	g/cm³	kg/m³	lb/in³	lb/ft³	lb/gal (US)
1 gram per cubic centimeter	1	1000	0.03613	62.43	8.345
1 kilogram per cubic meter	0.001	1	36.13×10^{-6}	0.06243	83.45×10^{-4}
1 pound per cubic inch	27.68	27680	1	1728	231
1 pound per cubic foot	0.01602	16.02	57.87×10^{-5}	1	0.1337
1 pound per gallon (US)	0.1198	119.8	43.29×10^{-4}	7.481	1

f. Linear velocities

Velocity	cm/s	m/s	km/h	in/s	ft/s	mile/h
1 centimeter per second	1	0.01	0.036	0.3937	0.03281	0.02237
1 meter per second	100	1	3.6	39.37	3.281	2.237
1 kilometer per hour	27.78	0.2778	1	10.937	0.9113	0.6214
1 inch per second	2.54	0.0254	0.09143	1	0.08333	0.0568
1 foot per second	30.48	0.3048	1.097	12	1	0.6818
1 mile per hour	44.70	0.447	1.609	17.604	1.467	1

TABLE A12.2 Conversion Factors *(Continued)*

g. Heat fluxes

	g-cal/(s)(cm²)	kcal/(h)(m²)	th/(d)(m²)	W/cm²	Btu/(h)(ft²)	Btu/(d)(ft²)
1 gram-calorie per second per square centimeter	1	36,000	864	4.183	13,276	31.86×10^4
1 kilocalorie per hour per square meter	27.78×10^{-6}	1	0.024	116.2×10^{-6}	0.36877	8.85
1 therm per day per square meter	115.74×10^{-5}	41.666	1	48.4×10^{-4}	15.365	368.77
1 watt per square centimeter	0.239	8,806	206.6	1	3,176.6	7.62×10^4
1 Btu per hour per square foot	75.32×10^{-6}	2.7117	0.06508	314.8×10^{-6}	1	24
1 Btu per day per square foot	3.14×10^{-6}	0.113	2.71×10^{-3}	13.12×10^{-6}	0.04167	1

h. Heat exchange

Heat Transfer Unit	kcal/(m²)(h)(°C)	Btu/(ft²)(h)(°F)
1 kilocalorie per square meter per hour per degree centigrade	1	0.2049
1 Btu per square foot per hour per degree Fahrenheit	4.88	1

Unit of Thermal Conductivity	kcal/(m²)(h)(°C/m)	Btu/(ft²)(h)(°F/ft)
1 kilocalorie per square meter per hour per (degree centigrade per meter)	1	0.6725
1 Btu per square foot per hour per (degree Fahrenheit per foot)	1.487	1

TABLE A12.2 Conversion Factors *(Continued)*

i. Viscosities

Dynamic Viscosity Unit	cP	kg/(m)(h)	lb/(ft)(h)
1 centipoise*	1	3.6	2.42
1 kilogram per meter per hour	0.2778	1	0.67209
1 pound per foot per hour	0.413	1.488	1

Kinematic Viscosity Unit	cSt	m^2/h	ft^2/h
1 centistokes†	1	0.0036	0.0387
1 square meter per hour	277.8	1	10.76
1 square foot per hour	25.81	0.0929	1

j. Pressure

Pressure Unit	kg/cm^2	bar	atm	psi	psf	Column of Hg at 0°C: mm	Column of Hg at 0°C: in	Water Column at 15°C: cm	Water Column at 15°C: in
1 kilogram per square centimeter	1	0.981	0.9678	14.22	2,048	735.5	28.96	1,001	394.05
1 bar‡	1.0197	1	0.987	14.50	2,088	749.79	29.522	1,020.72	401.85
1 atmosphere	1.033	1.013	1	14.70	2,116	760	29.92	1,034	407.14
1 pound per square inch	0.07031	0.06896	0.06804	1	144	51.71	2.036	70.37	27.70
1 pound per square foot	48.82×10^{-5}	47.89×10^{-5}	47.26×10^{-5}	69.44×10^{-4}	1	0.3591	141.38×10^{-4}	0.48873	0.19242
Hg column: millimeter	135.96×10^{-5}	13.34×10^{-4}	13.16×10^{-4}	19.34×10^{-3}	2.7845	1	0.03937	1.361	0.5357
inch	0.03453	0.03387	0.03342	0.4912	70.7266	25.40	1	34.56	13.61
Water column: centimeter	9.99×10^{-4}	97.97×10^{-5}	9.67×10^{-4}	0.01421	2.0463	0.7349	0.02893	1	0.3937
inch	25.38×10^{-4}	248.85×10^{-5}	24.56×10^{-4}	0.0361	5.1977	1.867	0.07349	2.54	1

TABLE A12.2 Conversion Factors *(Continued)*

k. Work, Energy, and Heat

Unit§	J	kWh	ch(h)	kgm	kcal	ft.lb	Btu	CHU	hp(h)
1 joule	1	27.78×10^{-8}	37.76×10^{-8}	0.10197	23.90×10^{-5}	0.7376	94.85×10^{-5}	52.70×10^{-5}	37.2×10^{-8}
1 kilowatthour	3.6×10^{6}	1	1.36	36.71×10^{4}	860.42	26.55×10^{5}	3414.37	1,896.87	1.341
1 chevalhour	2.65×10^{6}	0.736	1	2.70×10^{5}	634	1.95×10^{6}	2.51×10^{3}	1.39×10^{3}	0.986×10^{3}
1 kilogram-meter	9.807	2.72×10^{-6}	3.7×10^{-6}	1	2.34×10^{-3}	7.233	92.96×10^{-4}	51.64×10^{-4}	3.65×10^{-6}
1 kilocalorie	4.18×10^{3}	11.62×10^{-4}	1.58×10^{-3}	427	1	3,086	3.968	2.2046	1.56×10^{-3}
1 foot pound	1.356	37.66×10^{-8}	51.2×10^{-8}	0.1383	32.41×10^{-5}	1	12.86×10^{-4}	71.41×10^{-5}	50.50×10^{-8}
1 Btu	1,055	29.29×10^{-5}	3.98×10^{-4}	107.58	0.252	777.5	1	0.5555	3.93×10^{-4}
1 centigrade heat unit	1898	52.70×10^{-5}	7.16×10^{-4}	193.64	0.4536	1400.4	1.8	1	7.07×10^{-4}
1 horsepower-hour	2.68×10^{6}	0.7457	1.014	2.73×10^{5}	640.34	1.97×10^{6}	2.54×10^{3}	1.414×10^{3}	1

l. Power

Power unit¶	cal/s	kcal/s	kW	ch	ft.lb/s	ft.lb/min	Btu/h	hp	t of refrig.
1 calorie-gram per second	1	0.001	41.83×10^{-4}	5.69×10^{-3}	3.086	185.1	14.28	5.61×10^{-3}	11.91×10^{-4}
1 kilocalorie per second	1000	1	4.183	5.69	3,086	185.1×10^{3}	14,280	5.61	1.191
1 kilowatt	239	0.239	1	1.3596	737.6	44.25×10^{3}	3415	1.341	0.2845
1 cheval	175.7	0.1757	0.736	1	542.3	32,538	2511	0.986	0.2092
1 foot-pound per second	0.3241	32.41×10^{-5}	13.56×10^{-4}	1.84×10^{-3}	1	60	4.63	1.82×10^{-3}	38.58×10^{-5}
1 foot-pound per minute	54.02×10^{-4}	54.02×10^{-7}	22.60×10^{-6}	30.63×10^{-6}	0.01667	1	0.0771	30.33×10^{-6}	64.31×10^{-7}
1 Btu per hour	69.99×10^{-3}	69.99×10^{-6}	29.28×10^{-5}	39.67×10^{-5}	0.216	12.96	1	39.28×10^{-5}	83.33×10^{-6}
1 horsepower	178.2	0.1782	0.7457	1.014	550	33,000	2547	1	0.2122
1 ton of refrigeration	840	0.84	3.514	4.759	2592	15.55×10^{4}	12000	4.712	1

TABLE A12.2 Conversion Factors *(Continued)*

m. Flowrates

Volume flowrate	l/s	m^3/h	gal/min	gal/h	ft^3/s	ft^3/h	bbl/h	bbl/d
1 liter per second	1	3.6	15.85	951.2	0.03532	127.1	22.66	543.8
1 cubic meter per hour	0.2778	1	4.4031	264.2	98.10×10^{-4}	35.30	6.292	151
1 gallon per minute	0.06308	0.22711	1	60	22.28×10^{-4}	8.019	1.429	34.30
1 gallon per hour	10.52×10^{-4}	37.85×10^{-4}	0.01667	1	37.13×10^{-6}	0.1337	0.02382	0.5716
1 cubic foot per second	28.32	101.95	448.9	26,930	1	3,600	641.1	15,386
1 cubic foot per hour	78.67×10^{-4}	283.20×10^{-4}	0.1246	7.481	27.78×10^{-5}	1	0.1781	4.272
1 barrel per hour	0.04415	0.15894	0.700	42	15.6×10^{-4}	5.615	1	24
1 barrel per day	18.40×10^{-4}	66.24×10^{-4}	0.02917	1.75	64.98×10^{-6}	0.234	0.04167	1

n. Mass Flux

Flux Rate	$kg/(cm^2)(s)$	$kg/(m^2)(h)$	$lb/(ft^2)(h)$
1 kilogram per square centimeter per second	1	$3,600 \times 10^4$	7376.26×10^3
1 kilogram per square meter per hour	27.78×10^{-9}	1	0.2048
1 pound per square foot per hour	135.57×10^{-9}	4.882	1

o. Perfect-Gas Constants for the Equation $PV = nRT$

n	V	T	P	R
lb/mole	ft^3	°R	lb/in^2	10.73
			lb/ft^2	1,545.12
			atm	0.7302
		K	lb/in^2	19.314
			atm	1.314
g-mole	cm^3	K	atm	82.05
			bar	83.12
	L	K	atm	0.08205
			mmHg	62.36
kg-mole	L	K	atm	82.05
	m^3		bar	0.0831

TABLE A12.2 Conversion Factors *(Continued)*

p. Molecular Volume at 1 atm

Mole	Temperature	Volume
lb-mole	0 °C	359 ft^3
	60 °F	379 ft^3
g-mole	0 °C	22.4 liters

q. Heat Content

Unit	kcal/kg	Btu/lb
1 kilocalorie per kilogram	1	1.8
1 Btu per pound	0.555	1

r. Acceleration Due to the Force of Gravity

	cm/s^2	m/h^2	ft/s^2	ft/h^2
$g =$	981	1.271×10^8	32.18	4.17×10^8

* 1 poiseuille (Pl) = 10 poises = 1,000 centipoises.

†1 Stokes (St) = $10^{-4} m^2/s$.

‡1 bar = 1 hectopièzes = 100 pièzes = 10^6 baryes = 10^5 pascals.

§1 joule = 10^7 ergs; 1 therm = 10^3 kilocalories = 10^6 gram-calories; 1 CHU = 1 PCU (pound centigrade unit).

¶1 kilowatt = 10^{10} ergs per second; 1 cheval = 75 kilogram-meters per second.

TABLE A12.3 Comparison of International Standards for Steels

	Standard and Code Designation					
Type	*ASTM-AISI*	*AFNOR*	*BSI*	*SIS*	*VDEh*	Other
Mild Steel	SA 201 A-B					
	SA 212 A-B					
	SA 285 C	A37C NFA36205	BS 1449		DIN 17155 typ H1	
Low-Alloy steel	SA 203 A	NFA 36208			WI 5639 16 Ni 14	
	SA 203 D	—				
	SA 302 A-B					
	SA 357					
	SA 387 C					
Stainless and High-alloy steel	SA 240-304	Z6 CN 18-09	EN 58 E	142333	W 4301	
	SA 240-304 L	Z2 CN 18-10	—	—	W 4306	
	SA 240-310 S	Z12 CNS 25-20			W 4842	
	SA 240-316	Z6 CND 17-12	EN 58 J	142342	W 4401	
	SA 240-316 L	Z3 CND 17-12	—	—	W 4404	
	SA 240-316 (Ti)	Z3 CNDT 17-12	EN 58 J	142343	W 4571	
	SA 240-317					
	SA 240-321	Z6 CNT 18-11	EN 58 C	142334	W 4541	
	SA 240-347		EN 58 G	142334	W 4550	
	SA 240-410	Z12 C 13	EN 56 A	142302	W 4001	
Austen-itic/fer-ritic		Z5 CNDU 21-08				Uranus 50
		Z1 NCDU 25-20				Uranus B 6
Nickel-based						Hastelloy
						Monel 400
						Inconel 600
						Inconel 625

AISI: American Iron and Steel Institute.
BSI: British Standards Institution.
SIS: Sveriges Standardiseringskomission.
VDEh: Verein Deutscher Eisenhüttenleute.
ASTM: American Society for Testing Materials

Alloying Element and Concentration, %									
Fe	Cr	Ni	Mo	Si max	Mn max	S max	P max	Cu max	Miscellaneous
Balance	16–18	3.5–5.5	—	—	5.5–7.5	—	—	—	C = 0.15 max
Balance									N = 0.25 max
Balance									
Balance	17–19	8–10	0.5						C = 0.08–0.20
Balance	5		0.5						
Balance	1.25		0.5						
Balance	17–20	9–12	—	1	2	0.03	0.045	1	C = 0.08 max
Balance	17–20	9–12	—	1	2	0.03	0.045	1	C = 0.04 max
Balance	25	20	0.2	1	2	0.03	0.045	1	C = 0.05 max
Balance	16–19	10–14	2–3	1	2	0.03	0.045	1	C = 0.08 max
Balance	16–19	10–14	2–3	1	2	0.03	0.045	1	C = 0.04 max
Balance	16–19	10–14	2–3	1	2	0.03	0.045	1	C = 0.04 max-Ti
Balance	18–20	11–15	3–4	1	2	0.03	0.045	1	C = 0.08 max
Balance	17–20	9–13	—	1	2	0.03	0.045	1	C = 0.12 max
Balance	17–20	9–13	—	1	2	0.03	0.045	1	C = 0.12 max
Balance	12–14	—	—	1	1	0.03	0.04	0.4	C = 0.15 max
	21	8	2.5	—	—	—	—	1.5	C = 0.08 max
	20	25	4.5	—	—	—	—	1.5	C = 0.02 max
18–21	21–23.5	45	5.5–7.5	—	—	—	—	1.5–2.5	C = 0.05 max
									Nb + Ta = 2
1.25	—	66.5	—	0.15	0.9	—	—	31.5	C = 0.15 max
7	16.5	76	—	0.20	0.20	0.007	—	0.10	C = 0.04 max
12.5	20	63		0.35	0.55	0.007	—	0.05	C = 0.05 max,
									Al = 0.15
									Ti = 0.3
									Nb + Ta = 4

TABLE A12.4 General Corrosion Properties of Some Metals and Alloys*

Ratings: 0 unsuitable. Not available in form required or not suitable for fabrication requirements or not suitable for corrosion conditions.
1 poor to fair.
2 fair. For mild conditions or where periodic replacement is possible. Restricted use.
3 fair to good.
4 good. Suitable when superior alternatives are uneconomic.
5 good to excellent.
6 normally excellent.
Small variations in service conditions may appreciably affect corrosion resistance. Choice of materials is therefore guided wherever possible by a combination of experience and laboratory and site tests.

	Liquids											Gases				
	Nonoxidizing or Reducing Media				Oxidizing media			Natural waters				Common industrial media				
			Alkaline solutions, e.g.,					Fresh-water supplies		Sea water		Steam		Furnace gases with incidental sulfur content		
Material	Acid solutions, excluding hydrochloric, e.g., phosphoric, sulfuric, most conditions, many organics	Neutral solutions, e.g., many non oxidizing salt solutions, chlorides, sulfates	Caustic and mild alkalies, excluding ammonium hydroxide	Ammonium hydroxide and amines	Acid solu- tions, e.g., nitric	Neutral or alkaline solutions, e.g., persulfates, peroxides, chromates	Pitting media,† acid ferric chloride solutions	Static or slow- moving	Turbu- lent	Static or slow- moving	Turbu- lent	Moist, conden- sate	Dry at high temp., promoting slight dissociation	Reducing, e.g., heat- treatment furnace gases	Oxidizing, e.g., flue gases	Ambient air, city or industrial
Cast iron, flake graphite, plain or low alloy	1	3	4	5	0	4	0	4	3	4	2	4	4	1	1	3
Ductile iron (higher strength and hardness may be attained by composition and heat-treatment or both	1	3	4	5	0	4	0	4	4	4	3	4	4	1	1	3
Ni-Resist corrosion-resistant cast iron, type 1(14 Ni; 7 Cu; 2 Cr; bal. Fe)	4	5	5	5	0	5	0	5	5	5	5	5	5	3	2	4
Ni-Resist corrosion-resistant cast iron, type 2 Cu free (20-30 Ni; 2-3 Cr; bal. Fe)	4	5	5	6	0	5	0	5	5	5	5	5	5	3	2	4
Ni-Resist corrosion-resistant cast iron, ductile (24 Ni; bal. Fe)	4	5	5	6	0	5	0	5	5	5	5	5	5	3	2	4
14% silicon iron	6	6	2	5	6	6	3	5	5	5	5	6	4	4	3	6
Mild steel, also low-alloy irons and steels	1	3	4	5	0	4	0	4	3	4	2	4	4	1	1	3
Stainless steel, ferritc 17% Cr type	2	4	4	6	5	6	0	4	6	1	4	5	6	3	2	4
Stainless steel, austenitic 18 Cr; 8 Ni type	3	4	5	6	6	6	0	6	6	2	5	6	6	2	3	5
Stainless steel, austenitic 18 Cr; 12 Ni;2.5 Mo type	4	5	5	6	5	6	1	6	6	3	5	6	6	2	4	6
Stainless steel, austenitic 20 Cr; 29 Ni; 2.5 Mo; 3.5 Cu type	5	6	5	6	5	6	2	6	6	4	6	6	6	2	4	6
Incoloy 825 nickel-iron-chromium alloy(40 Ni; 21 Cr; 3 Mo; 1.5 Cu; bal. Fe)	6	6	5	6	5	6	2	6	6	4	6	6	6	2	5	6

TABLE A12.4 General Corrosion Properties of Some Metals and Alloys* *(Continued)*

Ratings: 0 unsuitable. Not available in form required or not suitable for fabrication requirements or not suitable for corrosion conditions.
1 poor to fair.
2 fair. For mild conditions or where periodic replacement is possible. Restricted use.
3 fair to good.
4 good. Suitable when superior alternatives are uneconomic.
5 good to excellent.
6 normally excellent.
Small variations in service conditions may appreciably affect corrosion resistance. Choice of materials is therefore guided wherever possible by a combination of experience and laboratory and site tests.

	Liquids											Gases				
	Nonoxidizing or Reducing Media				Oxidizing media			Natural waters				Common industrial media				
			Alkaline solutions, e.g.,					Fresh-water supplies		Sea water		Steam		Furnace gases with incidental sulfur content		
Material	Acid solutions, excluding hydrochloric, e.g., phosphoric, sulfuric, most conditions, many organics	Neutral solutions, e.g., many non oxidizing salt solutions, chlorides, sulfates	Caustic and mild alkalies, excluding ammonium hydroxide	Ammonium hydroxide and amines	Acid solutions, e.g., nitric	Neutral or alkaline solutions, e.g., persulfates, peroxides, chromates	Pitting media,† acid ferric chloride solutions	Static or slow-moving	Turbulent	Static or slow-moving	Turbulent	Moist, condensate	Dry at high temp., promoting slight dissociation	Reducing, e.g., heat-treatment furnace gases	Oxidizing, e.g., flue gases	Ambient air, city or industrial
Hastelloy alloy C-276 (55 Ni; 17 Mo; 16 Cr; 6 Fe; 4 W)	5	6	5	6	4	6	5	6	6	6	6	6	6	3	4	6
Hastelloy alloy B-2 (61 Ni; 28 Mo; 6 Fe	6	5	4	4	0	3	0	6	6	4	4	6	5	3	2	5
Inconel nickel-chromium alloy 600 (78 Ni; 15 Cr; 7 Fe)	3	6	6	6	3	6	1	6	6	4	6	6	6	2	4	6
Copper-nickel alloys up to 30% nickel	4	5	5	0	0	4	1	6	6	6	6	6	5	2	2	5
Monel 400 nickel-copper alloy (66Ni; 30 Cu; 2 Fe)	5	6	6	1	0	5	1	6	6	4	6	6	6	2	3	5
Alloy 505 nickel-copper cast alloy (66 Ni; 30 Cu; 4 Si)	5	6	6	1	0	5	1	6	6	4	6	6	6	2	3	5
Monel K-500 age hardenable Ni-Cu alloy (67 Ni; 30 Cu; 3 Al)	5	6	6	1	0	5	1	6	6	4	6	6	6	2	3	5
A nickel—commercial (99.4 Ni)	4	5	6	1	0	5	0	6	6	3	5	6	6	2	2	4
Copper and silicon bronze	4	4	4	0	0	4	0	6	5	4	1	6	5	2	2	5
Aluminum brass (76 Cu; 22 Zn; 2 Al)	3	4	2	0	0	3	0	6	6	4	5	6	5	2	2	5
Nickel-aluminum-bronze (80 cu; 10 Al; 5 Ni; 5 Fe)	4	4	2	0	0	3	0	6	6	4	5	6	5	2	3	5
Bronze, type A (88 Cu; 5 Sn; 5 Ni; 2 Zn)	4	5	4	0	0	4	0	6	6	5	5	6	5	2	2	5
Aluminum and its alloys	1	3	0	6	0–5	0–4	0	4	5	0–5	4	5	2	5	4	5
Lead, chemical or antimonial	5	5	2	2	0	2	0	6	5	5	3	2	0	4	3	5
Silver	4	6	6	0	0	2	0	6	6	5	5	6	5	4	4	4
Titanium	3	6	2	6	6	6	6	6	6	6	6	6	5	3	5	6
Zirconium	3	6	2	6	6	6	2	6	6	6	6	6	6	3	5	6

TABLE A12.4 General Corrosion Properties of Some Metals and Alloys* *(Continued)*

Ratings: 0 unsuitable. Not available in form required or not suitable for fabrication requirements or not suitable for corrosion conditions.
1 poor to fair.
2 fair. For mild conditions or where periodic replacement is possible. Restricted use.
3 fair to good.
4 good. Suitable when superior alternatives are uneconomic.
5 good to excellent.
6 normally excellent.
Small variations in service conditions may appreciably affect corrosion resistance. Choice of materials is therefore guided wherever possible by a combination of experience and laboratory and site tests.

	Gases *(Cont'd.)*									
	Halogens and derivatives									
	Halogens									
Material	Moist, *e.g.,* chlorine below dew point	Dry, *e.g.,* fluorine above dew point	Halide acids moist, *e.g.,* hydrochloric hydrolysis products of organic halides	Hydrogen halides, dry,‡ *e.g.,* dry hydrogen chloride, °F.	Available forms	Cold formability in wrought and clad form	Weld-ability	Max. strength annealed condition × 1000 lb/in²	Coeff. of thermal expansion, millionths per °F. 70–212°F	Remarks¶
Cast iron, flake graphite, plain or low alloy	0	2	0	$2 < 400$ $1 < 750$	Cast	No	Fair§	45	6.7	
Ductile iron (higher strength and hardness be attained by composition	0	2	0	$2 < 400$ $1 < 750$	Cast	No	Good§	67	7.5	
Ni-Resist corrosion-resistant cast iron, type 1 (14 Ni; 7 Cu; 2 Cr; bal. Fe)	0	2	3	$3 < 400$ $2 < 750$	Cast	No	Good§	22–31	10.3	
Ni-Resist corrosion-resistant cast iron, type 2 Cu free (20–30 Ni; 2–3 Cr; bal. Fe)	0	2	3	$3 < 400$ $2 < 750$	Cast	No	Good§	22–31	9.6	Type 3 Ni-Resist has same corrosion resistance
Ni-Resist corosion-resistantcast iron, ductile (24 Ni; bal. Fe)	0	2	3	$3 < 400$ $2 < 750$	Cast	No	Good§	56	10.4	
14% silicon iron	0	0	4	$1 < 400$	Cast	No	No	22	7.4	Very brittle, susceptible to cracking by mechanical and thermal shock
Mild steel, also low-alloy irons and steels	0	3	0	$3 < 400$ $1 < 750$	Wrought, cast	Good	Good	67	6.7	High strengths obtainable by alloying, also improved atmospheric corrosion resistance. See A.S.T.M. specifications for particular grade
Stainless steel, ferritic 17% Cr type	0	2	0	$2 < 400$	Wrought, cast clad	Good	Good§	78	6.0	A.I.S.I. type 430 A.S.T.M. corrosion- and heat-resisting steels
Stainless steel, austenitic 18 Cr; 8 Ni types	0	2	0	$3 < 400$	Wrought, cast, clad	Good	Good	90	9.6	A.I.S.I. type 304 A.S.T.M. corrosion- and heat-resisting steels. Stabilized or ELC types used for welding
Stainless steel, austenitic 18 Cr; 12 Ni; 2.5 Mo type	0	3	2	$4 < 400$ $3 < 750$	Wrought, cast, clad	Good	Good	90	8.9	A.I.S.I. type 316 A.S.T.M. corrosion- and heat-resisting steel. ELC type used for welding
Stainless steel, austenitic 20 Cr; 29 Ni; 2.5 Mo; 3.5 Cu type	1	3	3	$4 < 400$ $3 < 750$	Wrought, cast	Good	Good	90	9.4	A.C.I. CH-7M. Good resistance to sulfuric, phosphoric, and fatty acids at elevated temperatures

Incoloy 825 nickel-iron-chromium alloy (40 Ni; 21 Cr; 3 Mo; 1.5 Cu; bal. Fe)	2	3	3	4 < 400 3 < 750	Wrought, cast, clad	Good	Good	100	7.3	Special alloy with good resistance to sulfuric, phos-pENphoric, and fatty acids. Resistant to chlorides in some environments
Hastelloy alloy C-276 (55Ni; 17 Mo; 16 Cr; 6 Fe; 4 W)	5	4	4	4 < 750 3 < 900	Wrought, cast, clad	Fair	Good	145	6.3	Excellent resistance to wet chlorine gas and sodium hypochlorite solutions
Hastelloy alloy B-2(61 Ni; 28 Mo; 6 Fe)	1	3	5	4 < 750 3 < 900	Wrought, cast, clad	Fair	Good	135	5.6	Resistant to solutions of hydrochloric and sulfuric acids
Inconel nickel-chromium alloy 600 (78 Ni; 15 Cr; 7 Fe	2	5	3	5 < 400 4 < 900	Wrought, cast, clad	Good	Good	90	8.9	Wide application in food and pharmaceutical industries
Copper-nickel alloys up to 30% nickel	1	5	2	4 < 400 3 < 750	Wrought, cast, clad	Good	Good	38–62	9.3–8.5	High-iron types excellent for resisting high-velocity effects in condenser tubes
Monel 400 nickel-copper alloy (66 Ni 30 Cu; 2 Fe)	2	6	3	6 < 400 3 < 750 2 < 900	Wrought, cast, clad	Good	Good	77	7.5	Widely used for sulfuric acid pickling equipment. Also for propeller shafts in motor boats. Take precautions to avoid sulfur attack during fabrication
Alloy 505 nickel-copper cast alloy (66 Ni; 30 Cu; 4 Si)	2	4	3	6 < 400 3 < 750 2 < 900	Cast	No	No	100	8.8	Non-galling characteristics. Excellent for bearings or bushings. High strength developed by heat-treatment
Monel K-500 age hardenable Ni-Cu alloy (67 Ni; 30 Cu; 3 A1)	2	6	3	6 < 400 3 < 750 2 < 900	Wrought, cast	Fair	Good	99–155	7.4	High strength obtainable by heat-treatment. Take precautions to avoid sulfur attack during fabrication
A nickelommercial (99.4 Ni)	2	6	2	6 < 400 5 < 750 4 < 900	Wrought, cast, clad	Good	Good	54	6.6	Widely used for hot concentrated caustic solutions. Take precautions to avoid sulfur attack during fabrication
Copper and silicon bronze	0	5	2	3 < 400 2 < 750	Wrought, cast, clad	Excellent	Fair	29	9.3–9.5	Unsuitable for hot concentrated mineral acids or for high-velocity HF
Aluminum brass (76 Cu; 22 Zn; 2 A1)	0	4	2	2 < 400	Wrought, cast	Good	Fair	60	10.3	May develop localized corrosion in sea water
Nickel-aluminum-bronze (80 Cu; 10 A1; 5 Ni; 5 Fe)	0	4	3	3 < 400 2 < 750	Wrought, cast	Good	Fair	60–80	9.4	Ship propellers an excellent application
Bronze, type A (88 Cu; 5 Sn; 5 Ni; 2 Zn)	0	4	3	3 < 400 2 < 750	Cast	No	§	45	11.0	High strengths obtainable by heat-treatment. Not susceptible to dezincification
Aluminum and its alloys	0	6	0	3 < 400 1 < 750	Wrought, cast, clad	Good	Good	9–90	11.5–13.7	Extent of corrosion dependent upon type and concentration of acidicions. Wide range of mechanical properties obtainable by alloying and heat-treatment
Lead, chemical or antimonial	0	1	3	0	Wrought, cast, clad	Excellent	Good	2	16.4–15.1	High purity 'chemical lead' preferred for most applications
Silver	5	5	3	4 < 400 2 < 750	Wrought, cast, clad	Excellent	Good	21	10.6	Used as a lining
Titanium	6	0	1	0	Wrought, cast	Fair	Good§	6–90	5.0	Red fuming HNO_3 may initiate explosions. Good resistance to solutions containing chlorides
Zirconium	6	1	6	0	Wrought, cast	Fair	Good§			

[a]Also Chlorimet 3. [b]Also Chlorimet 2.

Data courtesy of the International Nickel Company, Inc.

†On unsuitable materials these media may promote potentially dangerous pitting.

‡Temperatures are approximate.

§Special precautions required.

¶Many of these materials are suitable for resisting dry corrosion at elevated temperatures.

Appendix 13

Calculation Sheets for Equipment Items

TABLE A13.1 Process Design Estimate: Towers and Tanks (Sec. A1.4)

a. Determining Wall Thicknesses

		Item No.					
	Temperature, max., °C						
	Pressure, max. P, bar						
	Maximum allowable stress t, bar (Table A1.8)						
	Welding coefficient, α						
	Shell radius, mm						
	Required shell thickness e_b mm, $= \frac{PR}{\alpha t - 0.6P}$						
	Corrosion allowance Δe, mm						
	Shell thickness $e = e_b + \Delta e$, mm						

b. Cost of Shell, Heads, and Skirt, $1,000

		Item No.					
	Diameter D', m						
	Height H, m						
1	Thickness e, mm						
	Weight in kilograms						
2	Shell: $24.7D'He$						
3	Heads: (Fig. A1.8) =						
4	**(1)** × **(3)**						
5	Shell + heads = **(2)** + **(4)**						
	Price of shell and heads:						
6	Base price, $1,000/kg (Fig. A1.9) × 10^{-3}						
	Correction factors						
7	f_e (Fig. A1.10)						
8	f_m (Table A1.9)						

TABLE A13.1 Process Design Estimate: Towers and Tanks (Sec. A1.4) *(Continued)*

	b. Cost of Shell, Heads, and Skirt, \$1,000 *(Continued)*						
		Item No.					
9	Corrected price = **(6)** × **(7)** × **(8)**						
10	Price of shell and heads = **(5)** × **(9)**						
	Price of the skirt (towers only)						
	Height H', m						
	Thicknesses e', mm						
11	Weight = $24.7D'H'e'$						
12	Correction factor f_e (Fig. A1.10)						
13	Price of skirt = **(6)** × **(11)** × **(12)**						

	c. Cost of Cladding or Lining, \$1,000						
		Item No.					
	Diameter D', m						
	Height H, m						
1	Shell surface, m^2 = $3.14D'H$						
2	Surface of heads, m^2 = (Fig. A.I.8)						
3	Total surface, m^2 = **(1)** + **(2)**						
	Cladding (6–20 mm thick):						
4	Base cost, \$1,000 = 0.109 × **(3)**						
	Correction factors:						
5	f'_e (Fig. A1.11)						
6	f'_m (Table A1.10)						
7	f'_c (Table A1.10)						
8	Price of cladding, \$1,000 = **(4)** × **(5)** × **(6)** × **(7)**						
	Lining (over 20 mm thick):						
9	Base price, \$1,000 = 0.182 × **(3)**						

TABLE A13.1 Process Design Estimate: Towers and Tanks (Sec. A1.4) *(Continued)*

	c. Cost of Cladding or Lining, $1,000 *(Continued)*						
		Item No.					
	Correction factors						
10	f''_e (Fig. A1.11)						
11	f''_m (Table A1.11)						
12	Price of lining, $1,000 = **(9)** × **(10)** × **(11)**						

	d. Cost of Internals, $ 1,000						
		Item No.					
	Trays						
	Diameter *D′*, m						
1	Number,						
	Material,						
2	Base price, $ (Fig. A1.13)						
	Correction factors (Table A1.13):						
3	Type f_{pl}						
4	Number of passes f_{pa}						
5	Thickness f_{pe}						
6	Number f_{pn}						
7	Cost, $1,000 = **(1)** × **(2)** × **(3)** × **(4)** × **(5)** × **(6)** × 10^{-3}						
	Packing						
	Type						
	Dimension						
8	Volume, m^3						
9	Base price (Table A1.14)						
10	Base-price reduction (10% for 50–100m^3, 15% for more)						

TABLE A13.1 Process Design Estimate: Towers and Tanks (Sec. A1.4) *(Continued)*

	d. Cost of Internals, $ 1,000 *(Continued)*						
		Item No.					
11	Price of packing, $1,000 = **(8)** × **(9)** × **(10)** × 10^{-3}						
12	Cost of internals, $1,000 = **(11)** + **(7)**						

	e. Final Estimated Current Cost, $ 1,000						
		Item No.					
1	Shell and heads (sheet b)						
2	Skirt (sheet b)						
3	Skirt, shell, and heads = **(1)** + **(2)**						
4	Cladding or lining (sheet c)						
5	Base price for accessories, [Fig. A1.12 applied to **(3)**]						
6	Correction factor, f_{am} (Table A1.12)						
7	Price of accessories = **(5)** × **(6)**						
8	Price of vessel = **(3)** + **(4)** + **(7)**						
9	Price of internals, (sheet d)						
10	Price = **(8)** + **(9)**						
11	Correction factors: For towers, 1.15 For tanks, 1.10						
12	Final cost in 1975 = **(10)** × **(11)**						
13	Cost index reference for pressure vessels (Fig. 4.7)						
14	Current cost index for pressure vessels						
	Final current cost = **(12)** × **(14)**/**(13)**						

TABLE A13.2 Process Design Estimate: Shell-and-Tube Heat Exchangers (Sec. A3.1)

		Item No.		Item No.	
		Hot Side	Cold Side	Hot Side	Cold Side
	Fluid:				
	Nature				
	Flow rate, t/h				
	Temperature,				
	Entering, °C	$T_1=$	$t_1=$	$T_1=$	$t_1=$
	Leaving, °C	$T_2=$	$t_2=$	$T_2=$	t_2
	Operating pressure, bar				
	Materials required, tubes, shell				
	Corrected LMTD				
	$T_1 - t_2$				
	$T_2 - t_1$				
	LMTD (Fig. A3.1)				
	$E = (t_2 - t_1)/(T_1 - t_1)$				
	$R = (T_1 - T_2)/(t_2 - t_1)$				
	f (from Fig. A3.2)				
	Corrected LMTD = $f \times$ LMTD				
	Required no. of shells				
	Heat exchange duty, Q, million kcal/h				
	Overall heat transfer coefficient U, kcal/(h)(m^2)(°C): (Table A3.1 or Fig. A3.3)				

TABLE A13.2 Process Design Estimate: Shell-and-Tube Heat Exchangers (Sec. A3.1) *(Continued)*

		Item No.		Item No.	
		Hot Side	Cold Side	Hot Side	Cold Side
	Type of exchanger required (Fig. A3.4)				
	Total estimated surface, m^2 = $Q/U \times$ LMTD corr.				
1	Base price, \$1,000 (Figs. A3.5 and A3.6)				
	Correction factors, (Tables A3.6 and A3.7)				
	For type f_d				
	For tube diameter and pitch $f\phi$				
	For length of bundle f_l				
	For number of tube passes f_{np}				
	For pressure f_p				
	For temperature f_t				
	For materials f_m				
2	Combined correction factor				
3	Corrected price for 1975 = **(1)** $\times$ **(2)**				
4	Cost index reference for exchangers (Fig. 4.7)				
5	Current cost index for exchangers				
6	Current estimated cost, \$1,000 = **(2)** $\times$ **(1)** $\times$ **(5)/(4)**				

TABLE A13.3 Process Design Estimate: Air Coolers (Sec. A3.2)

		Item No.					
	Process fluid						
	Nature						
	Inlet temperature T_1, °C						
	Outlet temperature T_2, °C						
1	Temperature difference $T_2 - T_1$, °C						
	Air Inlet temperature t_a, °C						
2	Inlet temperature difference $T_1 - t_a$, °C						
3	Heat exchange duty Q, million kcal/h						
4	Temperature-difference ratio $R = (T_2 - T_1)/(T_1 - t_a) =$ **(1)/(2)**						
5	Reduced heat exchange duty S $= Q10^{-3}/(T_1 - t_a) =$ **(3)/(2)**						
6	Overall resistance to heat transfer r, (h)(m²)(°C)/kcal: (Table A3.8)						
	Power absorbed from fans:						
	(Fig. A3.9) for K						
7	Power in cv $= P_{cv} = KS = K \times$ **(5)**						
	Width of cooler, m $= 0.112P_{cv} = 0.112 \times$ **(7)**						
	Number of tube rows, N						
	$r_m = 0.00015$						
	$r_a =$ (from Table A3.9)						
8	$1/U = r + r_m + r_a =$ **(6)** $+ r_m + r_a$						
	(2) × **(8)**						
	Corrected r_a (Table A3.9)						
	N (Table A3.9)						
	Air exit temperature, t_h, °C:						
9	V_f (Table A3.9)						
10	$t_h = t_a + Q/(1{,}061V_f + P) = t_a +$ **(3)**/[1.061 × **(9)** + **(7)**]						

TABLE A13.3 Process Design Estimate: Air Coolers (Sec. A3.2) *(Continued)*

		Item No.					
	Bare tube surface, m^2:						
11	$1/U$ with corrected r_a						
	LMTD corrected:						
	$(T_1 - t_h)$, °C						
	$(T_2 - t_a)$, °C						
12	LMTD (Fig. A3.1)						
	$(t_h - t_a)/(T_1 - T_2) = p =$ $(t_h - t_a)$/**(1)**						
	Number of passes: (Fig. A3.8)						
13	Correction factor, f (Fig. A3.8)						
14	Corr. LMTD = $f \times$ LMTD						
15	Surface $S = Q/(U)$(LMTD corr.) = **(3)** × **(11)/(14)**						
	Price						
	Base cost C, \$1,000 (Fig. A3.10)						
16	Price = $C \times S = C \times$ **(5)**						
	Correction factors: (Table A3.10)						
17	f_e, thickness						
18	f_p, pressure						
19	f_l, length						
20	f_n, no. of tube rows						
21	f_m, materials						
22	Corrected price in 1975, \$1,000 = **(16)** × **(17)** × **(18)** × **(19)** × **(20)** × **(21)**						
23	Cost index reference for exchangers (Fig. 4.7)						
24	Current cost index for exchangers						
25	Current estimated cost, \$1,000 = **(22)** × **(24)/(23)**						

TABLE A13.4 Process Design Estimate: Centrifugal Pumps (Sec. A4.1.3)

a. Estimating Cost							
		Item No.					
	Operating conditions						
	Fluid pumped						
	Mass flow, t/h						
	Temperature, °C						
	Density at						
	Ambient temp.						
	Operating temp., d_t						
1	Flow rate, m^3/h						
	Design conditions						
2	Safety factor (see Sec A4.1.2.1)						
3	Design flow, m^3/h = **(1)** × **(2)**						
	Pressure drop:						
	Differential head H, m						
	Pressure head, bar = $H/10d_t$						
	Upstream pressure P_1, bar						
	Downstream pressure P_2, bar						
	Friction loss F, bar						
	Total pressure differential ΔP,						
4	bar = $H/10d_t + (P_1 - P_2) + F$						
	Pressure head ΔH, m = $10\Delta P d_t$						
	Type of centrifugal pump						
	Material						
	Price						
5	Base price: (Fig. A4.5*a* and A4.5*b*)						
	Correction factors: (Tables A4.2)						
6	Type, f_d						
7	Material, f_m						
8	Temperature, f_t						
9	Suction pressure, f_p						

TABLE A13.4 Process Design Estimate: Centrifugal Pumps (Sec. A4.1.3) *(Continued)*

a. Estimating Cost *(Continued)*

		Item No.					
10	Corrected price for 1975: **(5)** × **(6)** × **(7)** × **(8)** × **(9)**						
11	Cost index reference for centrifugal pumps (Fig. 4.7)						
12	Current cost index for centrifugal pumps						
13	Current estimated cost, **(10)** × **(12)/(11)**						
14	If an operating spare is required: 2 × **(13)**						

b. Estimating the Drive Motor and Electrical Consumption (Sec. A5.1)

		Item No.					
	Operating conditions (See Table A13.4a)						
1	Operating flow, m^3/hr						
2	Design flow, m^3/h						
3	Total pressure differential, bar						
	Power						
4	Hydraulic horsepower, cv 0.038 × **(2)** × **(3)**						
5	Efficiency (Fig. A4.2)						
6	Power = **(4)/(5)**						
	Motor						
	Power, cv (Table A5.1)						
	Type						
7	Base price (Fig. A5.1)						
8	Correction factor (Fig. A5.2)						
9	Corrected price in 1975 = **(7)** × **(8)**						

TABLE A13.4 Process Design Estimate: Centrifugal Pumps (Sec. A4.1.3) ***(Continued)***

	b. Estimating the Drive Motor and Electrical Consumption (Sec. A5.1) *(Continued)*						
		Item No.					
10	Cost index reference for motors (Fig. 4.7)						
11	Current cost index for motors						
12	Current estimated cost, **(9)** × **(11)/(12)**						
13	If an operating spare is required: 2 × **(12)**						
	Electrical consumption						
14	Operating power, cv: 0.038 × **(1)** × **(3)/(5)**						
	Use percent for motor						
15	Efficiency of motor (Table A5.1)						
	Electrical consumption, kWh: 0.735 × **(14)/(15)**						

TABLE A13.5 Process Design Estimate: Compressors (Sec. A4.2.2)

	a. Simplified Calculation						
		Item No.					
	Number of stages *s*:						
1	Suction pressure P_1, bar						
2	Discharge pressure P_2, bar						
3	Stages *s* are the next highest whole number after 1.431og (P_2/P_1)						
	Compression ratio per stage, ρ						
4	Interstage pressure drop, bar: $(s - 1)0.350$						
5	Total pressure developed, bar: $P_2 + (s - 1)0.350 = \mathbf{(2)} + \mathbf{(4)}$						
	Overall compression ratio *R*: **(5)/(1)**						
	Compression ratio per stage ρ: $\rho = R^{1/s}$						
	Suction conditions						
	Flow, kmol/h, *D*						
6	Temperature T_1, K						
7	Volume V_1, m³/h: $0.082DT_1/P_1$						
	Theoretical power per stage PTA_s, cv						
	Ratio of specific heats, γ (Table A4.5)						
8	$(\gamma - 1)/\gamma$						
9	$\gamma/(\gamma - 1)$						
10	$\rho^{(\gamma - 1)/\gamma} = \rho^{\mathbf{(8)}}$						
11	**(10)** − 1						
12	$P_1V_1/26.5 = \mathbf{(1)} \times \mathbf{(7)}/26.5$						
13	$PTA_s = \mathbf{(12)} \times \mathbf{(9)} \times \mathbf{(11)} = cv$						
14	Overall theoretical horsepower *PTA*, cv						
	$PTA = (PTA_s) \times s = \mathbf{(3)} \times \mathbf{(13)}$						
	Note: This calculation assumes that the inlet temperature to each stage is the same as T_1.						

TABLE A13.5 Process Design Estimate: Compressors (Sec. A4.2.2) ***(Continued)***

	a. Simplified Calculation *(Continued)*						
		Item No.					
	Reciprocating compressor						
15	Mechanical efficiency (Fig. A4.12)						
16	Power with gas engine drive, cv **(14)/(15)**						
17	Power with electric motor drive, cv = **(16)**/0.95						
18	Discharge temperature, K, T_2 $T_2 = T_1\rho^{(\gamma - 1)/\gamma}$ = **(6)** × **(10)**						
	Centrifugal compressor						
19	Polytropic efficiency E_p						
20	Correction factor (Fig. A4.10)						
21	E_p = **(19)** × **(20)**						
	Isentropic efficiency E_i						
22	Ratio of E_i/E_p (Fig. A4.11)						
23	E_i = **(21)** × **(22)**						
24	Losses p, cv (See Sec. A4.2.2.6)						
25	Actual power, cv = $p + PTA/E_i$ = **(24)** + **(14)/(23)**						
	Discharge temperature						
26	Temperature rise, °C = $(T_2 - T_1)/E_i$ = [**(18)** − **(6)**]/**(22)**						
27	$T_2 = T_1$ + rise = **(6)** + **(26)**						

	b. Estimating Cost and Electrical Consumption						
		Item No.					
	Compressors						
1	Calculated horsepower, cv						
2	Base cost (Figs. A4.13 to A4.15)						

TABLE A13.5 Process Design Estimate: Compressors (Sec. A4.2.2) *(Continued)*

	b. Estimating Cost and Electrical Consumption *(Continued)*						
		Item No.					
	Correction factors (Table A4.6):						
3	f_{pmax}						
4	f_m						
5	Other factors (see Section A.4.2.3.3)						
6	Corrected cost for 1975 = **(2)** × **(3)** × **(4)** × **(5)**						
7	Cost index reference for pumps and compressors (Fig. 4.7)						
8	Current cost index for pumps and compressors						
9	Final current cost = **(6)** × **(8)/(7)**						
	Motors						
	Power, cv (Table A5.1)						
	Type (Table A5.1)						
10	Base cost (Fig. A5.1)						
11	Correction factor (Fig. A5.2)						
12	Corrected price in 1975 = **(10)** × **(11)**						
13	Reference cost index for motors for pumps and compressors (Fig. 4.7)						
14	Current cost index for motors for pumps and compressors						
15	Final current cost = **(12)** × **(14)/(13)**						
	Cost of compressor train:						
	Single unit only: **(9)** + **(15)**						
	Electrical consumption						
	Use percentage of motor						
17	Efficiency of motor (Table A5.1)						
18	Electrical consumption, kWh/h [0.735 × **(1)**]/**(17)**						

TABLE A13.6 Process Design Estimate: Steam Turbines (Sec. A5.2)

		Item No.					
	Type						
	Speed, rpm						
	Power consumption						
	Required power P_0, cv						
	Base efficiency ρ_0 (Fig. A5.3)						
	Steam inlet pressure, bars absolute						
	Steam outlet pressure, bars absolute						
	Steam correction factor f_v (Fig. A5.4)						
	Corrected efficiency $\rho = \rho_0 f_v$						
	Turbine power consumption, cv $= P_0/\rho$						
	Steam consumption						
	Available energy, kcal/kg						
	Correction for superheat:						
	Temperature, °C						
	Superheat Δt, °C						
	Correction: $1 + 0.08\Delta t/50$						
	Corrected available energy, kcal/kg						
	Correction for humidity:						
	% saturation						
	Correction: $1 - 0.02(\%)$						
	Corrected available energy, kcal/kg						
	Steam consumption, kg/(cv)(h) (Figs. A5.5 and A5.6)						
	Steam consumption, kg/h						
	Price						
1	Base price in 1975 (Fig. A5.7)						
2	Cost index reference (Fig. 4.7)						
3	Current cost index						
4	Final current price = **(1)** × **(3)/(2)**						

TABLE A13.7 Process Design Estimate: Furnaces (Sec. A6.2)

		Item No.					
	Type						
	Service pressure, bar						
	Materials						
	Duty						
	Demanded, million kcal/h						
	Fuel consumption						
	Efficiency, %						
	Low heating value, kcal/kg						
	Consumption, kg/h						
	Price						
1	Base price (Fig. A6.1)						
	Correction factors (Table A6.1):						
	f_d for type						
	f_m for materials						
	f_p for pressure						
2	Corrected price in 1975: **(1)** $\times (1 + f_d + f_m + f_p)$						
3	Reference cost index (Fig. 4.7)						
4	Current cost index						
5	Final current price: **(2)** $\times$ **(4)/(3)**						

TABLE A13.8 Process Design Estimate: Steam Ejectors (Sec. A7.1)

		Item No.					
	Fluid evacuated						
	Nature						
	Composition						
	Maximum flow at inlet						
	Inlet pressure, mmHg						
	Number of stages						
	Volume of equipment evacuated, m^3						
1	Inlet flow, kg/h (Fig. A7.1)						
2	Maximum inlet flow, kg/h **(1)** × 2						
	Flow of condensables						
3	Partial pressure of the condensables $P_v = P_i$						
4	Partial pressure of the noncondensables $P_{nc} = \Pi - P_v$						
5	Average molecular weight of the condensables M_c						
6	Average molecular weight of the noncondensables M_{nc}						
7	Kilomoles of noncondensables $N_{nc} = D_m / M_{nc}$ = **(2)/(6)**						
8	Kilomoles of condensables $N_v = N_{nc}(P_v / P_{nc})$ = **(7)** × **(3)/(4)**						
9	Flow of condensables $D_v = N_v \times M_v$ = **(8)** × **(5)**						

TABLE A13.8 Process Design Estimate: Steam Ejectors (Sec. A7.1) *(Continued)*

		Item No.					
	Flow equivalent of dry air at 20°C						
	Average entrainment coefficient:						
10	Temperature f_t						
	Air or gas						
	Water vapor						
11	Molecular weight f_M (Fig. A7.2)						
12	Flow equivalent to dry air $= D_m/f_1 f_M =$ **(2)/(10) × (11)**						
	Utility consumption per kg of air						
	Without condensers:						
	Steam *t*, (Table A7.2)						
	Correction factor (Fig. A7.3)						
	With condensers:						
	% noncondensables						
	Steam *t*, (Table A7.3)						
	Correction factor (Fig. A7.3)						
	Water, m^3, (Table A7.3)						
	Consumption per hour						
	Steam, t/h						
	Cooling water, m^3/h						
	Price						
13	Base price in 1975 (Figs. A7.4*a* to A7.4*c*)						
14	Reference cost index (Fig. 4.7)						
15	Current cost index						
16	Final current price: **(13) × (15)/(14)**						

Bibliography

1. GENERAL

Aries, R. S., and R. D. Newton: *Chemical Engineering Cost Estimation,* McGraw-Hill, New York, 1965.

Bauman, H. C.: *Fundamentals of Cost Engineering in the Chemical Industry,* Reinhold, New York, 1964.

Chilton, C. H.: *Cost Engineering in the Process Industries,* McGraw-Hill, New York, 1960.

"Cost Engineers' Notebook," *AACE Bulletin,* American Association of Cost Engineers 17th national meeting, St. Louis, Mo., 1973.

Depallens, G.: *Gestion Financière de l'Entreprise,* Sirey, Paris, 1970.

Givaudon, J., P. Massot, and R. Bensimon: *Précis de Génie Chimique,* Berger-Levrault, Paris, 1960.

Guthrie, K. M.: *Process Plant Estimating Evaluation and Control,* Craftsman Book Company of America, Solana Beach, Calif., 1974.

Jelen, F. C.: *Cost and Optimization Engineering,* McGraw-Hill, New York, 1970.

Jordan, D. G.: *Chemical Process Development,* Wiley Interscience, New York, 1968.

Letullier, A.: *Petit Lexique de Termes Financiers Américains,* Mena Press, SCM, Publications, Paris, 1975.

Ludwig, E. E.: *Applied Process Design for Chemical and Petrochemical Plants,* Gulf, Houston, 1964.

Mercier, C.: *L'Industrie Pétrochimique et Ses Possibilités d'Implantation dans les Pays en Voie de Développement,* Editions Technip, Paris, 1966.

Page, J. S.: *Estimator's Manual of Equipment and Installation Costs,* Gulf, Houston, 1963.

Perry, J. H.: *Chemical Engineer's Handbook,* 4th ed., McGraw-Hill, New York, 1963.

Peters, M. S., and K. D. Timmerhaus: *Plant Design and Economics for Chemical Engineers,* McGraw-Hill, New York, 1968.

Popper, H.: *Modern Cost Engineering Techniques,* McGraw-Hill, New York, 1970.

Rase, H. F., and M. H. Barrow: *Project Engineering of Process Plants,* Wiley, New York, 1966.

Transactions of the American Association of Cost Engineers, 17th national meeting, St. Louis, Mo., 1973.

Wuithier, P.: *Le pétrole: Raffinage et Génie Chimique,* 2d ed. entièrement mise à jour, Éditions Technip, Paris, 1972.

2. MARKET RESEARCH

a. General

Balaceanu, J. C., and P. Leprince: "La Pétrochimie d'Aujourd'hui à l'An 2000," *Rev. de l'Inst. Franc. du Petrole,* vol. 27, no. 3 (May–June 1972), p. 453.

Chemical Pricing Patterns, Schnell.

Chemicals Economics Handbook, Standford Research Institute, Current, Menlo Park, Calif.

Corrigan, T. E., and M.J. Dean: "Determining Optimum Plant Size," *Chem. Eng.,* 74 (17), p. 152, 1967.

Duchesnes, D.: "Le Rôle de l'Étude de Marché dans l'Entreprise Chimique," *Annales de la Société Belge de l'Industrie du Pétrole,* 14, p. 27, 1964.

Europe and the Europeans: An Overview, Hudson Institute, HI-1429/4-CC, New York, 25, 1971.

Eurostat: *Commerce Extérieur de la CEE* (Nimexe), Office Central des Statistiques de la CEE, Luxembourg.

L'Industrie Chimique 1969–1970, OCDE, Paris, 1971.

Saglin, I.: "How To Price New Products," *Modern Cost Engineering Techniques,* McGraw-Hill, New York, 1970.

Statistiques de Base de la Communauté, Office Statistique des Communautés Européennes, 1968–1969.

Stobaugh, R. B.: "Chemical Marketing Research," *Modern Cost Engineering Techniques,* McGraw-Hill, New York, 1970.

Synthetic Organic Chemicals, United States Production and Sales, 1970, U.S. Tariff Commission, Washington, 1972.

b. Existing Production Capacities

Chemical Economics Handbook, Stanford Research Institute, Current, Menlo Park, Calif.

"Chemical profiles," *The Chemical Marketing Newspaper,* Schnell, New York.

Lefebvre, G., L. Castex and N. Delavault: *Les Unités Pétrochimiques en Europe de l'Ouest,* Les Informations Technico-économiques de l'Institut Français du Pétrole.

Abstracts

Chemical Market Abstracts, Predicasts, Cleveland.

Industrial Notes.

S.E.A.R.C.H. (Systemized excerpts abstracts and reviews of chemical headlines), Compendium Publishers International Corp., New Jersey.

Periodicals *Informations Chimie, Hydrocarbon Processing, The Oil and Gas Journal, European Chemical News, Chemical Age International, Process Engineering, Chimie Actualités, Chemical Week, Japan Chemical Week, Kunststoffe, Gummi Asbest. Kunstoffe, Rubber Statistical Bulletin, L'industrie des Corps Gras.*

c. Prices of Chemicals

Malloy, J. B.: "Projecting Chemical Product Prices," *Chem. Eng. Progr.,* 70 (9), pp. 77–83, 1974.

Minerals Yearbook, Bureau of Mines, United States Department of the Interior, Washington, 1972.

Synthetic Organic Chemicals, United States Production and Sales, 1970, United States Tariff Commission, Tc Publication 479, Washington, 1972.

Chemical Pricing Patterns Comparisons of annual high and low prices for 1,250 key chemicals and related process materials for the years of 1952 through 1970, *The Chemical Marketing Newspaper,* Schnell, New York, 1971.

Current Price Data *Chemical Marketing, European Chemical News, Chimie Actualités, L'Usine Nouvelle, Japan Chemical Week, Informations Chimie, Petroleum Press Service.*

d. Market Trends

Badley, S. R., and C. M. Bromley: *Prices and Demand for Plastics and Rubber,* European Petrochemical Association, Venice, Oct. 1973.

Kaup, H. H.: *Trends in Man-Made Fibers,* The European Petrochemical Association, Venice, Oct. 1973.

Kirkley, T. A.: *Supply Trends for Hydrocarbon Raw Materials,* American Institute of Chemical Engineers, Houston, Mar. 1971.

Leprince, P. and A. Hahn: *World Energy Situation and Its Effects on Olefins and Aromatics in Europe,* The European Petrochemical Association, Venice, Oct. 1973.

Lunde, K. E.: *The Present and Future Petrochemical Situation in the United States in Relation to Western Europe,* European Petrochemical Association, Monte Carlo, Oct. 15–18, 1972.

Twaddle, W. W., and J. B. Malloy: *Evaluating and Sizing New Chemical Plants in a Dynamic Economy,* American Association of Cost Engineers 10th national meeting, Philadelphia, Pa., June 20, 21, and 22, 1966.

Waggoner, J. V.: *Marketing Trends of the 70's in Petrochemical Intermediates,* American Institute of Chemical Engineers, Houston, Mar. 1971.

3. INDEXES

a. Cost Indexes

Chemical Engineering Cost Index

Caldwell, D. W., and J. H. Ortego: "A Method for Forecasting the CE Plant Cost Index," *Chem. Eng.,* 82 (14), pp. 83–85, 1975.

"CE's Plant Cost Index Shows the Least Rate of Growth," *Chem. Eng.,* 70 (4), pp. 143–152, 1963.

Norden, R. B.: "CE Cost Indexes: A Sharp Rise Since 1965," *Chem. Eng.,* 76 (10), pp. 134–138, 1969.

Ricci, L. J.: "CE Cost Indexes Accelerate 10-Year Climb," *Chem. Eng.,* 82 (9), pp. 117–118, 1975.

Engineering News Record Index

Engineering News Record, 143 (9), p. 398, 1949.

Marshall and Swift or Marshall and Stevens Cost Index

Stevens, R. W.: "Equipment Cost Indexes for Process Industries," *Chem. Eng.,* 54 (11), pp. 124–126, 1947.

———, "Costs Recover from Downward Dip," *Chem. Eng.,* 69 (5), pp. 125–126, 1962.

Nelson Cost Index

Nelson, W. L.: "A Refinery Cost Index," *Oil Gas J.,* p. 91, Dec. 15, 1969.

———, "How the Nelson Index Is Computed," *Oil Gas J.,* p. 110, Oct. 1, 1956.

———, "Construction Costs," *Oil Gas J.,* 55 (32), 1957.

———, "How To Construct Cost Indexes from Components and Labor Productivity," *Oil Gas J.,* 60 (46), pp. 153–154, 1962.

———, "New Cost Indexes for This Journal Feature," *Oil Gas J.,* 61 (14), pp. 119–120, 1963.

———, "Tabulated Values of Nelson Refinery, Construction Cost Index," *Oil Gas J.,* 63 (14), p. 185, 1965.

———, "Where To Find Yearly Indexes," *Oil Gas J.,* 63 (27), p. 117, 1965.

———, "True Cost Indexes of Refinery Construction," *Oil Gas J.,* 64 (2), p. 76, 1966.

———, "How Nelson Index Is Computed," *Oil Gas J.,* 65 (19), p. 97. 1967.

———, "Here Are 'True' and Other Refinery Cost Indexes," *Oil Gas J.,* 68 (46), p. 165, 1970.

———, "Labor Costs Remain Key Factor in Determining Inflation Index," *Oil Gas J.,* 69 (5), pp. 74–75, 1971.

———, "Explaining the Nelson Indexes," *Oil Gas J.,* 69 (34), p. 78, 1971.

———, "How To Use the Nelson Productivities," *Oil Gas J.,* 70 (38), p. 84, 1972.

———, "Refinery Construction Costs Continue Upward," *Oil Gas J.,* 71 (5), p. 114, 1973.

———, "What Are the Nelson Refinery Indexes," *Oil Gas J.,* 72(44), pp. 105–106, 1974.

Process Engineering Index

"Productivity and Construction Costs," *Process Eng.*, p. 13, Jan. 1972.

"Cost indexes," *Process Eng.*, p. 18, Jan. 1973.

"Plant Construction Cost Indexes: How To Compare with the U.S.," *Process Eng.*, pp. 108–109, Mar. 1973.

Others

Heller, K. B.: *A Cost Index For Plant Construction in the Netherlands* (a WEBCI study), Third International Cost Engineering Symposium, Ref. F.5., London, Oct. 6–9 1974.

Schulze, J.: *Development of a Cost Index for a Chemical Plant in Western Germany,* Third International Cost Engineering Symposium, Ref. F.6., London, Oct. 6–9 1974.

b. Location

Bauman, H. C.: "Costs for Chemical Process Plants Abroad," *Eng. Chem.*, 54 (9), pp. 40–43, 1962.

Callagher, J. T.: "Efficient Estimating of Worldwide Plant Costs," *Chem. Eng.*, 76 (12), p. 196, 1969.

Catry, J. P.: "La Localisation Industrielle," conférence donnée 5/30/75 au Centre de Perfectionnement des Industries Chimiques de Nancy, Séminaire "Evaluation des Avant-Projets."

Cran, J.: "Location Index Compares Costs of Building Process Plants Overseas," *Process Eng.*, pp. 109–111, Apr. 1973.

Johnson, R. J.: "Costs of Overseas Plants," *Chem. Eng.*, 76 (5) p. 146, 1969.

Nelson, W. L.: "Comparing Construction Costs Abroad," *Oil Gas J.*, 67 (50), p. 76, 1969.

Terris, Ph.: "Rôle des Pays Producteurs de Pétrole dans l'Industrie Pétrochimique," *Revue de l'Association Française des Techniciens du Pétrole,* pp. 186–189, Nov.–Dec. 1973.

Yen-Chen, Yen: "Estimating Plant Costs in the Developping Countries," *Chem. Eng.*, 79 (15), pp. 89–92, 1972.

c. Miscellaneous

Alonso, J. R. F.: "Estimating the Cost of Gas Cleaning Plants," *Chem. Eng.*, 78 (28), pp. 86–96, 1971.

Hirschmann, W. B.: "Has the Cost of Building New Refineries Really Gone Up?" *Chem. Eng. Progr.*, 67 (8), pp. 39–47, 1971.

4. METHODS FOR ESTIMATING INVESTMENT COSTS

a. Accuracy of Estimating Methods

Hackney, J. W.: "Estimating Methods for Process Industry Capital Costs," *Chem. Eng.*, 67 (5), pp. 113–131, 1960 and 67 (7), pp. 119–134, 1960.

Nichols, W. T.: "Capital Cost Estimating," *Ind. Eng. Chem.*, 43 (10), pp. 2295–2298, 1951.

b. Extrapolating Investments

Chase, J. D.: "Plant Cost vs. Capacity, New Way To Use Exponents," *Chem. Eng.*, 77 (7), pp. 113–118, 1970.

Guthrie, K. M.: "Capital and Operating Costs for 54 Chemical Processes," *Chem., Eng.*, 77 (10), pp. 140–156, 1970.

Nelson, W. L.: "Overall Plant Costs Cracking, *Oil Gas J.*, 48 (29), pp. 149, 1949.

———, "How To Scale Plant Costs to Other Sizes," *Oil Gas J.*, 62 (39), pp. 84–85, 1964.

Williams, R., Jr.: "Six-Tenths Factor Aids in Approximating Costs," *Chem. Eng.*, 54 (12), pp. 124–125, 1947.

Woodier, A. B., and J. W. Voolcock: "The ABC of the 0.6 Scale Up Factor," *European Chem. News,* large plant supplement, pp. 7–9, Sept. 10, 1965.

c. Calculating Investment Costs

Allen, D. H., and R. C. Page: "Revised Technique for Predesign Cost Estimating," *Chem. Eng.*, 82 (3), pp. 142–150, 1975.

Bach, N. G.: "More Accurate Plant Cost Estimates," *Chem. Eng.*, 65 (19), pp. 155–159, 1958.

Bridgwater, A. V.: "Rapid Cost Estimation in the Chemical Process Industries," Third International Cost Engineering Symposium, Ref. A6, London, Oct. 6–9 1974.

Callagher, J. T.: "Rapid Estimation of Plant Costs," *Chem. Eng.*, 74 (26), pp. 89–96, 1967.

Carmichael, A., B. M. Sood, and R. G. Muller: "Capital Cost Estimates for Process Screening," 164th ACS meeting, Washington, Aug. 28–Sept. 2 1972.

Chilton, C. H.: "Cost Data Correlated," *Chem. Eng.*, 54 (6), pp. 97–106, 1949.

Clerk, J.: "Multiplying Factors Give Installed Costs of Process Equipment," *Chem. Eng.*, 70 (4), pp. 182–184, 1963.

Guthrie, D. M.: " 'Rapid Calc' Chart," *Chem. Eng.*, 76 (1), pp. 138–144, 1969.

———, "Capital Cost Estimating," *Chem. Eng.*, 76 (6), pp. 114–142, 1969.

Hackney, J. W.: "Capital Cost Estimates for Process Industries," *Chem. Eng.*, 67 (5), pp. 113–130, 1960.

Hand, W. E.: "From Flow Sheet to Cost Estimate," *Petrol. Ref.*, 37 (9), pp. 331–334, 1958.

Haselbarth, J. E., and J. M. Berk: "Chemical Plant Cost Breakdown," *Chem. Eng.*, 67 (10), p. 158, 1960.

Hellenach, L. J., et al.: "Module Estimating Technique as an Aid in Developing Plant Capital Cost, *AACE Bulletin,* 6 (4), pp. 116–121, 1964.

Hill, R. D.: "What Petrochemical Plants Cost," *Petrol. Ref.*, 35 (8), pp. 106–110, 1956.

Hirsch, J. H., and E. M. Glazier: "Estimating Plant Investment Costs," *Chem. Eng. Progr.*, 56 (12), pp. 37–43, 1960.

Lang, H. J.: "Cost Relationships In Preliminary Cost Estimation," *Chem. Eng.*, 54 (10), pp. 117–121, 1947.

———, "Simplified Approach to Preliminary Estimates," *Chem. Eng.*, 55 (6), pp. 112–113, 1948.

Miller, C. A.: "New Cost Factors Give Quick, Accurate Estimates," *Chem. Eng.*, 72 (19), pp. 226–236, 1965.

O'Donnell, J. P.: "New Correlation of Engineering and Other Indirect Project Costs," *Chem. Eng.*, 60 (1), pp. 188–190, 1953.

Stallworthy, E. A.: "The Viewpoint of a Large Chemical Manufacturing Company," *Chem. Eng.*, pp. 182–189, juin 1970.

Walas, S. M.: "Plant Investment Costs by the Factor Method," *Chem. Eng. Progr.*, 57 (6), 1961.

Wilson, G. T.: "Capital Investment for Chemical Plant," *Brit. Chem. Eng. Process. Tech.*, 16 (10), pp. 931–934, 1971.

Zevnik, F. C., and R. L. Buchanan: "Generalized Correlation of Process Investment," *Chem. Eng. Progr.*, 59 (2), pp. 70–77, 1963.

5. METHODS FOR CALCULATING PROFITABILITY OF A PROJECT

Allen, D. H.: "Two New Tools for Project Evaluation," *Chem. Eng.*, 74(15), pp. 75–78, 1967.

———, "Economic Evaluation and Taking Decision," *Brit. Chem. Eng.*, 14 (6), pp. 790–793, 1969.

———, *A Guide to The Economic Evaluation of Projects,* Department of Chemical Engineering, University of Nottingham, The Institution of Chemical Engineers, London, 1972.

Babusiaux, D.: "Introduction au Calcul Économique," *Rapport IFP,* 21,013, Feb. 1973.

Chilton, T. H.: "Investment Return via the Engineer's Method," *Chem. Eng. Progr.*, 65 (7), pp. 29–34, 1969.

Clark, R. W.: "Consider These Factors Affecting Plant Costs," *Petrochem. Eng.*, C8–C44, Aug. 1960.

Earley, W. E.: "How Process Companies Evaluate Capital Investments," *Chem. Eng.*, 71 (6), pp. 111–116, 1964.

Hackney, J. W.: "How To Appraise Capital Investments," *Chem. Eng.*, 67 (11), pp. 145–164, 1961.

Happel, J.: "Economic Evaluation," *Petrochem. Eng.*, C8–C44, Aug. 1960.

——— and W. H. Kapfer: "Evaluate Plant Design with This Simple Index," *Petrochem. Eng.*, C8–C44, Aug. 1960.

Herron, D. P.: "Comparing Investment Evaluation Methods," *Chem. Eng.*, 74 (2), pp. 125–132, 1967.

Holland, F. A., F. A. Watson, and J. K. Wilkinson:
Part 1: "Engineering Economics for Chemical Engineers," *Chem. Eng.,* 80 (15), pp. 103–107, 1973.
Part 2: "Capital Costs and Depreciation," *Chem. Eng.,* 80 (17), pp. 118–121, 1973.
Part 3: "Profitability of Invested Capital," *Chem. Eng.,* 80 (19), pp. 139–144, 1973.
Part 4: "Time Value of Money," *Chem. Eng.,* 80 (21), pp. 123–126, 1973.
Part 5: "Methods of Estimating Project Profitability," *Chem. Eng.,* 80 (22), pp. 80–86, 1973.
Part 6: "Sensitivity Analysis of Project Profitabilities," *Chem. Eng.,* 80 (25), pp. 115–119, 1973.
Part 7: "Time, Capital, and Interest Affect Choice of Project," *Chem. Eng.,* 80 (27), pp. 83–89, 1973.
Part 8: "Statistical Techniques Improve Decision Making," *Chem. Eng.,* 80 (29), pp. 61–66, 1973.
Part 9: "Probability Techniques for Estimates of Profitability," *Chem. Eng.,* 81 (1), pp. 105–110, 1974.
Part 10: "Estimating Profitability When Uncertainties Exist," *Chem. Eng.,* 81 (3), pp. 73–79, 1974.
Part 11: "Numerical Measures of Risk," *Chem. Eng.,* 81 (5), pp. 119–125, 1974.
Part 12: "How To Estimate Capital Costs," *Chem. Eng.,* 81 (7), pp. 71–76, 1974.
Part 13: "Manufacturing Costs and How To Estimate Them," *Chem. Eng.,* 81 (8), pp. 91–96, 1974.
Part 14: "How To Budget and Control Manufacturing Costs," *Chem. Eng.,* 81 (10), pp. 105–110, 1974.
Part 15: "How To Allocate Overhead Cost and Appraise Inventory," *Chem. Eng.,* 81 (12), pp. 83–87, 1974.
Part 16: "Principles of Accounting," *Chem. Eng.,* 81 (14), pp. 93–98, 1974
Part 17: "How To Evaluate Working Capital for a Company," *Chem. Eng.,* 81 (16), pp. 101–106, 1974.
Part 18: "Financing Assets by Equity and Debt," *Chem. Eng.,* 81 (18), pp. 62–66, 1974.
Part 19: "How To Assess Your Company's Progress," *Chem. Eng.,* 81 (19), pp. 119–124, 1974.
Part 20: "Inflation and Its Impact On Costs and Prices," *Chem. Eng.,* 81 (23), pp. 107–112, 1974.

Jelen, F. C., and M. S. Cole: "Methods for Economic Analysis":
Part 1: *Hydr. Process.,* pp. 133–139, July 1974.
Part 2: ibid., pp. 227–233, Sept. 1974.
Part 3: ibid., pp. 161–163, Oct. 1974.

Leibson, I., and C. A. Trischman, Jr.:
Part 1: "Spotlight on Operating Cost," *Chem. Eng.,* 78 (12), pp. 69–74, 1971.
Part 2: "How to Cut Operating Costs: Evaluate Your Feedstocks," *Chem. Eng.,* 78 (13), pp. 92–95, 1971.
Part 3: "How To Get Approval of Capital Projects," *Chem. Eng.,* 78 (15), pp. 95–102, 1971.
Part 4: "Avoiding Pitfalls in Developing a Major Capital Project," *Chem. Eng.,* 78 (18), pp. 103–110, 1971.

Part 5: "A Realistic Project Development Case Study," *Chem. Eng.,* 78 (20), pp. 86–92, 1971.

Part 6: "Case Study Shows Project Development in Action," *Chem. Eng.,* 78 (22), pp. 85–92, 1971.

Part 7: "Final Stage in the Project," *Chem. Eng.,* 78 (25), pp. 78–85, 1971.

Part 8: "When and How To Apply Discounted Cash Flow and Present Worth," *Chem. Eng.,* 78 (28), pp. 97–105, 1971.

Part 9: "Decision Trees: A Rapid Evaluation of Investment Risk," *Chem. Eng.,* 79 (2), pp. 99–105, 1972.

Part 10: "Should You Make or Buy Your Major Raw Material?" *Chem. Eng.,* 79 (4), pp. 76–83, 1972.

Part 11: "Zeroing in on "Make or Buy" Decisions," *Chem. Eng.,* 79 (6), pp. 113–118, 1972.

Part 12: "How To Profit from Product Improvement and Development," *Chem. Eng.,* 79 (8), pp. 103–112, 1972.

Malloy, J. B.: "Instant Economic Evaluation," *Chem. Eng. Progr.,* 65 (11), pp. 47–54, 1969.

Mapstone, G. E.: "The Present Value of Plant Allowances," *Chem. Process Eng.* pp. 66–69, April 1971.

Martin, J. C.: "How To Measure Project Profitability," *Petrochem. Eng.,* C8–C44, Aug. 1960.

McLean, J. G.: "Techniques for the Appraisal of New Investments," 6th Congrès Mondial du Pétrole, Frankfurt, Sect. VIII, pp. 47–63, June 1963.

Newton, R. D.: "Selection of Plant Capacity," *Petrochem. Eng.,* C8–C44, Aug. 1960.

Nitchie, E. B.: "Accounting Data and Methods Help Control Costs and Evaluate Profits," *Chem. Eng.,* 74 (1), pp. 87–92, 1967.

Ohsol, E. O.: "Commercial Evaluation of New Projects," American Chemical Society 164th meeting, New York, Aug.–Sept. 1972.

Schraishuhn, E. A., and J. Kellett: "How To Obtain Speed and Accuracy in Preliminary Appraisals," *Petrochem Eng.,* C8–C44, Aug. 1960.

Sinclair, C. G.: "Measurement of Investment Worth," *Chem. Process Eng.,* 47, pp. 137–139, 1966.

Sniffen, T. J.: "The Right Cos Can Reduce Pumping Costs," *Petrochem. Eng.,* C8–C44, Aug. 1960.

Twaddle, W. W., and J. B. Malloy: "Evaluating and Sizing New Chemical Plants," *Chem. Eng. Progr.,* 62 (7), pp. 90–96, 1966.

Urhan, W. J., and F. A. Holland: "How to Determine Optimum Plant Size," *Chem. Eng.,* 73 (6), pp. 103–108, 1966.

Walton, P. R.: "Cost Estimating at the Research Level," American Association of Cost Engineers 10th national meeting, Philadelphia, June 20–22, 1966.

Weinberger, A. J.: "Improving R&D's Batting Average," "How To Estimate Required Investment", "Calculating Manufacturing Costs", "Estimating Sales and Markets", "Methods for Estimating Profitability", "Post Audits and Qualitative Factors," *Chem. Eng.,* 71 (9), pp. 123–168, 1964.

6. PROCESS DESIGN ESTIMATION OF EQUIPMENT

a. General

Physical Data

"Azeotropic data," *Advances in Chemistry Series,* American Chemical Society, Washington, vol. 1, 1952, vol. 2, 1962, vol. 3, 1973.

Gallant, R. W.: *Physical Properties of Hydrocarbons,* Gulf, Houston, 1968.

Maxwell, J. B.: *Data Book on Hydrocarbons,* D. Van Nostrand, New York, 1960.

"Physical Properties of Chemical Compounds," *Advances in Chemistry Series,* American Chemical Society, Washington, vol. 1, 1955, vol. 2, 1959, vol. 3, 1961.

Thermodynamic Data

Cox, J. D., and G. Pilcher: *Thermochemistry of Organic and Organometallic Compounds,* Academic Press, London, 1970.

Engineering Data Book, Natural Gas Processors Suppliers Association, Tulsa, 1967, 8th ed., pp. 225–293.

Noddings, C. R., and G. M. Mullet: *Handbook of Compositions at Thermodynamic Equilibrium,* Wiley Interscience, New York, 1965.

"Selected Values of Chemical Thermodynamic Properties," Circular 500, National Bureau of Standards, U.S. Government Printing Office, Washington, 1961.

Stull, D. R., E. F. Westrum, Jr., and G. C. Sinke: *The Chemical Thermodynamics of Organic Compounds,* Wiley, New York, 1969.

Materials of Construction

Adams, L.: "Relative Metal Economy of Pressure-Vessel Steels," *Chem. Eng.,* 76 (27), pp. 150–151, 1969.

"Chemical Resistance of Tank Lining Materials," *Petrochem. Eng.,* pp. 42–49, May 1966.

Le Gouic, Y.: "Corrosion," *Inf. Chimie,* 87, Aug.–Sept. 1970, pp. 100–114.

McDowell, D. W., Jr.: "Materials of Construction Report," *Chem. Eng.,* 81 (24), pp. 118–146, 1974.

Tyson, S. E.: "Sure Shortcut to Stainless Steel Specification," *Chem. Eng.,* 76 (21), pp. 188–192, 1969.

b. Primary Equipment

General

Callagher, J. T.: "Rapid Estimation of Plant Costs, *Chem. Eng.,* 74 (27), pp. 89–96, 1967.

Drew, J. W., and A. F. Ginder: "How To Estimate the Cost of Pilot Plant Equipment," *Chem. Eng.,* 77 (3), pp. 100–110, 1970.

Guthrie, K. M.: "Capital Cost Estimating," *Chem. Eng.,* 76 (6), pp. 114–142, 1969.

Maristany, B. A.: "Figure LREP for Batch Plants," *Hydr. Process,* 45 (12), pp. 123–126, 1966.

Miller, R., Jr.: "Process Energy Systems," *Chem. Eng.,* 75 (11), pp. 130–148, 1968.

Mills, H. E.: "Cost of Process Equipment," *Chem. Eng.,* 71 (6), pp. 133–156, 1964.

Mouchel, J. C.: *Méthode d'Estimation Rapide des Devis: Planification et Contrôle des Constructions Industrielles pour l'Exploitation des Gisements de Pétrole,* Editions Technip, Collection Colloques et Séminaires, Paris, 1968.

Documents internes à l'Institut Français du Pétrole:

Rifflet, M.: "Estimation de Prix," Technip Engineering, Cours ENSPM.

Pigeyre, A.: Notions d'Estimation de Prix des Installations Pétrolières et de Pétrochimie, BEICIP, cours ENSPM (formation maîtrise), 1972.

Pressure Vessels

Badhwar, R. K.: "Quick Sizing of Distillation Columns," *Chem. Eng. Progr.,* 66 (3), pp. 56–61, 1970.

———, "Shortcut Design Methods for Piping, Exchangers, Towers," *Chem. Eng.,* 78 (24), pp. 112–122, 1971.

Barona, N., and H. W. Prengle, Jr.: "Design Reactors This Way for Liquid Phase Processes":

Part 1: "Types, Design Factors, Selection, and Scale Up," *Hydr. Process,* 52 (3), pp. 63–79, 1973.

Part 2: "Selected Processes, Design, Procedures, and Examples," *Hydr. Process,* 52 (12), pp. 73–89, 1973.

Callagher, J. T.: "Estimate Column Size by Nomogram," *Hydr. Process,* Petro. Ref., 42 (6), pp. 151–156, 1963.

Maccary, R. R.: "How To Select Pressure-Vessel Size," *Chem. Eng.,* 67 (23), pp. 187–190, 1960.

Sommerville, R. F.: "New Method Gives Quick, Accurate Estimate of Distillation Costs," *Chem. Eng.,* 79 (9), pp. 71–76, 1972.

Heat Exchangers

Badhwar, R. K.: "Shortcut Design Methods for Piping, Exhangers, Towers," *Chem. Eng.,* 78 (23), pp. 112–122, 1971.

Clerk, J.: "Costs of Air vs. Water Cooling," *Chem. Eng.,* 72 (1), pp. 100–102, 1965.

Grange, M.: "Technip Engineering," Cours de l'ENSPM, Institut Français du Pétrole, Rueil-Malmaison, 1973 (publication interne).

Lerner, J. E.: "Simplified Air Cooler Estimating," *Hydr. Process.,* 51 (2), pp. 93–100, 1972.

Maze, R. W.: "Air vs. Water Cooling":

Part 1: "How To Decide Which System To Choose," *Oil Gas J.,* 72 (46), pp. 74–78, 1974.

Part 2: "How To Make the Choice," *Oil Gas J.,* 72 (47), pp. 125–128, 1974.

"New Tema Standards: What's New, What's Changed," *Hydr. Process,* 47 (9), pp. 277–281, 1968.

Palmer, A. J., N. G. Boyd, and E. A. D. Saunders: "Prices of Shell and Tube Heat Exchangers," *Chem. Process Eng.,* Nov. 1963.

Retoret, J. P.: "Comment Spécifier un Échangeur de Chaleur," *L'Industrie du Pétrole en Europe,* 426, pp. 73–92, Dec. 1971.

Russel, C. M. B., and J. Tiley: "Air Cooler Estimation," *Chem. Process. Eng.* (Heat Transfer Survey), pp. 70–72, 1969.

"Welded and Seamless Heat Exchanger Tubing," *Chem. Eng.*, 67 (7), pp. 156, 1960.

Worsham, H. N.: "Analysis Can Reduce Air Cooler Design Uncertainty," *Oil Gas J.*, 67 (2), pp. 75–77, 1969.

Pumps, Compressors, and Drivers

Bresler, S. A., and J. H. Smith: "Guide to Trouble-Free Compressor," *Chem. Eng.*, 77 (12), pp. 161–170, 1970.

Guthrie, K. M.: "Costs," *Chem. Eng.* (deskbook issue), 76 (8), pp. 201–216, 1969.

Guthrie, K. M.: "Pump and Valve Costs," *Chem. Eng.*, 78 (22), pp. 151–159, 1971.

Hallock, D.: "Quick Method for Centrifugal Compressor Estimates," *Hydr. Process.*, 44 (10), pp. 115–121, 1965.

Kirk, M. M.: "Chemical Feed Pumps," *Chem. Eng.*, 72 (5), p. 112, 1965.

Neerken, R. F.: "Pump Selection for the Chemical Process Industries," *Chem. Eng.*, 81 (4), pp. 104–115, 1974.

Retoret, J. P., and M. Moreau: "Les Compresseurs Alternatifs," *L'Industrie du Pétrole en Europe,* 432, pp. 69–75, July 1972, and 433, pp. 53–63, Sept. 1972.

Stindt, W. H.: "Pump Selection," *Chem. Eng.*, 78 (22), pp. 43–49, 1971.

Thurlow, C.: "Centrifugal Pump," *Chem. Eng.*, 78 (22), pp. 29–35, 1971.

Vachez, J.: "Les Pompes dans l'Industrie du Pétrole et de la Chimie," *Ind. Pétrole,* 399, pp. 63–68, June 1969, and 400, pp. 41–45, July–Aug. 1969.

c. Utilities

Balou, D. F., T. A. Lyons, and J. R. Tacquard, Jr.: "Mechanical Refrigeration Systems," *Hydr. Process,* 46 (6), pp. 119–135, 1967.

Bourbillion, P.: *Les utilités: Planification et Contrôle des Constructions Industrielles pour l'Exploitation des Gisements de Pétrole,* Editions Technip, Collection Colloques et Séminaires, Paris, 1968.

"Process Plant Utilities," *Chem. Eng.*, 77 (27), pp. 130–146, 1970:
Monroe, L. R.: "Steam."
Brooke, M.: "Water."
McAllister, D. G., Jr.: "Air."
Price, H. A., and D. G. McAllister, Jr.: "Inert Gas."
Yuhen, M. H.: "Electricity."

Spencer, E.: "Estimating the Size and Cost of Steam Vacuum Refrigeration," *Hydr. Process,* 46 (6), pp. 136–140, 1967.

d. Miscellaneous Costs

Bauman, H. C.: "Estimate design engineering costs," *Hydr. Process,* 43 (10), pp. 141–144, 1964.

———: "Engineering Cost," American Association of Cost Engineers annual convention, New York, 1964.

Feldman, R. P.: "Economics of Plant Startups," *Chem. Eng.,* 76 (24), pp. 87–90, 1969.

McCabe, W. P., and E. V. Rymer: "Estimating Engineering Costs," American Association of Cost Engineers 10th national meeting, Philadelphia, June 1966.

O'Donnel, J. P.: "New Correlation of Engineering and Other Indirect Project Costs," *Chem. Eng.,* 60 (1), pp. 188–190, 1953.

Staudt, W. E.: "Overhead Costs in Economic Evaluations," *Chem. Eng. Progr.,* 62 (5), pp. 39–47, 1966.

Index

NOTE: Page numbers in *italics* refer to material included in a table or chart.

Index